AF380631

From Insect Pheromones to Mating Disruption: Theory and Practice

From Insect Pheromones to Mating Disruption: Theory and Practice

Editors

Giovanni Benelli
Andrea Lucchi

MDPI • Basel • Beijing • Wuhan • Barcelona • Belgrade • Manchester • Tokyo • Cluj • Tianjin

Editors
Giovanni Benelli
Agriculture, Food
and Environment
University of Pisa
Pisa
Italy

Andrea Lucchi
Agriculture, Food
and Environment
University of Pisa
Pisa
Italy

Editorial Office
MDPI
St. Alban-Anlage 66
4052 Basel, Switzerland

This is a reprint of articles from the Special Issue published online in the open access journal *Insects* (ISSN 2075-4450) (available at: www.mdpi.com/journal/insects/special_issues/pheromone_mating_disruption).

For citation purposes, cite each article independently as indicated on the article page online and as indicated below:

LastName, A.A.; LastName, B.B.; LastName, C.C. Article Title. *Journal Name* **Year**, *Volume Number*, Page Range.

ISBN 978-3-0365-3179-3 (Hbk)
ISBN 978-3-0365-3178-6 (PDF)

Contents

About the Editors

Giovanni Benelli

Giovanni Benelli serves as an Associate Professor of General and Applied Entomology at the Department of Agriculture, Food and Environment, University of Pisa, Italy. He teaches Agrarian Zoology, Biological Control, Trends and Challenges in the Management of Vineyard Pests, and Biotechnologies for Managing Animal Parasites.

He achieved an International Ph.D. in Agrarian and Veterinary Sciences at the University of Pisa and Sant' Anna School of Advanced Studies. Giovanni has worked in several international institutions, including the University of Hawaii at Manoa (USA) and the University of Jaén (Spain).

Giovanni's research focuses on insect behaviour, biological control, chemical ecology (with special reference to sex pheromones and mating disruption), and insect-inspired robotics, covering agricultural pests as well as vectors of medical and veterinary importance.

He has cooperated with a wide number of researchers worldwide on various research projects (e.g., iGuess-MED PRIMA, and STRADIOL). He is actively engaged in Third Mission activities, through agricultural extension services focused on IPM and the biological control of olive and vineyard insect pests.

Giovanni serves as Editor-in-Chief/Associate Editor/Editorial Board Member for many top-ranked international journals in the field of general and applied entomology.

He has been awarded various research prizes from international and national organizations, including the Odile Bain Memorial Prize 2018 (Parasites and Vectors & Boehringer Animal Health) and the Antico Fattore Prize 2016 (Accademia dei Georgofili, Firenze). He was appointed as a Member of Accademia dei Georgofili in December 2021.

Andrea Lucchi

Andrea Lucchi is a Full Professor at the University of Pisa, where he teaches Agricultural Entomology, Viticultural Entomology and "New strategies in the control of grape pests". Since September 2021, he has been President of the University Course "Viticulture and Enology" at the Dept of Agriculture, Food & Environment, in Pisa. He has co-authored 300 publications (86 Scopus), 6 chapters in international books, and 1 university book, *Note di Entomologia viticola*. He has been the Vice-President of IOBC/wprs for the last 8 years, and is now a member of the Audit Committee of the same organization.

Member of the USDA-APHIS Technical Working Group (TWG) for the management of the European Grapevine Moth (EGVM) in USA. Active cooperation with SAG (Chile), EMBRAPA (Brazil), COSAVE (South America) and ADVID (Portugal) for the management of grape moths. Since 2018 Prof. Lucchi is an EFSA expert on alternative pest control methods and High Risk Plants.

Preface to "From Insect Pheromones to Mating Disruption: Theory and Practice"

The study of insect chemical ecology with special reference to their pheromones is a fascinating field of research. Pheromone-mediated mating disruption (MD) represents an effective and eco-friendly biocontrol technique to manage insect pests of agricultural importance. It is currently used across more than 800,000 ha globally. This eco-friendly pest management approach is based on the release of synthetic sex pheromones from dispensers in crops, which interfere with mate-finding and reproduction of the pest through competitive and non-competitive mechanisms. To date, MD is still applied on a narrow amount of insect pests, with a major focus on moths. However, the MD potential is huge and urgently needs to be explored further.

In this framework, the present book, a reprint of the successful *Insects* Special Issue "From Insect Pheromones to Mating Disruption: Theory and Practice", includes laboratory and field studies dealing with insect pheromones, as well as on mating disruption efficacy against insect species of economic importance, with special reference to the development and optimization of mating disruption approaches, their mechanisms of action, and possible non-target effects.

Giovanni Benelli, Andrea Lucchi
Editors

Editorial

From Insect Pheromones to Mating Disruption: Theory and Practice

Giovanni Benelli and Andrea Lucchi *

Department of Agriculture, Food and Environment, University of Pisa, Via del Borghetto 80, 56124 Pisa, Italy; giovanni.benelli@unipi.it
* Correspondence: andrea.lucchi@unipi.it; Tel.: +39-050-221-6119

Citation: Benelli, G.; Lucchi, A. From Insect Pheromones to Mating Disruption: Theory and Practice. *Insects* **2021**, *12*, 698. https://doi.org/10.3390/insects12080698

Received: 28 July 2021
Accepted: 29 July 2021
Published: 3 August 2021

Publisher's Note: MDPI stays neutral with regard to jurisdictional claims in published maps and institutional affiliations.

1. Introduction

Insects perceive and integrate a hierarchy of visual, chemical and tactile cues for feeding and reproductive purposes, as well as for predator and parasitoid avoidance. Among semiochemicals routing insect's decisions, pheromones play a key role in mediating intraspecific communication [1,2]. The study of insect chemical ecology with special reference to their pheromones is a fascinating research field. Pheromone-mediated mating disruption represents an effective and eco-friendly biocontrol technique to manage insect pests of agricultural importance. Worldwide, agricultural pests affecting more than 800,000 hectares of crops are managed by mating disruption [3]. This technique relies on the release of synthetic sex pheromones from dispensers in crops, interfering with mate finding and reproduction of the pest through both competitive and non-competitive mechanisms [4,5]. Unfortunately, the use of mating disruption is still restricted to a rather limited number of crop pests, with special efforts being directed toward moths [3]. However, the mating disruption potential is huge and urgently needs to be explored further.

2. Insect Chemical Ecology: From the Laboratory to the Field

In this framework, the Special Issue "From Insect Pheromones to Mating Disruption: Theory and Practice" includes both laboratory and field studies on insect pheromones, as well as on mating disruption efficacy against insect pests of economic importance.

Herein, the following topics have been covered:

(a) **Pheromone biology**, with special reference to the various factors, often overlooked, affecting pheromone production in species of economic importance, including wood boring beetles, e.g., *Lyctus africanus* Lesne (Coleoptera: Bostrichidae) [6]; a further focus has been provided on the citrophilous mealybug, *Pseudococcus calceolariae* (Maskell) (Hemiptera, Pseudococcidae), assessing the potential negative impact of male multiple matings on mass trapping and mating disruption [7].

(b) **Development of novel mating disruption tools and approaches.** Several novel mating disruption approaches have been developed and/or optimized to manage a broadly diverse number of insect pests. A major focus has been devoted to moth pests, showing carefully conducted mating disruption experiments on grape (*Cryptoblabes gnidiella* (Millière, 1867) (Lepidoptera: Pyralidae)) [8], plum (*Grapholita funebrana* Treitschke (Lepidoptera: Tortricidae)) [9], almond (*Amyelois transitella* (Walker) (Lepidoptera: Pyralidae)) [10], and tomato (*Helicoverpa armigera* (Hubner) (Lepidoptera: Noctuidae)) [11], as well as on polyphagous leafrollers, such as *Proeulia auraria* (Clarke) (Lepidoptera: Tortricidae) [12]. Concerning mealybugs, further research efforts have been directed to evaluating the potential of *P. calceolariae* mating disruption in apple and tangerine orchards [13].

(c) **Reviews on insect chemical ecology.** The Special Issue ends with two broad reviews. The first summarizes current knowledge on insect sex pheromone research and its application in Integrated Pest Management [14]. The second review offers an updated analysis of what we really know on tephritid fruit fly semiochemicals, their real-world applications and the related research challenges [15].

3. The Future

Overall, we enjoyed organizing the Special Issue "From Insect Pheromones to Mating Disruption: Theory and Practice" greatly, and sincerely thank all the authors for their fine contributions. On the other hand, we are very aware that the present Special Issue cannot reflect the wide diversity of topics and challenges currently characterizing the pheromone and MD research. In particular, further research efforts are still needed to develop novel approaches to combat emerging pests, with special reference to invasive species, to fully clarify the mechanisms of action of MD-based control tools, to shed light on potential non-target effects, as well as to boost the efficacy of MD programs through the optimization of release geometries.

Author Contributions: Conceptualization, G.B. and A.L.; writing—original draft preparation, G.B. and A.L.; writing—review and editing, G.B. and A.L. Both authors have read and agreed to the published version of the manuscript.

Funding: This manuscript received no external funding.

Institutional Review Board Statement: Not applicable.

Acknowledgments: We are grateful to Insects Editorial Office for their kind assistance in the preparation on the present Special Issue.

Conflicts of Interest: The authors declare no conflict of interest.

References

1. Stökl, J.; Steiger, S. Evolutionary origin of insect pheromones. *Curr. Opin. Insect Sci.* **2017**, *24*, 36–42. [CrossRef] [PubMed]
2. Henneken, J.; Jones, T.M. Pheromones-based sexual selection in a rapidly changing world. *Curr. Opin. Insect Sci.* **2017**, *24*, 84–88. [CrossRef] [PubMed]
3. Benelli, G.; Lucchi, A.; Thomson, D.; Ioriatti, C. Sex Pheromone Aerosol Devices for Mating Disruption: Challenges for a Brighter Future. *Insects* **2019**, *10*, 308. [CrossRef] [PubMed]
4. McGhee, P.S.; Gut, L.J.; Miller, J.R. Aerosol emitters disrupt codling moth, *Cydia pomonella*, competitively. *Pest Manag. Sci.* **2014**, *70*, 1859–1862. [CrossRef] [PubMed]
5. Miller, J.R.; Gut, L.J. Mating Disruption for the 21st Century: Matching Technology with Mechanism. *Environ. Entomol.* **2015**, *44*, 427–453. [CrossRef] [PubMed]
6. Kartika, T.; Shimizu, N.; Himmi, S.K.; Guswenrivo, I.; Tarmadi, D.; Yusuf, S.; Yoshimura, T. Influence of Age and Mating Status on Pheromone Production in a Powderpost Beetle *Lyctus africanus* (Coleoptera: Lyctinae). *Insects* **2020**, *12*, 8. [CrossRef] [PubMed]
7. Ricciardi, R.; Lucchi, A.; Benelli, G.; Suckling, D.M. Multiple Mating in the Citrophilous Mealybug *Pseudococcus calceolariae*: Implications for Mating Disruption. *Insects* **2019**, *10*, 285. [CrossRef] [PubMed]
8. Ricciardi, R.; Di Giovanni, F.; Cosci, F.; Ladurner, E.; Savino, F.; Iodice, A.; Benelli, G.; Lucchi, A. Mating Disruption for Managing the Honeydew Moth, *Cryptoblabes gnidiella* (Millière), in Mediterranean Vineyards. *Insects* **2021**, *12*, 390. [CrossRef] [PubMed]
9. Verde, G.L.; Guarino, S.; Barone, S.; Rizzo, R. Can Mating Disruption Be a Possible Route to Control Plum Fruit Moth in Mediterranean Environments? *Insects* **2020**, *11*, 589. [CrossRef] [PubMed]
10. Higbee, B.; Burks, C. Individual and Additive Effects of Insecticide and Mating Disruption in Integrated Management of Navel Orangeworm in Almonds. *Insects* **2021**, *12*, 188. [CrossRef]
11. Burgio, G.; Ravaglia, F.; Maini, S.; Bazzocchi, G.G.; Masetti, A.; Lanzoni, A. Mating Disruption of *Helicoverpa armigera* (Lepidoptera: Noctuidae) on Processing Tomato: First Applications in Northern Italy. *Insects* **2020**, *11*, 206. [CrossRef]
12. Flores, M.; Bergmann, J.; Ballesteros, C.; Arraztio, D.; Curkovic, T. Development of Monitoring and Mating Disruption against the Chilean Leafroller *Proeulia auraria* (Lepidoptera: Tortricidae) in Orchards. *Insects* **2021**, *12*, 625. [CrossRef]
13. Ballesteros, C.; Romero, A.; Castro, M.; Miranda, S.; Bergmann, J.; Zaviezo, T. Mating Disruption of *Pseudococcus calceolariae* (Maskell) (Hemiptera, Pseudococcidae) in Fruit Crops. *Insects* **2021**, *12*, 343. [CrossRef] [PubMed]
14. Rizvi, S.; George, J.; Reddy, G.; Zeng, X.; Guerrero, A. Latest Developments in Insect Sex Pheromone Research and Its Application in Agricultural Pest Management. *Insects* **2021**, *12*, 484. [CrossRef] [PubMed]
15. Scolari, F.; Valerio, F.; Benelli, G.; Papadopoulos, N.; Vaníčková, L. Tephritid Fruit Fly Semiochemicals: Current Knowledge and Future Perspectives. *Insects* **2021**, *12*, 408. [CrossRef] [PubMed]

 *insects*

Article

Multiple Mating in the Citrophilous Mealybug *Pseudococcus calceolariae*: Implications for Mating Disruption

Renato Ricciardi [1], Andrea Lucchi [1], Giovanni Benelli [1,*] and David Maxwell Suckling [2,3]

1   Department of Agriculture, Food and Environment, University of Pisa, via del Borghetto 80, 56124 Pisa, Italy
2   The New Zealand Institute for Plant & Food Research Limited, PB 4704, Christchurch 8140, New Zealand
3   School of Biological Sciences, University of Auckland, Tamaki Campus, PB 92019, Auckland 1142, New Zealand
*   Correspondence: giovanni.benelli@unipi.it; Tel.: +39-050-221-6141

Received: 24 July 2019; Accepted: 27 August 2019; Published: 5 September 2019

Abstract: The citrophilous mealybug *Pseudococcus calceolariae* (Maskell) (Hemiptera, Pseudococcidae) is a primary pest of various crops, including grapevines. The use of insecticides against this species is difficult in most cases because its life cycle includes an extended duration of eggs, juveniles, and adults under the bark and on the roots. Pheromone-based control strategies can present new eco-friendly opportunities to manage this species, as in the case of *Planococcus ficus* (Signoret) and *Planococcus citri* (Risso). With this aim it is critical to understand behavioral aspects that may influence pheromone-based control strategies. Herein, the capability of males to fertilize multiple females was investigated, trying to understand whether this behavior could negatively impact the efficacy of mass trapping, mating disruption, or the lure and kill technique. Results showed that a *P. calceolariae* male can successfully mate and fertilize up to 13 females. The copulation time in subsequent mating events and the time between copulations did not change over time but the number of matings per day significantly decreased. In a further experiment, we investigated the mate location strategy of *P. calceolariae* males, testing the attractiveness of different loadings of sex pheromone on males in a flight tunnel. Males constantly exposed to 16 rubber septa loaded with the sex pheromone showed a significant decrease in female detection at 1 and 30 µg loadings (0.18 and 0.74 visits per female for each visit per septum, respectively), whereas in the control about 9.2-fold more of the released males successfully detected the female in the center of the array of 16 septa without pheromone. Male location of females in the control (45%) was significantly higher than in the arrays with surrounding pheromone (5% and 20% at 1 and 30 µg loadings, respectively). Mating only occurred in the control arrays (45%). This study represents a useful first step to developing pheromone-based strategies for the control of citrophilous mealybugs.

Keywords: sex pheromone; biological control; flight tunnel; Integrated Pest Management; mealybug monitoring

1. Introduction

Vineyards host a great variety of pests worldwide which impact grapevine health in different ways [1]. Many of these pests belong to the order Hemiptera and especially to the family Pseudococcidae, commonly known as mealybugs [2]. They compromise grape quality by contaminating bunches with honeydew, which leads to the development of sooty mold fungi. Furthermore, they can transmit several important grapevine viral diseases [3]. Farmers usually rely on insecticide applications to manage these pests. However, mealybugs can develop resistance to currently marketed insecticides and new reports are still emerging [4]. Several species of mealybugs have become resistant to insecticides, which is unsustainable as this negatively affects their control [5–11]. However, to boost grape production under an environmentally-safe agricultural system, it is necessary to find alternative solutions to effectively manage crop pests while reducing insecticide overuse [12–14]. Hence, it is critical to develop more sustainable Integrated Pest Management (IPM) systems [15,16]. An alternative is presented in synthetic sex pheromones, which disrupt the chemical communication between male and female of a given insect species, thereby preventing mate location and mating [17]. This approach is currently known as sex pheromone-based mating disruption (MD). Each mealybug species relies on unique pheromone(s) for sexual communication, with structures characterized by an irregular non-head-to-tail monoterpenoid structure [18]. To date, the pheromones of 19 mealybugs species belonging to seven genera have been identified [19]. This approach can be very effective. For example, mating disruption of *Planococcus ficus* (Signoret) (Hemiptera, Pseudococcidae) usually leads to a significant reduction of the abundance of ovipositing females, an increase of the pre-oviposition period, lower damage rates on grape bunches, and thus an overall decrease of the pest population densities [20–22].

The citrophilous mealybug, *Pseudococcus calceolariae* (Maskell) (Hemiptera, Pseudococcidae) impacts upon grape production through the abundant emission of honeydew, with subsequent development of sooty mold and transmission of two closteroviruses, GLRaV-1 and GLRaV-3 [23]. Moreover, *P. calceolariae* is a cosmopolitan and polyphagous pest on different crops (orange, lemon, and avocado, as well as other vegetable, fruit, and floricultural plants) [24–26].

Notably, comprehensive studies on the basic biology of *P. calceolariae* have been carried out, also shedding light on its chemoecology. Rotundo and Tremblay [27] and Rotundo et al. [28] focused on the male flight activity, providing evidence of the release of sex pheromones. Later, Silva et al. [29] studied its mating behavior, showing that this mealybug is obliged to mate for reproduction [30]. The sex pheromone of *P. calceolariae* was identified [25], allowing for the synthesis of a synthetic pheromone. No evidence of habituation was detected in *P. calceolariae* males exposed to the sex pheromone (1 mg for 24 h); indeed, they were able to promptly locate the sex pheromone source despite earlier pre-exposure to the chemical [31]. However, even though Silva et al. [32] studied multiple mating in this mealybug species, little is currently known about the potential effect of this factor on monitoring and control.

Multiple mating occurs in many orders of insects, including moths such as *Epiphyas postvittana* (Walker) [33] and *Cydia pomonella* (Linnaeus) (Lepidoptera, Tortricidae) [34], as well in other mealybug species such as *P. ficus*, *Pseudococcus longispinus* (Targioni Tozzetti) and *Pseudococcus viburni* (Signoret) (Hemiptera, Pseudococcidae) [35]. Knowledge about the presence and frequency of multiple mating in a given pest species is timely and important. Indeed, a key issue to watch for is that the males are able to mate with several females, thus reducing the efficacy of several control tools, including mating disruption, mass trapping, and the lure and kill technique, which all have the goal of preventing mating or reducing it by removing as many males from the population as possible [26].

In this scenario, the present study provides insights into the reproductive biology of *P. calceolariae* by having investigated how many females can be fertilized by a single male within a day as well as during the whole lifespan. The effect of subsequent mating events on copulation duration and time elapsed between copulations has also been evaluated. By simulating mating disruption in the laboratory, we tried to assess the potential effectiveness of this technique to manage *P. calceolariae*, following recent successful attempts on another mealybug species (*P. ficus*) [14,21,22,36,37].

2. Materials and Methods

2.1. Insect Rearing

Pseudococcus calceolariae mealybugs tested here were obtained from a colony held at The New Zealand Institute for Plant & Food Research Ltd. (Lincoln, New Zealand). The new colony was maintained inside ventilated plastic containers (30 × 25 × 10 cm) and periodically fed with new potato sprouts (*Solanum tuberosum* L.). We did not use individuals from the field to avoid contamination of other species. Crawlers were collected every 3–4 days and separated in different boxes (10 × 7 × 7 cm) to obtain mealybug cohorts of different ages. After 15–20 days, mealybugs were sexed by separating male cocoons before they hatched to avoid possible mating (males were easily identified by the presence of a cocoon). Immature individuals were reared on fresh potato sprouts in separated ventilated plastic containers (10 × 7 × 7 cm). Each box contained 20–40 virgin females. We tested 24–32 h-old males; their sexual maturation was indicated by complete growth of the wax tail, as detailed also for *Planococcus citri* (Risso) [38,39].

2.2. Multiple Mating Experiment

To estimate the number of females that could be fertilized by a single virgin male, we followed the method proposed by Waterworth et al. [35] with minor modifications. Ten *P. calceolariae* females and a virgin male were kept close to each other for 6 h in a sterile Petri dish (35 × 10 mm). All replicates were recorded using a high definition webcam (HD C525, Logitech, Lausanne, Switzerland) placed above the arena and connected to a computer (OptiPlex 745, DELL, Round Rock, TX, U.S.). Recordings were analyzed to measure (i) the copulation duration (s), (ii) the copulation number per day, and (iii) the time interval between matings. Then, each female was moved to a Petri dish (35 × 10 mm) and provided ad libitum with potato sprouts. Oviposition activity was monitored daily to measure the time required to produce an ovisac, following Silva et al. [29]. The following day, surviving males were introduced into a new arena with a new group of virgin females; this procedure was replicated with 30 males.

All tests were conducted in a room (22.5 ± 1 °C, 40 ± 2.5% RH; 16:8 (L:D) photoperiod) illuminated with daylight fluorescent tubes to obtain a light intensity in the proximity of the arena of 1000 lux. Each trial was carried out between 8:30 and 14:30.

2.3. Flight Tunnel Experiments

Controlled tests were conducted in a laminar airflow flight tunnel as described by El-Sayed et al. [40]. For each trial, two *P. calceolariae* males were set in a plastic vial (20 mL). Before the start of the experiment, the vial was placed in the center of the floor of the flight tunnel and males were allowed to acclimatize to the flight tunnel conditions (22.5 ± 1 °C, 40 ± 2.5% R.H, wind speed 0.4 ± 0.1 m/s) for 2 min. Each trial lasted 60 min. The different loadings of pheromone used were chosen based on pre-exposure tests reported by Suckling et al. [31]. Three experiments were carried out, as detailed below.

Firstly, we measured the flight tunnel response of *P. calceolariae* males to increasing loadings of sex pheromone in a 4 × 4 grid composed by 16 lures baited with synthetic sex pheromone; in this way, we replicated, on a reduced scale, what happens in the field when MD is attempted. We determined whether the male ability to locate a female could be disrupted using an amount of sex pheromone that was much higher than that naturally released by the females in the field, having observed much higher trap catches to synthetic lures when compared with mealybug females [25]. This information, considering the knowledge acquired on the extent of multiple mating, is critical to fully understand the potential and effectiveness of pheromone-based control strategies, with special reference to MD.

2.3.1. Trial 1: Male Response to a Single Loading of Sex Pheromone

To simulate the short distance between males and females in a field colony, we released two males close to a sex pheromone-loaded array of lures. The goal was to see which and how many lures were visited and whether they preferred the upwind or downwind lures. Herein, a 4 × 4 array of 16 rubber lures (Ø = 0.5 mm, h = 0.2 mm, 2 cm apart) was placed in the flight tunnel, with each lure loaded with 1 µg of synthetic pheromone; the pheromone was synthesized by the New Zealand Institute for Plant & Food Research laboratories according to the method described by El-Sayed et al. [25]. An A4 white sheet with a grid of 16 unloaded lures was used as a control. Herein, we tested both downwind and upwind directions (using different males every time) to evaluate the mealybug preference in terms of number of visits per lure. After confirming that the preferred direction was upwind, we compared all treatments upwind. With each new couple of males released 10–15 cm from the grid, the grid was rotated clockwise 90° to avoid positional effects [41]. We tested 30 male pairs.

2.3.2. Trial 2: Male Response to Four Different Loadings of Sex Pheromone

In this experiment, we used the same grid used in trial 1; however, it was simultaneously baited with four increasing sex pheromone loadings from downwind to upwind. The aim was to understand whether males preferred certain loadings over others. As in the previous experiment, *P. calceolariae* males were released close to the grid. An A4 white sheet with 16 rubber septa (4 lines × 4 lines as above) was used with four different loadings of synthetic sex pheromone for each row: 1, 3, 10, and 30 µg, respectively, increasing upwind. A control with 16 unloaded lures was used. The odor sources were renewed after every 10 males tested. During each replicate, we observed which and how many lures were visited by *P. calceolariae* males. A total of 30 replicates were carried out.

2.3.3. Trial 3: Mimicking Mating Disruption

Trial 3 followed the array setup of trial 1; attraction of males to a central virgin female surrounded with a control or either of two loadings of synthetic sex pheromone (each septum was loaded with 0, 1, or 30 µg of sex pheromone) was tested. By placing one virgin female in the middle of the array, it was possible to observe whether the lures were able to mask the presence of the female. The female was held with a small piece of double-sided tape.

During each replicate, we recorded: (i) which septa were visited by *P. calceolariae* males, (ii) the number of male visits per septum, (iii) the duration of each visit, and (iv) the mate location success of males attempting to locate a female in each treatment and control.

2.4. Statistical Analysis

In the multiple mating experiment, data about copulation duration, time between copulations and number of copulations per day were analyzed by JMP 11 [42] with a general linear mixed model (GLMM) with one fixed factor, i.e., the day of the observation [43]: $y_{i,w} = \mu + D_i + ID_w + e_{iw}$, where $y_{i,w}$ is the observation, μ is the overall mean, D_i is the i-th fixed effect of the day of the observation ($i = 1$–3), ID_w is the w-th random effect of the individual over repeated testing phases ($w = 1$–30), and e_{iw} is the residual error. Means of treatments were separated by the Tukey's HSD test. A p-value of 0.05 was selected as the threshold to assess significant differences.

In flight tunnel trial 1, the possible presence of significant differences among male visits to different rows of pheromone-baited lures was studied using a contingency analysis, highlighting that males randomly visited pheromone-baited lures in the flight tunnel. Therefore, data testing different pheromone loadings in the flight tunnel experiments (i.e., male visits per lure (n), female location events (n), and time spent on each lure (s)) were not normally distributed and it was not possible to normalize the distribution to homogenize the variance (Shapiro-Wilk test, goodness of fit $p < 0.001$). Therefore, all data were analyzed by Kruskal-Wallis test followed by Steel-Dwass test to make nonparametric comparisons between all pairs. A p-value of 0.05 was selected as the threshold to assess significant differences.

3. Results

3.1. Multiple Mating Experiment

Males with multiple mating events showed no significant differences in copula duration over three days of observations ($F_{2,172} = 2.466$; $p = 0.088$) (Figure 1a) while the copulation number per day was significantly affected ($F_{2,58} = 39.549$; $p < 0.0001$). The mean number of male successful copulation events per day was 4.73 the first day and then dropped to 1.73 and 0.47 on the second and third days, respectively (Figure 1b). Note that data from the fourth day of observations were not included in our analysis since only one male survived. The time between copulation events was not significantly affected by three subsequent days of repeated exposure to females ($F_{2,58} = 0.839$; $p = 0.434$) (Figure 1c).

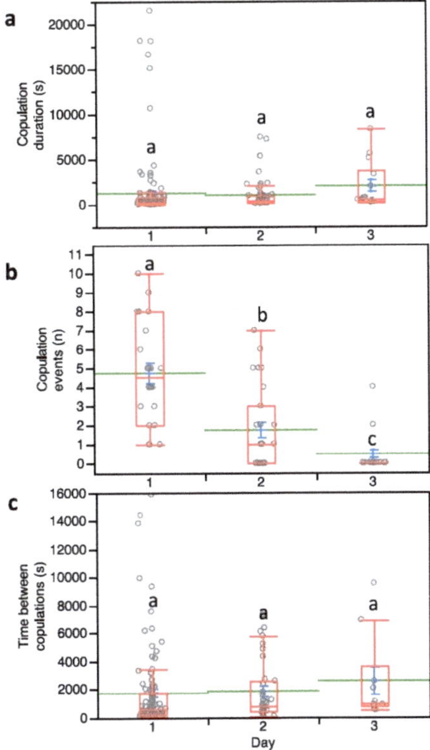

Figure 1. (**a**) Copulation duration, (**b**) number of copulation events and (**c**) time between copulations in *Pseudococcus calceolariae*. Box plots indicate the median (line) within each box and the range of dispersion (lower and upper quartiles and outliers). Green lines and blue T-bars indicate means and standard errors, respectively. Above each box plot, different letters indicate significant differences (general linear mixed model (GLMM), Tukey's HSD test, $p < 0.05$).

After the first day of mating, 56.7% of males survived and were then able to mate on the second day; 20% survived the second day and then mated during the third day. Just one male survived the third day and mated on the fourth day (Figure 2).

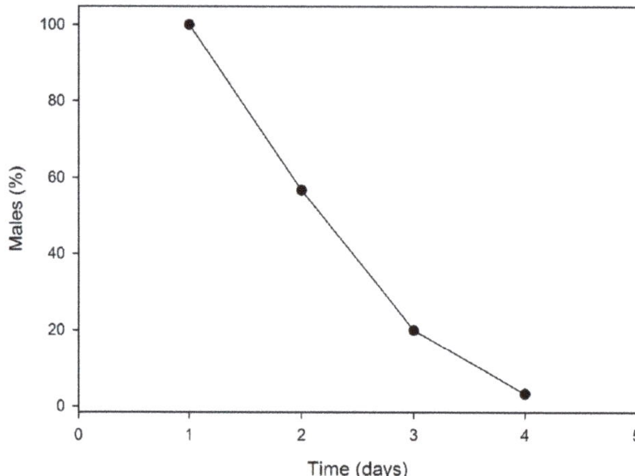

Figure 2. Longevity of *P. calceolariae* males subjected to multiple mating under laboratory conditions.

Overall, the maximum number of lifetime mating events was 13 and the daily number of mating events for each male ranged from 1 to 10, with a mean mating success of 92.8%. The mean number of copulation attempts during the whole life of *P. calceolariae* males was 7.0 ± 0.8 (mean ± SE), leading to 6.5 ± 0.7 fertilized females (n = 30).

3.2. Flight Tunnel Experiments

When exposed to an array of rubber septa baited with synthetic pheromone, *P. calceolariae* males were attracted by upwind lures. In trial 1, with all 16 lures loaded with the same quantity of pheromone (1 µg), males did not show preferences for a specific rubber septum, visiting different rows of septa in a random way ($\chi^2 = 0.9422$, *d.f.* = 3, $p = 0.815$) (Figure 3).

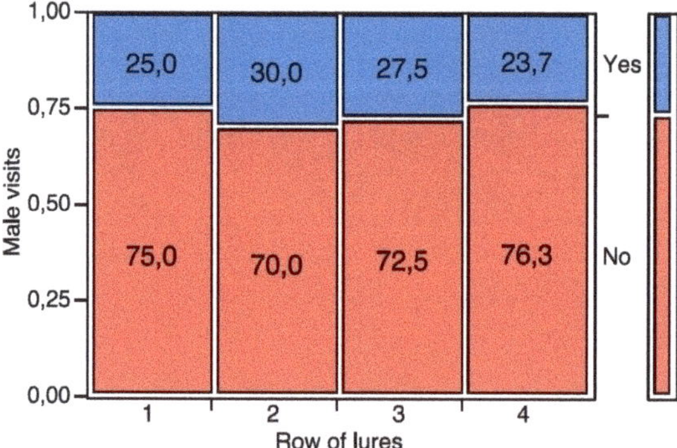

Figure 3. Contingency analysis between the rows of pheromone-baited lures (1 µg per lure, 16 lures, four per row across the wind) tested in the flight tunnel and the visits received by *P. calceolariae* males. The bar on the right represents the relative abundance (%) of male visits to a given row of lures over the total number of tested individuals. The number within each box shows the percentage of males visiting a given row of pheromone-baited lures. Yes = the male visited a given row of lures. No = the male did not visit a given row of lures. No significant differences were found among male visits to the different row of lures ($p > 0.05$).

In trial 2, where we used an array with four different sex pheromone loadings, a significant effect of the tested loading was present (χ^2 = 26.805, *d.f.* = 3, *p* < 0.0001). Most of the males showed positive chemotaxis towards the septa loaded with the highest quantities of pheromone, 10 and 30 µg (Figure 4). The highest number of visits per lure was achieved when testing 10 µg- and 30 µg-loaded septa (Figure 4); these septa showed significant differences over 3 µg- (Z = 2.677, *p* = 0.007; Z = 1.882, *p* = 0.05, respectively) and 1 µg-loaded lures (Z = 3.766, *p* = 0.001; Z = 4.790, *p* < 0.0001, respectively). Also, the number of male visits to 3 µg-loaded lures was significantly higher over 1 µg-loaded ones (Z = 2.426, *p* = 0.015) (Figure 4).

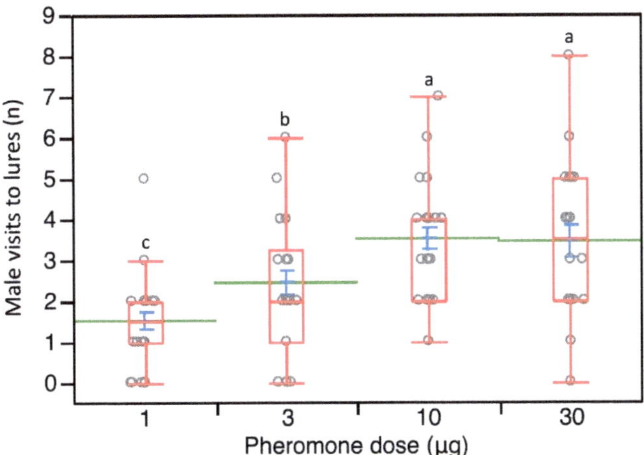

Figure 4. Distribution of *P. calceolariae* male visits to 16 lures organized in four rows with increasing loadings of sex pheromone going upwind (1, 3, 10, and 30 µg). Box plots indicate the median (line) within each box and the range of dispersion (lower and upper quartiles and outliers). Green lines and blue T-bars indicate means and standard errors, respectively. Above each box plot, different letters indicate significant differences (Steel-Dwass test, *p* < 0.05).

In trial 3, MD was mimicked at the flight tunnel scale by baiting all lures with 1 or 30 µg of *P. calceolariae* pheromone compared with unloaded controls. At 1 µg loading, the males successfully detected a female in the middle of the grid in just three cases out of 60 (5%) (Figure 5). No males attempted courtship or copulation. By increasing the loading of each bait to 30 µg, 12 out of 60 males detected the female (20%); none tried to fertilize her. In absence of the pheromone, 27 out of 60 males (45%) successfully detected the female and then mated and fertilized her.

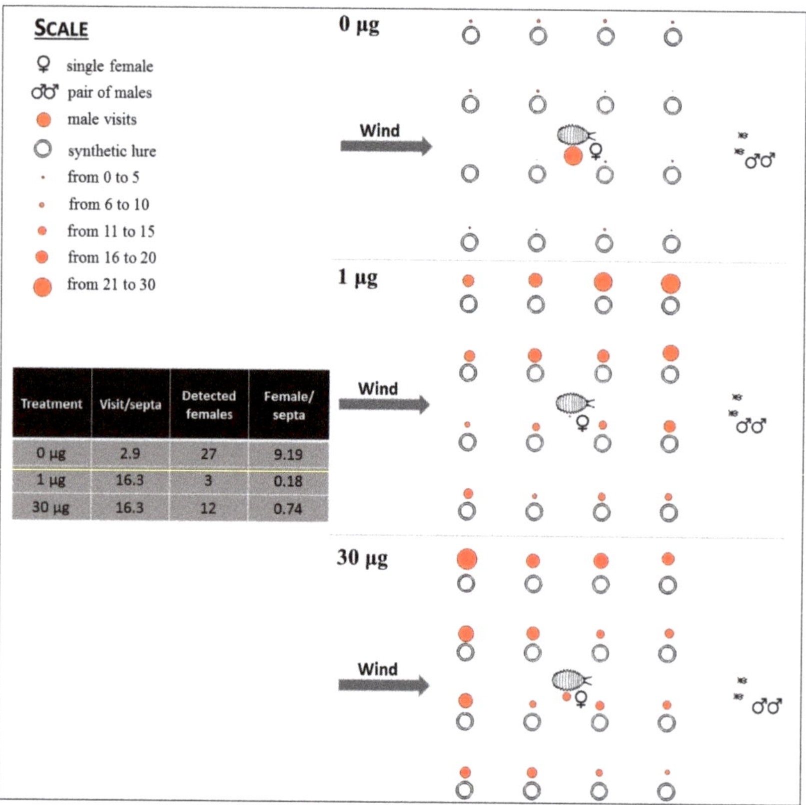

Treatment	Visit/septa	Detected females	Female/septa
0 µg	2.9	27	9.19
1 µg	16.3	3	0.18
30 µg	16.3	12	0.74

Figure 5. Distribution of visits of *P. calceolariae* males to the central female and to the 16 surrounding lures loaded with 0, 1, and 30 µg of synthetic sex pheromone. The diameter of spots represents the number of visits to each lure during 30 replicates, with two males released in each.

A significant effect of the sex pheromone loadings on successful mate location was detected ($x^2 = 22.317$, *d.f.* = 2, $p < 0.0001$) (Figure 6a). Successful mate detections were significantly reduced at both 1 and 30 µg per lure over the control (means: 0.10 and 0.40 versus 0.90 events; Z = 4.539, $p < 0.0001$ and Z = 2.647, $p = 0.008$, respectively). Successful mate detection events at 30 µg per lure were higher compared to those achieved at 1 µg loading per lure (Z = 2.436, $p = 0.015$) (Figure 6a).

Furthermore, the time spent by *P. calceolariae* males on the sex pheromone lures was significantly affected by the tested pheromone loading ($x^2 = 235.964$; *d.f.* = 2; $p < 0.0001$) (Figure 6b).

Indeed, the time spent on the lure was significantly higher at 1 µg of pheromone (421.38 ± 41.29 s, mean ± SE) when compared to the control (28.17 ± 5.70 s) (Z = −9.599; *d.f.* = 1; $p < 0.0001$) as well as to when 30 µg was used (56.39 ± 7.05 s) (Z = −14.202, $p < 0.0001$). Also, the time spent on the lure was significantly higher at 30 µg of sex pheromone over the control (Z = −2.813, $p = 0.0049$) (Figure 6b). In the control, the mean duration of copulation attempts was 1663.33 ± 262.73 s, while males did not attempt copulation when exposed to either sex pheromone loading.

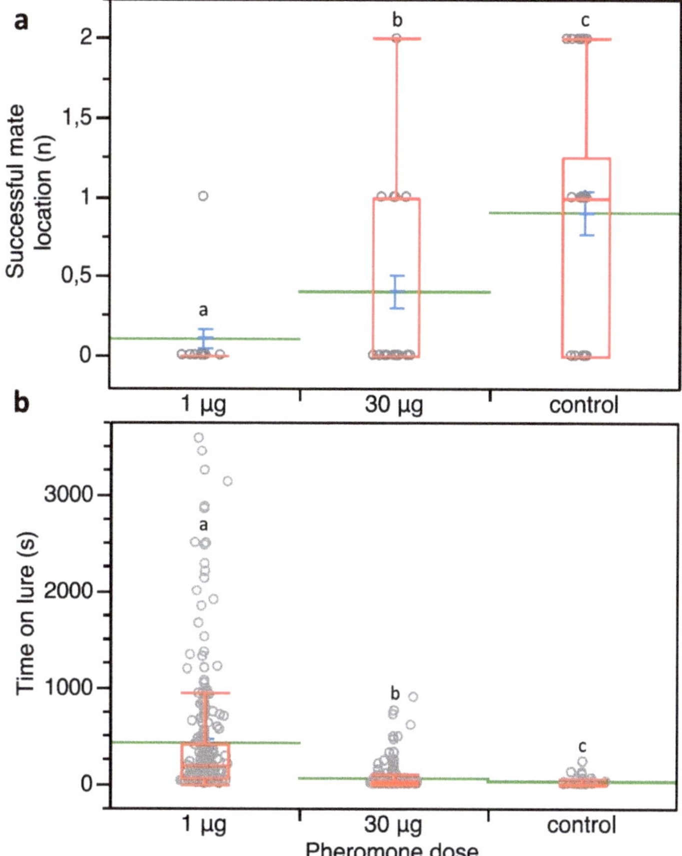

Figure 6. (**a**) Number of successful mate locations by *P. calceolariae* males surrounded by two different loadings of sex pheromone on 16 rubber septa at 2 cm spacing. (**b**) Time spent by *P. calceolariae* males on lures loaded with 1 or 30 µg of synthetic sex pheromone and control ones. Box plots indicate the median (line) within each box and the range of dispersion (lower and upper quartiles and outliers). Green lines and blue T-bars indicate means and standard errors, respectively. Above each box plot, different letters indicate significant differences (Steel-Dwass test, $p < 0.05$).

4. Discussion

In mealybugs, the number of nymphs in a population depends on the number of mated females, which in turn depends on the number of males able to fertilize them. Multiple mating is common in *P. longispinus*, *P. ficus*, *P. viburni* [35], and *P. calceolariae* [32]. In the present research, we investigated the reproductive potential of *P. calceolariae* males in the context of simulated reduced-scale mating disruption.

Our research provides new insights into the number of copulations occurring during the male lifetime, their duration, and the time elapsing between copulations. The life expectancy and consequently the potential mating capability of *P. calceolariae* males was lower than that reported by Silva et al. [32]. The different results are probably due to various factors, i.e., number of males tested (100 against 30), RH conditions (50–70% in Silva et al. against 40% in our trial), and male life span (5–6 days in Silva et al. against 3–4 days in our case). In the days following the first mating, the number of copulation events strongly decreased. Several authors [29,35] have suggested that mated males need less time to successfully conclude forthcoming matings, putatively because they have gained experience after the first copulation event. Males may acquire more experience about how to approach a female, but, since they do not feed, they could potentially have less energy to devote to mating approaches [44]. By contrast, we did not find a significant difference in the daily mating duration nor in the time between copulations.

The multiple mating capacity of this mealybug potentially represents a limit to pheromone-based control techniques. Indeed, in mass trapping and lure and kill techniques, even if we eliminate a high number of males, the remaining ones would be able to fertilize several females, allowing the perpetuation and increase of the population [26]. Studies carried out in citrus orchards infested by *P. citri* have demonstrated that mass-trapping using pheromone-baited traps can produce a significant reduction of males, but this was not enough to decrease infestation on the plants [45]. Considering other examples on extremely different insect species, such as *Calliphora vicina* Robineau-Desvoidy (Diptera, Calliphoridae), mass trapping has proved to be an effective technique, probably because the traps were baited with an attractant that allowed for the catching of males (20%) but especially many females (80%) [46]. Moreover, the costs related to attractive products and the labour required for the installation of numerous traps can be very high [47]. Another potential control system of this pest could be MD. Indeed, the latter has been found effective for the management of the vine mealybug, *P. ficus* [16,20–22,36,37]. Earlier tests revealed that habituation to the sex pheromone did not occur in *P. calceolariae* [31]. Our flight tunnel data pointed to the potential use of MD against this mealybug species by analyzing the male searching behavior towards females in the presence of multiple sources of synthetic sex pheromone.

The flight tunnel has been used to study several facets of insect behavior and in chemical ecology [48–51]. The flight tunnel can also be used to assess MD in Lepidoptera [52,53]. In this framework, our study used a flight tunnel to investigate MD in Hemiptera. We reproduced, on a small scale, conditions that are essentially comparable with mating disruption in the field, where the male emerges at a close range to the females in the colony. Results were promising from the perspective of reducing mating in the presence of an overabundance of *P. calceolariae* sex pheromone. In fact, as demonstrated by Lentini et al. [54] on *P. ficus*, even a mating delay has positive effects if mating is delayed >7 days. Field trapping trials by Unelius et al. [55] have revealed that synthetic lures are much more attractive than calling females to *P. calceolariae* males. Our results from flight tunnel trial 2 support this observation, showing that searching males prefer to arrest on the lure loaded with the highest loading of pheromone.

The MD tests using synthetic sex pheromone have shown success in obscuring the presence of the female compared to the control, in which the female was detected and fertilized by males. Testing both sex pheromone loadings, no females were approached or fertilized, while in the control group males performed prolonged copulation attempts and all the females were fertilized. Under MD conditions, a few males found the females and this could raise some doubts about performance in the field, since spacings were very close (2 cm). However, after finding the female, the males changed direction, and thus the observed events may be due to male random movements. In the trial testing the effect of 30 µg of synthetic sex pheromone, more males detected the female, showing a significant difference with male performances in trials testing 1 µg as well as the negative control. Probably the greater disruption, generated by a high pheromone quantity in the air, caused this difference by stimulating the males to walk much more instead of dwelling on the lures. Indeed, they spent an average of 56 s on the lure. The above-cited increased male activity could boost the chance of finding a female. On the contrary, under MD with 1 µg, males spent about 420 s per lure, eight times more if compared to time spent in trials testing 30 µg. Furthermore, comparing the male copulation attempts, it was evident that in the control males spent a lot of time trying to approach females, while in presence of both pheromone loadings, males did not perform copulation attempts, spending all the time above lures or walking. This confirms the observations carried out by Silva et al. [29] on the mating behavior of this species. After finding the female during the courtship phase, the male explores the female body by drumming it with the antennae; in response, the consenting female raises her abdomen to accept mating. Starting from this assumption, we can suppose that the loadings were not subjected to copulation attempts because they were not recognized as females. Comparable results have been achieved on *Grapholita molesta* Busck (Lepidoptera, Tortricidae) males in presence of increasing loadings of the sex pheromone [56]. Further research is still needed to investigate the behavior of *P. calceolariae* in response to the increasing amount of pheromone to identify the most appropriate loading to be used for MD purposes. At the same time, it would be useful to increase knowledge on the use of potential biological control agents of this species [57] as already done for other mealybugs [58].

Our results on *P. calceolariae*, as well as the evidence from field tests outlining the effectiveness of MD on *P. ficus* in California, Italy, Israel, and Tunisia [20–22,36,37,59], constitute a basis for undertaking further investigations into the potential of MD or male removal for *P. calceolariae* management. While MD applied to another mealybug species has led to good results at a large scale, other techniques such as mass trapping and lure and kill have similar limitations [45,60,61] and remain to be investigated. The flight tunnel can represent a fundamental tool to assess the MD effectiveness at a small scale in a wider range of species beyond Lepidoptera, where it has been noted that very few mechanistic studies are usually undertaken [31]. This would provide rapid and relatively cheap preliminary results of MD efficacy on a given pest. Certainly, the field conditions are more complex than those in the flight tunnel, since a greater number of variables are involved in field trials, including immigration from upwind by crawlers, requiring very large plots and multiple measures of population size to demonstrate efficacy. Nevertheless, flight tunnel experiments can provide important information of the possible efficacy of MD, as well as assess the absence of habituation and identify the risk of overstimulating male searching from high pheromone loadings. Indeed, this may be a research topic to be considered in further MD research, involving evaluating whether MD formulations releasing plumes of pheromones, such as aerosol or sprayable technologies, might be more effective over hand-held dispensers releasing higher pheromone loadings per surface area unit. However, at present there does not seem to be any possible way to make the *P. calceolariae* pheromone in large amounts at a cost that would be competitive with insecticides. This is a major and potentially insurmountable problem that could easily prevent MD ever being used on a large scale for this insect and future research into field applications of the sex pheromone of this species should focus on male removal strategies [26] and monitoring.

Author Contributions: All authors designed the research; R.R. conducted the experiments; G.B., D.M.S., and R.R. analyzed the data; D.M.S. contributed laboratory materials and tools; all authors wrote and approved the manuscript.

Funding: This research received no external funding.

Acknowledgments: Three anonymous reviewers improved an earlier version of this manuscript. We would like to thank the Biosecurity Group for their kind assistance during the use of the flight tunnel employed in our experiments. We are grateful to Flore Mas (The New Zealand Institute for Plant & Food Protection Ltd., Lincoln) and Sabina Avosani (University of Trento, Italy) for their insightful comments on an earlier version of the manuscript.

Conflicts of Interest: The authors declare no conflict of interest.

References

1. Bostanian, N.J.; Vincent, C.; Isaacs, R. *Arthropod Management in Vineyards: Pests, Approaches, and Future Directions*; Springer: Dordrecht, The Netherlands, 2012; pp. 1–15.

2. Daane, K.M.; Almeida, R.P.; Bell, V.A.; Walker, J.T.; Botton, M.; Fallahzadeh, M.; Mani, M.; Miano, J.L.; Sforza, R.; Walton, V.M.; et al. Biology and management of mealybugs in vineyards. In *Arthropod Management in Vineyards*; Bostanian, N.J., Vincent, C., Isaacs, R., Eds.; Springer: Dordrecht, The Netherlands, 2012; pp. 271–307.

3. Tsai, C.W.; Rowhani, A.; Golino, D.A.; Daane, K.M.; Almeida, R.P.P. Mealybug transmission of grapevine leafroll viruses: An analysis of virus–vector specificity. *Phytopathology* **2010**, *100*, 830–834. [CrossRef]

4. Venkatesan, T.; Jalali, S.K.; Ramya, S.L.; Prathibha, M. Insecticide resistance and its management in mealybugs. In *Mealybugs and Their Management in Agricultural and Horticultural Crops*; Mani, M., Shivaraju, C., Eds.; Springer: New Delhi, India, 2016; pp. 223–229.

5. Charles, J.G.; Walker, J.T.S.; White, V. Resistance to chlorpyrifos in the mealybugs *Pseudococcus affinis* and *P. longispinus* in Hawkes Bay and Waikato pipfruit orchards. *Proc. N. Z. Plant Prot. Conf.* **1993**, *46*, 120–125.

6. Saddiq, B.; Shad, S.A.; Khan, H.A.A.; Aslam, M.; Ejaz, M.; Afzal, M.B.S. Resistance in the mealybug *Phenacoccus solenopsis* Tinsley (Homoptera: Pseudococcidae) in Pakistan to selected organophosphate and pyrethroid insecticides. *Crop Prot.* **2014**, *66*, 29–33. [CrossRef]

7. Ahmad, M.; Akhtar, S. Development of resistance to insecticides in the invasive mealybug *Phenacoccus solenopsis* (Hemiptera: Pseudococcidae) in Pakistan. *Crop Prot.* **2016**, *88*, 96–102. [CrossRef]

8. Grimes, E.W.; Cone, W.W. Control of the grape mealybug, *Pseudococcus maritimus* (Hom.: Pseudococcidae), on Concord grape in Washington. *J. Entomol. Soc. Br. Columbia* **1985**, *82*, 3–6.

9. Prabhaker, N.; Gispert, C.; Castle, S.J. Baseline susceptibility of *Planococcus ficus* (Hemiptera: Pseudococcidae) from California to select insecticides. *J. Econ. Entomol.* **2012**, *105*, 1392–1400. [CrossRef]

10. Anand, P.; Ayub, K. The effect of five insecticides on *Maconellicoccus hirsutus* (Green) (Homoptera: Pseudococcidae) and its natural enemies *Anagyrus kamali* Moursi (Hymenoptera: Encyrtidae), and *Cryptolaemus montrouzieri* Mulsant and *Scymnus coccivora* Aiyar (Coleoptera: Coccinellidae). *Int. Pest Control* **2000**, *42*, 170–173.

11. Zettler, J.L.; Follett, P.A.; Gill, R.F. Susceptibility of *Maconellicoccus hirsutus* (Homoptera: Pseudococcidae) to methyl bromide. *J. Econ. Entomol.* **2002**, *95*, 1169–1173. [CrossRef]

12. Ioriatti, C.; Lucchi, A. Semiochemical strategies for tortricid moth control in apple orchards and vineyards in Italy. *J. Chem. Ecol.* **2016**, *42*, 571–583. [CrossRef]

13. Pertot, I.; Caffi, T.; Mugnai, L.; Hoffmann, C.; Grando, M.S.; Gary, C.; Lafond, D.; Duso, C.; Thiéry, D.; Mazzoni, V.; et al. A critical review of plant protection tools for reducing pesticide use on grapevine and new perspectives for implementation of IPM in viticulture. *Crop Prot.* **2017**, *97*, 70–84. [CrossRef]

14. Lucchi, A.; Benelli, G. Towards pesticide-free farming? Sharing needs and knowledge promotes Integrated Pest Management. *Environ. Sci. Pollut. Res.* **2018**, *25*, 13439–13445. [CrossRef]

15. Daane, K.M.; Bentley, W.J.; Millar, J.G.; Walton, V.M.; Cooper, M.L.; Biscay, P.; Yokota, G.Y. Integrated management of mealybugs in California vineyards. In Proceedings of the International Symposium on Grape Production and Processing, Baramati (Pune), Maharashtra, India, 6–11 February 2006; Adsule, P.G., Indu, S., Sawant, I.S., Shikhamany, S.D., Eds.; International Society for Horticultural Science: Madison, WI, USA, 2008; pp. 235–252. [CrossRef]

16. Mansour, R.; Belzunces, L.; Suma, P.; Zappalà, L.; Mazzeo, G.; Grissa-Lebdi, K.; Russo, A.; Biondi, A. Vine and citrus mealybug pest control based on synthetic chemicals. *A review. Agron. Sustain. Dev.* **2018**, *38*, 37. [CrossRef]

17. Carde, R.T. Principles of mating disruption. In *Behavior-Modifying Chemicals for Pest Management: Applications of Pheromones and Other Attractants*; Ridgway, R., Silverstein, R.M., Inscoe, M., Eds.; Marcel Dekker: New York, NY, USA, 1990; pp. 47–71.

18. Millar, J.G.; Daane, K.M.; McElfresh, J.S.; Moreira, J.A.; Bentley, W.J. Chemistry and applications of mealybug sex pheromones. In *Semiochemicals in Pest and Weed Control*; Petroski, R.J., Maria, R., Tellez, M.R., Behle, R.W., Eds.; ACS Publications: Washington, DC, USA, 2005; Volume 906, pp. 11–27.

19. Pherobase. Available online: http://www.pherobase.com (accessed on 21 July 2019).

20. Walton, V.M.; Daane, K.M.; Bentley, W.J.; Millar, J.G.; Larsen, T.E.; Malakar-kuenen, R. Pheromone-based mating disruption of *Planococcus ficus* (Hemiptera: Pseudococcidae) in California vineyards. *J. Econ. Entomol.* **2006**, *99*, 1280–1290. [CrossRef]

21. Cocco, A.; Muscas, E.; Mura, A.; Iodice, A.; Savino, F.; Lentini, A. Influence of mating disruption on the reproductive biology of the vine mealybug, *Planococcus ficus* (Hemiptera: Pseudococcidae), under field conditions. *Pest Manag. Sci.* **2018**, *74*, 2806–2816. [CrossRef]

22. Lucchi, A.; Suma, P.; Ladurner, E.; Iodice, A.; Savino, F.; Ricciardi, R.; Cosci, F.; Marchesini, E.; Conte, G.; Benelli, G. Managing the vine mealybug, *Planococcus ficus*, through pheromone-mediated mating disruption. *Environ. Sci. Pollut. Res.* **2019**, *26*, 10708–10718. [CrossRef]

23. Petersen, C.L.; Charles, J.G. Transmission of grapevine leafroll-associated closteroviruses by *Pseudococcus longispinus* and *P. calceolariae*. *Plant Pathol.* **1997**, *46*, 509–515. [CrossRef]

24. Bartlett, B.R. *Introduced Parasites and Predators of Arthropod Pests and Weeds: A World Review*; Agric Res. Service, Agriculture Handbook nr 480; U.S. Dep. Agric.: Washington DC, USA, 1978.

25. El-Sayed, A.M.; Unelius, C.R.; Twidle, A.; Mitchell, V.; Manning, L.A.; Cole, L.; Suckling, D.M.; Flores, M.F.; Zavievo, T.; Bergamann, J. Chrysanthemyl 2-acetoxy-3-methylbutanoate: The sex pheromone of the citrophilous mealybug, *Pseudococcus calceolariae*. *Tetrahedron Lett.* **2010**, *51*, 1075–1078. [CrossRef]

26. Sullivan, N.J.; Butler, R.C.; Salehi, L.; Twidle, A.M.; Baker, G.; Suckling, D.M. Deployment of the sex pheromone of *Pseudococcus calceolariae* (Hemiptera: Pseudococcidae) as a potential new tool for mass trapping in citrus in South Australia. *N. Z. Entomol.* **2019**, 1–12. [CrossRef]

27. Rotundo, G.; Tremblay, E. Osservazioni sull'attivita di volo dei maschi di *Pseudococcus calceolariae* (Mask.)(Homoptera Coccoidea). *Boll. Lab. Entomol. Agrar. Filippo Silvestri* **1976**, *33*, 108–112.

28. Rotundo, G.; Tremblay, E.; Papa, P. Short-range orientation of *Pseudococcus calceolariae* (Mask.) males (Homoptera Coccoidea) in a wind tunnel. *Boll. Lab. Entomol. Agrar. Filippo Silvestri* **1980**, *37*, 31–37.

29. Silva, E.B.; Branco, M.; Mendel, Z.; Franco, J.C. Mating behavior and performance in the two cosmopolitan mealybug species *Planococcus citri* and *Pseudococcus calceolariae* (Hemiptera: Pseudococcidae). *J. Insect Behav.* **2013**, *26*, 304–320. [CrossRef]

30. Silva, E.B.; Mendel, Z.; Franco, J.C. Can facultative parthenogenesis occur in biparental mealybug species? *Phytoparasitica* **2010**, *38*, 19–21. [CrossRef]

31. Suckling, D.M.; Stringer, L.D.; Jimenez-Perez, A.; Walter, G.H.; Sullivan, N.; El-Sayed, A.M. With or without pheromone habituation: Possible differences between insect orders? *Pest Manag. Sci.* **2018**, *74*, 1259–1264. [CrossRef]

32. Silva, E.B.; Mourato, C.; Branco, M.; Mendel, Z.; Franco, J.C. Biparental mealybugs may be more promiscuous than we thought. *Bull. Entomol. Res.* **2018**, 1–9. [CrossRef]

33. Foster, S.P.; Ayers, R.H. Multiple mating and its effects in the lightbrown apple moth, *Epiphyas postvittana* (Walker). *J. Insect Physiol.* **1996**, *42*, 657–667. [CrossRef]

34. Knight, A.L. Multiple mating of male and female codling moth (Lepidoptera: Tortricidae) in apple orchards treated with sex pheromone. *Environ. Entomol.* **2007**, *36*, 157–164. [CrossRef]

35. Waterworth, R.A.; Wright, I.M.; Millar, J.G. Reproductive biology of three cosmopolitan mealybug (Hemiptera: Pseudococcidae) species, *Pseudococcus longispinus*, *Pseudococcus viburni*, and *Planococcus ficus*. *Ann. Entomol. Soc. Am.* **2011**, *104*, 249–260. [CrossRef]

36. Sharon, R.; Zahavi, T.; Sokolsky, T.; Sofer-Arad, C.; Tomer, M.; Kedoshim, R.; Harari, A.R. Mating disruption method against the vine mealybug, *Planococcus ficus*: Effect of sequential treatment on infested vines. *Entomol. Exp. Appl.* **2016**, *161*, 65–69. [CrossRef]

37. Cocco, A.; Lentini, A.; Serra, G. Mating disruption of *Planococcus ficus* (Hemiptera: Pseudococcidae) in vineyards using reservoir pheromone dispensers. *J. Insect Sci.* **2014**, *14*, 144. [CrossRef]

38. Silva, E.B.; Mouco, J.; Antunes, R.; Mendel, Z.; Franco, J.C. Mate location and sexual maturity of adult male mealybugs: Narrow window of opportunity in a short lifetime. *IOBC Wrps Bull.* **2009**, *41*, 3–9.

39. Mendel, Z.; Protasov, A.; Jasrotia, P.; Silva, E.B. Sexual maturation and aging of adult male mealybug (Hemiptera: Pseudococcidae). *Bull. Entomol. Res.* **2012**, *102*, 385–394. [CrossRef]

40. El-Sayed, A.M.; Gödde, J.; Witzgall, P.; Arn, H. Characterization of pheromone blend for grapevine moth, Lobesia botrana by using flight track recording. *J. Chem. Ecol.* **1999**, *25*, 389–400. [CrossRef]

41. Benelli, G.; Stefanini, C.; Giunti, G.; Geri, S.; Messing, R.H.; Canale, A. Associative learning for danger avoidance nullifies innate positive chemotaxis to host olfactory stimuli in a parasitic wasp. *Naturwissenschaften* **2014**, *10*, 753–757. [CrossRef]

42. SAS Institute. *JMP 11*; SAS Institute Inc.: Cary, NC, USA, 2013.

43. Benelli, G.; Desneux, N.; Romano, D.; Conte, G.; Messing, R.H.; Canale, A. Contest experience enhances aggressive behaviour in a fly: When losers learn to win. *Sci. Rep.* **2015**, *5*, 9347. [CrossRef]

44. Franco, J.C.; Zada, A.; Mendel, Z. Novel approaches for the management of mealybug pests. In *Biorational Control of Arthropod Pests*; Ishaaya, I., Horowitz, A.R., Eds.; Springer: Dordrecht, The Netherlands, 2009; pp. 233–278.

45. Franco, J.C.; Gross, S.; Silva, E.B.; Suma, P.; Russo, A.; Mendel, Z. Is mass-trapping a feasible management tactic of the citrus mealybug in citrus orchard? *Anais Inst. Super. Agron.* **2003**, *49*, 353–367.

46. Aak, A.; Birkemoe, T.; Knudsen, G.K. Efficient mass trapping: Catching the pest, *Calliphora vicina* (Diptera, Calliphoridae), of Norwegian stockfish production. *J. Chem. Ecol.* **2011**, *37*, 924–931. [CrossRef]

47. Navarro-Llopis, V.; Alfaro, F.; Domínguez, J.; Sanchis, J.; Primo, J. Evaluation of traps and lures for mass trapping of Mediterranean fruit fly in citrus groves. *J. Econ. Entomol.* **2008**, *101*, 126–131. [CrossRef]
48. Visser, J.H.; Avé, D.A. General green leaf volatiles in the olfactory orientation of the Colorado beetle, *Leptinotarsa decemlineata. Entomol. Exp. Appl.* **1978**, *24*, 738–749. [CrossRef]
49. Elzen, G.W.; Williams, H.J.; Vinson, S.B. Wind tunnel flight responses by hymenopterous parasitoid *Campoletis sonorensis* to cotton cultivars and lines. *Entomol. Exp. Appl.* **1986**, *42*, 285–289. [CrossRef]
50. Bruce, T.J.A.; Wadhams, L.J.; Woodcock, C.M. Insect host location: A volatile situation. *Trends Plant Sci.* **2005**, *10*, 269–274. [CrossRef]
51. Miller, J.R.; Roelofs, W.L. Sustained-flight tunnel for measuring insect responses to wind-borne sex pheromones. *J. Chem. Ecol.* **1978**, *4*, 187–198. [CrossRef]
52. Sanders, C.J. Disruption of male spruce budworm orientation to calling females in a wind tunnel by synthetic pheromone. *J. Chem. Ecol.* **1982**, *8*, 493–506. [CrossRef]
53. Sanders, C.J. Effects of prolonged exposure to different concentrations of synthetic pheromone on mating disruption of spruce budworm moths in a wind tunnel. *Can. Entomol.* **1996**, *128*, 57–66. [CrossRef]
54. Lentini, A.; Mura, A.; Muscas, E.; Nuvoli, M.T.; Cocco, A. Effects of delayed mating on the reproductive biology of the vine mealybug, *Planococcus ficus* (Hemiptera: Pseudococcidae). *Bull. Entomol. Res.* **2018**, *108*, 263–270. [CrossRef]
55. Unelius, C.R.; El-Sayed, A.M.; Twidle, A.; Bunn, B.; Zavievo, T.; Flores, M.F.; Bell, V.; Bergmann, J. The absolute configuration of the sex pheromone of the citrophilous mealybug, *Pseudococcus calceolariae. J. Chem. Ecol.* **2011**, *37*, 166–172. [CrossRef]
56. Figueredo, A.J.; Baker, T.C. Reduction of the response to sex pheromone in the oriental fruit moth, *Grapholita molesta* (Lepidoptera: Tortricidae) following successive pheromonal exposures. *J. Insect Behav.* **1992**, *5*, 347–363. [CrossRef]
57. Laudonia, S.; Viggiani, G. Natural enemies of the citrophilus mealybug (*Pseudococcus calceolariae* Mask.). *Boll. Lab. Entomol. Agrar. Filippo Silvestri* **1986**, *43*, 167–171.
58. Afifi, A.I.; Arnaouty, S.A.E.; Attia, A.R.; Alla, A.E.M.A. Biological Control of Citrus Mealybug, *Planococcus citri* (Risso.) using Coccinellid Predator, *Cryptolaemus montrouzieri* Muls. *Pak. J. Biol. Sci.* **2010**, *13*, 216–222. [CrossRef]
59. Mansour, R.; Grissa-Lebdi, K.; Khemakhem, M.; Chaari, I.; Trabelsi, I.; Sabri, A.; Marti, S. Pheromone-mediated mating disruption of *Planococcus ficus* (Hemiptera: Pseudococcideae) in Tunisian vineyards: Effect on insect population dynamics. *Biologia* **2017**, *72*, 333–341. [CrossRef]
60. El-Sayed, A.M.; Suckling, D.M.; Wearing, C.H.; Byers, J.A. Potential of mass trapping for long-term pest management and eradication of invasive species. *J. Econ. Entomol.* **2006**, *99*, 1550–1564. [CrossRef]
61. El-Sayed, A.M.; Suckling, D.M.; Byers, J.A.; Jang, E.B.; Wearing, C.H. Potential of "Lure and Kill" in long-term pest management and eradication of invasive species. *J. Econ. Entomol.* **2009**, *102*, 815–835. [CrossRef]

Article

Mating Disruption of *Pseudococcus calceolariae* (Maskell) (Hemiptera, Pseudococcidae) in Fruit Crops

Carolina Ballesteros [1], Alda Romero [1], María Colomba Castro [1], Sofía Miranda [1], Jan Bergmann [2] and Tania Zaviezo [1,*]

1   Facultad de Agronomía e Ingeniería Forestal, Pontificia Universidad Católica de Chile, Avda. Vicuña Mackenna 4860, Macul, Santiago 7820436, Chile; cballestero@uc.cl (C.B.); aromeroga@uc.cl (A.R.); mccastro3@uc.cl (M.C.C.); smiranda1@uc.cl (S.M.)
2   Instituto de Química, Pontificia Universidad Católica de Valparaíso, Avda. Universidad 330, Curauma, Valparaíso 2340000, Chile; jan.bergmann@pucv.cl
*   Correspondence: tzaviezo@uc.cl

Simple Summary: The citrophilous mealybug is an economically important pest that is mainly controlled using insecticides, not always successfully, and with unintended negative environmental side effects. In our research, we tested a specific and sustainable control tool using the mealybug sex pheromone. Mating disruption is a technique that aims to reduce mating between males and females by inundating the area with the synthetic sex pheromone of the species, thereby reducing reproduction and consequently populations over time and damage. For this purpose, the mealybug pheromone, incorporated into a polymeric substance for its release, was applied in a tangerine and an apple orchard, in two seasons (2017/2018 and 2019/2020). In all seasons, a reduction in the males catches in traps after deploying pheromone was observed, which would indicate a decrease in the probability of successful mating compared to control plots. The duration of this effect was around one year. Mealybug abundance on trees was extremely low throughout the trials, so it was not possible to observe a reduction of populations or damage. This research shows that the use of this pheromone-based technique has good potential for controlling the citrophilous mealybug, with the advantage of being environmentally friendly and non-toxic.

Citation: Ballesteros, C.; Romero, A.; Castro, M.C.; Miranda, S.; Bergmann, J.; Zaviezo, T. Mating Disruption of *Pseudococcus calceolariae* (Maskell) (Hemiptera, Pseudococcidae) in Fruit Crops. *Insects* **2021**, *12*, 343. https://doi.org/10.3390/insects12040343

Academic Editors: Giovanni Benelli and Andrea Lucchi

Received: 15 March 2021
Accepted: 9 April 2021
Published: 13 April 2021

Publisher's Note: MDPI stays neutral with regard to jurisdictional claims in published maps and institutional affiliations.

Abstract: *Pseudococcus calceolariae*, the citrophilous mealybug, is a species of economic importance. Mating disruption (MD) is a potential control tool. During 2017–2020, trials were conducted to evaluate the potential of *P. calceolariae* MD in an apple and a tangerine orchard. Two pheromone doses, 6.32 g/ha (2017–2018) and 9.45 g/ha (2019–2020), were tested. The intermediate season (2018–2019) was evaluated without pheromone renewal to study the persistence of the pheromone effect. Male captures in pheromone traps, mealybug population/plant, percentage of infested fruit at harvest and mating disruption index (MDI) were recorded regularly. In both orchards, in the first season, male captures were significantly lower in MD plots compared to control plots, with an MDI > 94% in the first month after pheromone deployment. During the second season, significantly lower male captures in MD plots were still observed, with an average MDI of 80%. At the third season, male captures were again significant lower in MD than control plots shortly after pheromone applications. In both orchards, population by visual inspection and infested fruits were very low, without differences between MD and control plots. These results show the potential use of mating disruption for the control of *P. calceolariae*.

Keywords: citrophilous mealybug; IPM; semiochemicals; sex pheromones; sustainable pest control

1. Introduction

Pheromones and other semiochemicals have been used in pest management worldwide for more than 50 years, becoming an important tool in the development of sustainable management strategies for agriculture and forest pests [1]. Their use in pest management can be for detection, monitoring or control [1–5]. Their use for pest control may be through mass trapping, "attract and kill", or mating disruption (MD). MD has been one of the most successfully used strategies for controlling various pests, with over 800,000 hectares treated worldwide [1–3,6]. MD has been mainly used for the control of Lepidoptera species attacking vegetables, orchards and forests [7], such as codling moth (*Cydia pomonella* L.), grapevine moth (*Lobesia botrana* Denis & Schiffermüller), the plum fruit moth (*Grapholita funebrana* Treitschke) [8] and gypsy moth (*Lymantria dispar* L.) [1,2,6,9]. Successful cases of mating disruption have also been reported in other insect orders such as the Oriental beetle (*Anomala orientalis* (Waterhouse)) and *Prionus californicus* Motschulsky, both Coleoptera [2,10–12]. Only two commercial formulations of MD have been developed so far for Hemiptera, one against the California red scale, *Aonidiella aurantii* (Maskell) (Diaspididae), the other for the vine mealybug, *Planococcus ficus* Signoret (Pseudococcidae) [3,13–15].

Although the first pheromones for Pseudococcidae were identified in the early 1980s from *Pseudococcus comstocki* (Kuwana) [16] and *Planococcus citri* (Risso) [17], the identification and synthesis of pheromones of new species resumed only from 2001 onwards, summing to date 21 species [17–32]. The most studied mealybug species worldwide has been *Pl. ficus*, given its great economic importance as a key pest in several crops, particularly grapes. Its sex pheromone was first described by Hinkens et al. [18], and later monitoring methods [33,34] and mating disruption studies were developed [15,34–38]; it is the first mealybug species for which the successful use of MD has been reported. An additional case of successful application of MD for a mealybug species has been reported for *Planococcus kraunhiae* (Kuwana) in Japanese persimmon orchards [39].

Pseudococcus calceolariae (Maskell), commonly known as citrophilous mealybug, is distributed in Australia, New Zealand, USA, South Africa, and several countries in South America and south-eastern Europe [40,41]. This mealybug is a polyphagous species attacking several fruit crops such as orange, lemon, apple, pear, and avocado. Its main economic impact is due to its quarantine status for many markets. Usually synthetic insecticides are used for its control, although with low effectiveness due to its cryptic habit [38,42]. The sex pheromone of *P. calceolariae* was identified as (1*R*,3*R*)-chrysanthemyl (*R*)-2-acetoxy-3-methylbutanoate [27,28,43]. The potential use of the *P. calceolariae* pheromone for monitoring its presence and abundance in fruit crops has been reported [43,44]. Recently, Sullivan et al. [41] studied the potential use of the pheromone for the control of *P. calceolariae* through a mass trapping approach, showing a 90% decrease in male catches in treated plots. The objectives of the present study were to determine the potential for using *P. calceolariae* pheromone for control through a mating disruption (MD) approach in fruit orchards and to study the persistence of the disruption effect under field conditions.

2. Materials and Methods

2.1. Experimental Sites and Treatment Applications

Field trials were conducted in two commercial orchards in central Chile: a conventionally managed tangerine orchard (6 years-old, cv. Orri, planted at 7 × 2 m) located near Pomaire ($-33.64°$ S, $-71.13°$ W) in the Metropolitan Region, and an organic apple orchard (10 years-old, cv. Fuji, planted at 2.7 × 2 m) near San Fernando ($-34.59°$ S, $-70.98°$ W) in the O'Higgins Region. Both orchards were infested with *P. calceolariae*. In September 2017, 10 plots of 0.1 ha each (experimental unit), separated by at least 50 m, were selected in each orchard. Five plots of these units were randomly selected to receive the mating disruption pheromone treatment (MD) and the other half were used as controls (Control). An isomeric mixture of chrysanthemyl 2-acetoxy-3-methylbutanoate was synthesized as described previously in El-Sayed et al. [27]. The proportion of the active isomer (*R, R, R*) in the mix was approximately 15%. Previous studies have shown that the non-natural isomers do not interfere with male attraction [43]. On 19–20 December 2017 the pheromone was applied in the tangerine and apple orchards using SPLAT® (Specialized Pheromone & Lure Application Technology; ISCA Technologies, Riverside, CA, USA) at a dose of

6.32 g/ha of isomeric mixture. This dose was based on a small-scale study that we carried out in 2015. SPLAT® is an emulsified microcrystalline wax matrix allowing controlled release of formulated volatile compounds [45] that has been successfully used for MD in lepidopteran and coleopteran pests [45–50]. It can be applied manually or mechanically to the crop, it is biodegradable and has a low production cost [45]. This matrix (1 g approx.) was placed on a piece of cardboard of 20 cm^2 that was hung on trees, with a density of 750 dispensers / ha (i.e., 75 in the 0.1 ha plots). Control plots received only SPLAT®.

To determine mating disruption effect over time, treatments were not reapplied in the 2018–2019 season, but dispensers were left hanging on the trees. In September 2019, treatments were deployed again in the same plots; three plots were relocated in the apple orchard because of very low populations. The pheromone dose was increased to 9.45 g/ha in this field season because it was deployed earlier in the season, and also because similar studies with *Pl. ficus* have used higher doses. Moreover, in this season the SPLAT® application was made directly to the tree branches in tangerines, to avoid the cardboard being broken and/or removed due to pruning.

2.2. Male Captures in Pheromone Traps

To determine if a mating disruption effect was occurring, males were monitored by placing one Delta sticky trap in the center of each 0.1 ha plot (Feromonas Chile LTDA, Santiago, Chile) with rubber septum lures loaded with 50 µg (isomeric mixture) of *P. calceolariae* sex pheromone. Traps were deployed in September 2017, 3 months before the first pheromone application, and from then on continuously monitored until March 2020 (30 months, 50–52 monitoring occasions). This comprised three entire apple productive seasons (September–March in the years 2017/2018; 2018/2019 and 2019/2020) and two and a partial third for tangerine (September–August 2017/2018 and 2018/2019, and September–March 2019/2020). The sticky floors of the traps were replaced every two weeks from August to April (spring-summer) of each year and monthly from May to July (autumn-winter) of each year. Lures were renewed in July 2018 and August 2019 in both orchards. Trapped males were counted using a stereomicroscope, and counts were transformed to males per trap per day.

Additionally, a mating disruption index (MDI) [35] was calculated after the pheromone application for each monitoring date when controls captured at least 0.4 males $\times$ trap^{-1} $\times$ day^{-1}. This index indicates the percentage reduction of male captures in pheromone traps in disruption plots (MCD, mean of 5 plots) in relation to control plots (MCC, mean of 5 plots), using the following equation:

$$\text{MDI (\%)} = [1 - (\text{MCD/MCC})] \times 100 \tag{1}$$

2.3. Mealybugs Population and Damage

To assess the abundance of *P. calceolariae* population on the plants, every two weeks (spring-summer) or monthly (autumn-winter), twelve plants were randomly selected in each plot and inspected for 3 min counting nymphs, ovisacs, and adult females present.

In season 2017–2018, the damage, i.e., presence of mealybugs on fruits at harvest (April for apples and May for tangerines), was estimated by inspecting 120 fruit per plot (total = 1200 per field), and counting the number of mealybugs present (ovisacs, nymphs, and adult females) and the percentage of infested fruits. In the season 2019–2020, damage at harvest was evaluated in the case of apples (April 2020) by inspecting 50 apples per plot (500 total for the field). Apples were cut and checked under a stereomicroscope observing the presence of individuals inside the fruit (calyx and peduncle).

2.4. Statistical Analysis

Male counts (males $\times$ trap^{-1} $\times$ day^{-1}) were analyzed for different periods using Generalized Linear Models (GLM) with Poisson error distribution and log link, with treatment and date as independent variables. When more than two means were compared, Tukey HSD test were used. For periods or dates with very low density and many zero counts, we used non-parametric Kruskal-Wallis ANOVA. The time periods compared were before pheromone application in the first season (September to December 2017), first season after treatment applications (January–April 2018), first season winter (May–July 2018), second growing season (September 2018–April 2019 for apples and only September–December 2018 for tangerines due to the very low counts after that), second season winter (May–July 2019), third growing season after 2nd pheromone application (September 2019–March 2020). We also compared male captures at the beginning of the third growing season, after renewing the dispensers of traps but before the second pheromone application, using GLM (Poisson error distribution and log link) with treatment as the independent variable.

Mealybug population density on plants (mealybugs $\times$ plant^{-1}) per season were compared by non-parametric Kruskal–Wallis, as populations were always very low or zero on many occasions. Mealybug damage was analyzed only for apples in the 2017–2018 season, using GLM with binomial error distribution and logit link for fruit infestation and Poisson error distribution and log link for insects per fruit. For the 2019–2020 season in apples and for tangerines only a few insects were found on fruits, so no statistical analyses were carried out. Means presented in figures and text are adjusted means of untransformed data $\pm$ 1 standard error. Statistical analyses were done with Infostat version 2019 [51] and R 4.0.4 [52].

3. Results

3.1. Mating Disruption Effect on Males Captures and Persistence in the Apple Orchard

In the first season, before deploying the treatment, male captures were low and increasing in time (Figure 1A). For this period (September–December 2017, 6 monitoring dates) there was no effect of treatment ($F_{1,48} = 0.45$, $p = 0.507$), with male captures before treatment being similar in the plots that later received the pheromone applications and the control plots (mean captures MD = 4.79 $\pm$ 0.52, control = 4.30 $\pm$ 0.52 males $\times$ trap^{-1} $\times$ day^{-1}). There was a date effect ($F_{5,48} = 108.02$, $p < 0.0001$), with the largest male captures right before the treatment applications (Figure 1A). No effect of treatment $\times$ date was observed ($F_{5,48} = 1.43$, $p = 0.232$).

In the rest of the first season (January–April 2018), after deploying the treatments, male captures were lower (Figure 1A). For this period (7 monitoring dates) there was an effect of treatment ($F_{1,66} = 45.76$, $p < 0.0001$), but no effect of date ($F_{6,66} = 0.43$, $p = 0.856$) or treatment $\times$ date ($F_{6,66} = 0.29$, $p = 0.938$). Male captures were more than tenfold larger in control than in MD plots, with the latter being very low ("trap shutdown") (mean January–April: MD = 0.22 $\pm$ 0.1, control = 3.06 $\pm$ 0.3 males $\times$ trap^{-1} $\times$ day^{-1}). In this period, one month after the pheromone application, MDI was 97% and remained around 90% for the rest of the season (Figure 2A). In autumn–winter of the first season, similar results were observed (May–July 2018, 4 monitoring dates; Figure 1A), with an effect of treatment ($F_{1,32} = 5.04$, $p = 0.032$), but no effect of date ($F_{3,32} = 1.08$, $p = 0.372$) or treatment $\times$ date ($F_{3,32} = 0.24$, $p = 0.865$). Male captures were very low, but still sevenfold larger in control than in MD plots (mean May–July: MD = 0.10 $\pm$ 0.1, control = 0.70 $\pm$ 0.2 males $\times$ trap^{-1} $\times$ day^{-1}).

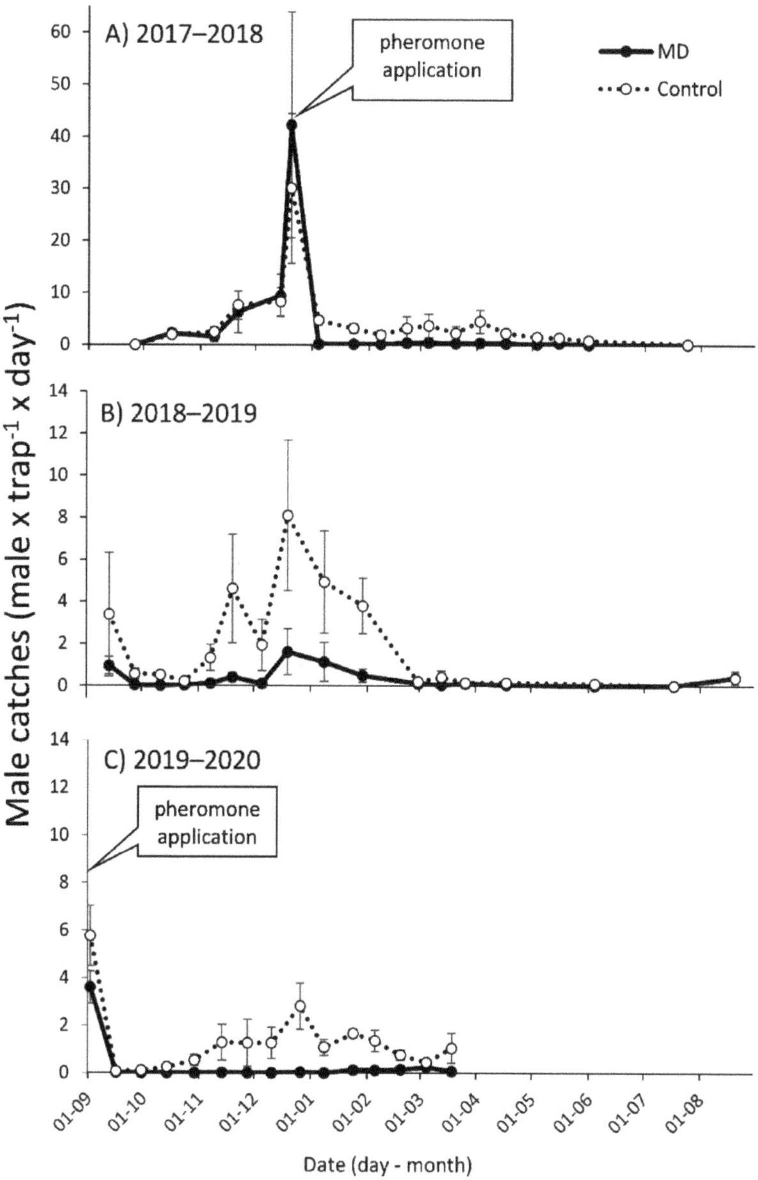

Figure 1. Season long adult male *P. calceolariae* (mean ± SEM) captures in pheromone baited traps in mating disruption (MD) and control plots in apple orchard from 1 September to 31 August of (**A**) 2017–2018, (**B**) 2018–2019 and (**C**) 2019–2020 seasons. Pheromone application dates: 19 December, 2017 and 29 August, 2019. Note that the scale for the Y axis in the first season differs from the rest.

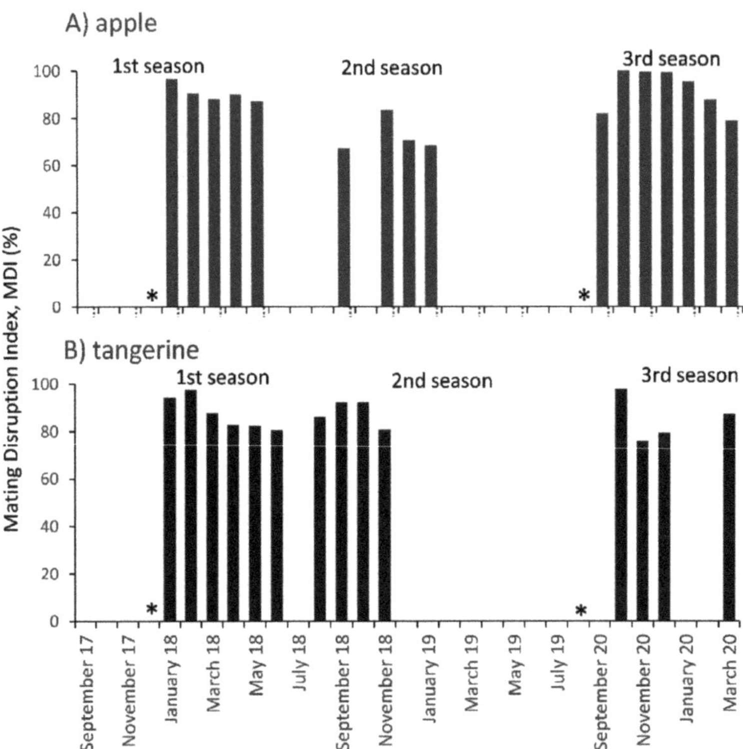

Figure 2. Mating Disruption Index (%), for the three seasons when male captures > 0.4 males × trap^{-1} × day^{-1} for the (**A**) apple and (**B**) tangerine orchards. * Date of pheromone application.

Although the pheromone was not renewed in the second season, there was a treatment effect (September 2018–April 2019, 14 monitoring dates, $F_{1,54}$ = 9.71, p = 0.003). Male captures in MD plots were still significantly lower than in control plots (mean September 2018–April 2019; MD = 0.16 ± 0.1, control 0.98 ± 0.2 males × trap^{-1} × day^{-1}, Figure 1B). There was also a date effect ($F_{13,54}$ = 3.18, p = 0.001) but not a treatment × date effect ($F_{13,54}$ = 0.21, p = 0.998). Male captures were larger in December, similar to the first season, and lower in October and April, although not significantly different (Figure 1B). The mean of MDI for this season was 73% (Figure 2A). During winter of the second season (June–July 2019, 2 monitoring dates) captures in both MD and control plots were very low and similar (mean captures: MD = 0.02 ± 0.1 and Control 0.04 ± 0.1 males × trap^{-1} × day^{-1}, Figure 1B). No effects of treatment, date or treatment × date were found (p > 0.8 for all).

At the end of August 2019, before the beginning of 2019–2020 season, after renewing the lures in the traps (20 August) but before the new pheromone applications (2 September), there was an increase in male captures, being similar for MD and control plots (MD = 3.60 ± 0.9, control = 5.77 ± 1.1 males × trap^{-1} × day^{-1}; $F_{1,8}$ = 2.47, p = 0.16; Figure 1C). Subsequently, after the second pheromone application in this third season, male captures in the MD plots were very low (0.0 ± 0.0 males × trap^{-1} × day^{-1}) while in the control plots, even when captures were also low (0.68 ± 0.1 males × trap^{-1} × day^{-1}), they followed the temporal dynamics observed in previous seasons, with a peak by the end of December (Figure 1C). Significant reduction of males captures in MD plots compared to control plots was detected for October and then from December through February (H > 4.36; p < 0.05 for all). Mean MDI for this period was 92% (Figure 2A).

3.2. Mating Disruption Effect on Males Captures and Persistence in the Tangerine Orchard

In the first season, before deploying the treatments, male captures were high but decreasing in time (Figure 3A). For this period (5 monitoring dates) there was an effect of treatment ($F_{1,39}$ = 33.65, p < 0.0001), date ($F_{4,39}$ = 26.16, p < 0.0001) and treatment × date ($F_{4,39}$ = 5.52, p = 0.001). Plots that would receive the control treatment had larger populations than those that would receive the MD (mean captures September–November 2017 MD = 18.5 ± 0.9; control = 26.8 ± 1.1 males × trap^{-1} × day^{-1}). Nevertheless, for the date previous to the treatment applications, male captures were lower and similar for both (MD = 9.4 ± 1.4; control = 10.9 ± 1.5 males × trap^{-1} × day^{-1}, LSD p > 0.05). Male captures significantly decreased in this period, with the least numbers in the last date.

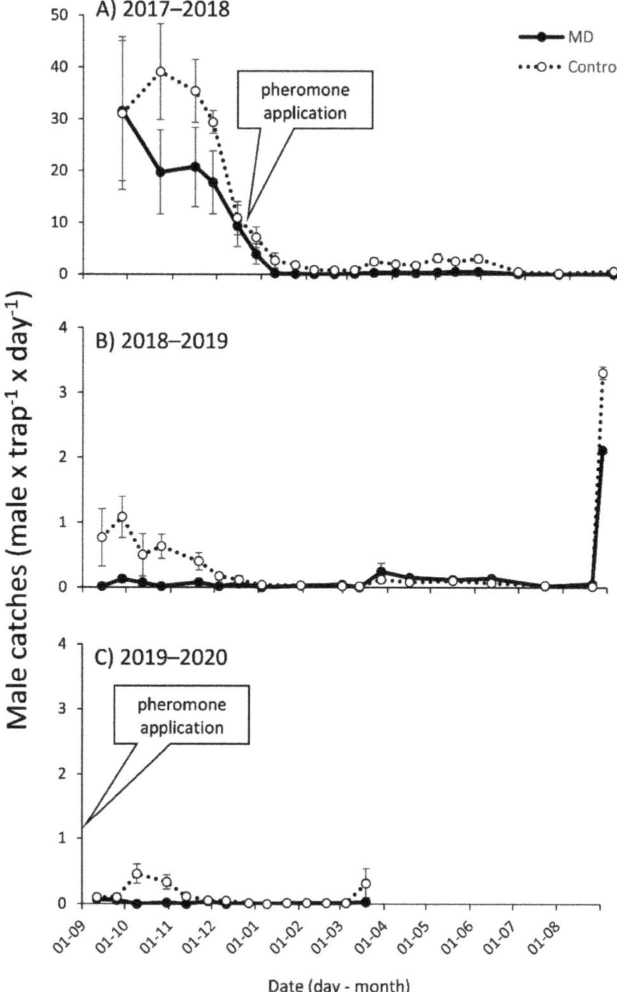

Figure 3. Season long adult male *P. calceolariae* (mean ± SEM) captures in pheromone baited traps in mating disruption (MD) and control plots in tangerine orchard from 1 September to 31 August of (**A**) 2017–2018, (**B**) 2018–2019 and (**C**) 2019–2020 seasons. Pheromone application dates: 20 December 2017 and 2 September 2019. Note that the scale for the Y axis in the first season differs from the rest.

During the rest of the first season (December 2017–April 2018), after deploying the treatments, male captures were lower (Figure 3A). For this period (9 monitoring dates) there was an effect of treatment ($F_{1,70}$ = 16.24, p = 0.0001), date ($F_{8,70}$ = 9.51, p < 0.0001) but not for treatment × date ($F_{8,70}$ = 1.41, p = 0.206). Male captures were tenfold larger in control than in MD plots, with the latter being very low ("trap shutdown") (mean December–April: MD = 0.17 ± 0.1, control = 1.78 ± 0.2 males × trap^{-1} × day^{-1}). Male captures were also significantly larger in the at the beginning of this period than the rest of the dates. In this period, one month after the pheromone application, MDI was 94% and then remained > 80% (Figure 2B). Similar results were observed for the fall-winter of the first season (May–August 2018, 6 monitoring dates; Figure 3A) with an effect of treatment ($F_{1,48}$ = 10.03, p = 0.003), date ($F_{5,48}$ = 2.43, p = 0.049) but not for treatment × date ($F_{5,48}$ = 0.07, p = 0.997). Male captures were fivefold larger in control than in MD plots, with the latter being very low (mean May–August: MD = 0.20 ± 0.1, control = 1.15 ± 0.3 males × trap^{-1} × day^{-1}). MDI in this period was also above 80% (Figure 2B).

In the second season (2018–2019) when pheromone application was not renewed but dispensers were left on the trees, male captures in the pheromone traps were much lower than the previous season, and mostly concentrated in the spring (Figure 3B, note the change in scales of Y axis). In this period (September–December 2018, 7 monitoring dates) an effect of treatment ($F_{1,56}$ = 4.93, p = 0.031), but not for date ($F_{6,56}$ = 0.39, p = 0.885) or treatment × date ($F_{6,56}$ = 0.15, p = 0.989) was observed. Male captures were very low, but nevertheless they were tenfold larger in control than in MD plots (mean September–December: MD = 0.04 ± 0.04, control = 0.41 ± 0.13 males × trap^{-1} × day^{-1}). During fall and winter (May–August 2019, 4 monitoring dates) captures in both, MD and control plots, were low and similar (mean May–August: MD = 0.07 ± 0.07, control = 0.04 ± 0.06 males × trap^{-1} × day^{-1}) (Figure 3B). No effects of treatment, date or treatment × date were found (p > 0.8 for all).

At the end of August 2019, before the beginning of 2019–2020 season, after renewing the lures in the traps (21 August) but before the new pheromone applications (29 August), there was an increase in male captures, being similar for MD and control plots (MD = 2.11 ± 0.7, control = 3.32 ± 0.8 males × trap^{-1} × day^{-1}; $F_{1,8}$ = 1.31, p = 0.29; Figure 3B). After that, male captures were low, with a small increase in captures in the control plots in October 2019 and March 2020 (Figure 3C). In this period (September 2019–March 2020, 14 monitoring dates), MD plots had significantly fewer male captures than control plots (H = 40.94; p = 0.0005; Figure 3C). For October 2019, when the small peak in male captures in the control plots was observed, MDI was 98% (Figure 2B).

3.3. Mealybugs Population and Fruit Damage in Tangerines and Apples

During the whole evaluation period, mealybug population levels on the plants (visual counts) were exceptionally low in both orchards, with no significant differences between MD and control for all seasons (Table 1).

In the first season, the number of mealybugs and percentage of infested fruit was very low in both crops. In apples, fruit infestation was 4 ± 1% in MD plots and 2 ± 1% in control plots, without significant differences ($F_{1,8}$ = 3.10, p = 0.12). Number of *P. calceolariae* per fruit were also similar and low (MD = 0.06 ± 0.11, control 0.05 ± 0.10, H = 0.39, p = 0.59). In the 2019–2020 season, out of the 500 apples harvested and inspected, only one presented two *P. calceolariae* nymphs.

In tangerines, only 4 fruits (out of the 1200 inspected) of the control plots were infested, with a total of 10 *P. calceolariae* individuals. This resulted in a 0.67% fruit infestation in the control plots and 0% infestation in the MD plots. Interestingly, a 9% fruit infestation with *P. longispinus* was found in both control and MD plots.

Table 1. Mean number of mealybugs (all life stages) per plant over the season for the three seasons studies, in mating disruption (MD) and control plots in tangerine and apple orchards. Twelve plants per experimental unit were monitored (60 plants per treatment in each date).

Fruit Crop and Season	Treatments		Kruskal-Wallis Test
	MD	Control	
	Mealybugs $\times$ Plant^{-1} $\pm$ SE		
Apple			
September 2017–April 2018	0.27 ± 0.09	0.04 ± 0.01	$H = 0.28; p = 0.59$
September 2018–April 2019	0.59 ± 0.14	0.30 ± 0.08	$H = 1.47; p = 0.22$
September 2019–March 2020	0.03 ± 0.01	0.13 ± 0.04	$H = 2.79; p = 0.09$
Tangerine			
September 2017–April 2018	0.30 ± 0.08	0.31 ± 0.09	$H * = 0.0; p = 0.96$
September 2018–April 2019	0.02 ± 0.009	0.001 ± 0.001	$H = 3.83; p = 0.05$
September 2019–March 2020	0.00 ± 0.0	0.00 ± 0.0	NA *

* not applicable.

4. Discussion

The results of our experiments provide valuable information about the potential of mating disruption for the control of *P. calceolariae* in two fruit crops of worldwide economic importance, such as apples and tangerines. Previously, studies showing the usefulness of this technique as a control tool for a mealybug species have been carried out only for *Pl. ficus* in grapes [15,34,35,37,38], leading to commercial formulations. Our results showed that in MD plots "trap shutdown" occurred, as male captures in pheromone traps were reduced by 97% to 100% in both fruit crops one to two months after the applications in two seasons.

The pheromone doses used in this study (6.32 and 9.45 g/ha of the stereoisomeric pheromone mixture) were much lower than those used in studies with *Pl. ficus*, which found effects with 62.5 and 93.8 g/ha [35]; 93 g/ha [36]; 54 to 90 g/ha [38] and 20.0 to 61.7 g/ha [15]. When lower doses were tested, such as 4.15 and 8.3 g/ha, male trap captures were similar in the MD and control treatment [15]. In our study with *P. calceolariae*, with ten times less pheromone than other studies with *Pl. ficus*, we obtained MDI between 80 and 100% in both crops and seasons. In addition to the dosage, the type of dispenser may play an important role in the successful implementation of mating disruption. In previous MD studies with *Pl. ficus*, other different types of dispensers and formulations have been used, including Checkmate VBM-XL (Suterra LLC., Bend, OR, USA), Isonet® PF (Shin-Etsu Chemical Co.Ltd., Tokyo, Japan) and a sprayable microencapsulated formulation [15,35–38]. Our study reports for the first time the use of SPLAT® as a dispenser for MD of mealybugs. The SPLAT® technology for mating disruption has been used and commercialized mainly in Lepidoptera [45]. The results presented here demonstrate the feasibility of using SPLAT® as a pheromone dispenser for MD in mealybugs.

Interestingly, MDIs calculated during the first season indicated that the mating disruption effect was maintained at least 12 months after the application of the pheromone treatment, with mean MDI values between 65 and 80%. Moreover, in the following season, 2018–2019, when no pheromone was applied (but dispensers were left in the orchard), MDI was around 80% during spring in tangerines (August to November, 2018) and during spring and summer (September 2018 to January 2019) in apples before trap captures declined. Low male captures were observed in both MD and control plots during the autumn-winter months (2018 and 2019), since there is a natural decrease of the populations due to the lower temperatures of the winter months. The amount of pheromone released or remaining in dispensers (SPLAT®) over time in the field was not measured in our study, so the results observed in the second season could be due to residual emission from the dispensers left from the previous season, because of the carry over effect of lower populations or some other factor. Therefore, to understand more comprehensively the eventual persistence of a MD pheromone release effect from dispensers under different conditions, it should be measured systematically through time. This is important, because for successful population control the pheromone not only should remain active for a long period in the field, but it is also key that it be consistently released in a constant amount [38]. The previous studies with *Pl. ficus* in grapes only evaluated the MD treatments up to six or seven months, maintaining a satisfactory disruptive effect in males [15,35,37]. However, in evergreen fruit trees such as tangerines, the MD effect needs to be maintained for a longer period of time to cover the productive period fully. Our results show that during the second season 2018–2019 (without pheromone renewal), the mean mating disruption index for the months of September to November was 88% in tangerines, when a male flight peak occurs. This long-lasting treatment effect, 6 to 12 months, in mealybug MD is desirable for farmers, because it makes this technique cost effective compared to insecticide applications. Pheromones can also be used in combination with insecticides or biological control, potentially generating a synergistic effect on mealybug control [15].

Trap shutdown is one of the main signs that mating disruption is working [35], but to measure control effectiveness it is desirable to demonstrate that populations and damage are reduced [2,46]. The abundance of mealybugs was extremely low throughout the duration of the trials in both crops in our study. Due to this low abundance, we were not able to observe a reduction of populations on trees or even of damage of fruits. Probably the pest control measures taken by the farmers in the experimental fields also impacted mealybug populations on the plants and did not allow us to evaluate properly the impact of MD through visual monitoring and fruit infestation. It is important to note that studies have shown that MD is more effective when used at low population levels [35,37,38], while at high mealybug levels at least two continuous seasons will be needed to produce an observable effect of MD [36], or it might even be necessary to combine the use of MD with insecticide applications [37]. Another factor to consider when dealing with high mealybug populations is that high doses of pheromone are needed, as otherwise the disruptive effect can decrease due to the high abundance of females as natural sources of pheromone [15]. Thus, studies in orchards with larger populations of *P. calceolariae* are desirable to test the potential of MD for control in a wider range of field situations.

This study shows that mating disruption has a good potential for controlling *P. calceolariae* in fruit crops, as previously demonstrated for *Pl. ficus* in grapes [34,37,38]. The final growers' decision to implement disruption will depend on several factors, such as the cost of the pheromone product, efficiency of the technique, pest population level and cost comparative to other control alternatives available (e.g. biological control, insecticides) [53]. Nevertheless, it is important to consider that in quarantine pests very low populations are required to avoid infestation of the products before harvest, and that MD is a more efficient technique at lower population levels. Moreover, it has been found that a low pheromone concentration is sufficient to provide good control at low pest population levels [54]. The combination of higher pheromone concentration and insecticide application may be necessary with high pest populations, which may increase the cost of implementing the technique, and thus not be economically sustainable [53,54]. However, the long-lasting effect of the pheromone would make it cost-effective compared to repetitive insecticide applications. The use of this pheromone-based technique instead of insecticides also has the advantage of being environmentally friendly and non-toxic.

Author Contributions: T.Z., A.R., C.B., and J.B. secured funding; T.Z., A.R., and C.B. conceived the article and supervised data collection; T.Z., A.R., C.B., S.M., and M.C.C. helped with data collection; A.R. and S.M. were in charge of data curation; T.Z. and C.B. analyzed the data and wrote the initial draft. All authors contributed to the final draft. All authors have read and agreed to the published version of the manuscript.

Funding: Funding was provided by Fundación para la Innovación Agraria (FIA Chile), project FIA PYT-2017-0140 and CONICYT Doctoral fellowship #2015.

Institutional Review Board Statement: Not applicable.

Data Availability Statement: The data presented in this study are available on request from the corresponding author.

Acknowledgments: The authors would like to thank Sebastian Morales for technical assistance in the pheromone synthesis, and Iván Osorio, Patricia Toro, Daniela Urrutia and Cristóbal Calvo for field and laboratory help, and growers for allowing us to carry out experiments in their fields.

Conflicts of Interest: The authors declare no conflict of interest.

References

1. Witzgall, P.; Kirsch, P.; Cork, A. Sex pheromones and their impact on pest management. *J. Chem. Ecol.* **2010**, *36*, 80–100. [CrossRef]
2. Miller, J.R.; Gut, L.J. Mating disruption for the 21st century: Matching technology with mechanism. *Environ. Entomol.* **2015**, *44*, 427–453. [CrossRef] [PubMed]
3. Benelli, G.; Lucchi, A.; Thomson, D.; Ioriatti, C. Sex pheromone aerosol devices for mating disruption: Challenges for a brighter future. *Insects* **2019**, *10*, 308. [CrossRef] [PubMed]
4. Reddy, G.V.P.; Guerrero, A. New pheromones and insect control strategies. *Vitam. Horm.* **2010**, *83*, 493–519. [CrossRef] [PubMed]
5. Smart, L.E.; Aradottir, G.I.; Bruce, T.J.A. *Role of Semiochemicals in Integrated Pest Management*; Elsevier Inc.: Amsterdam, The Netherlands, 2014; ISBN 9780124017092.
6. Ioriatti, C.; Lucchi, A. Semiochemical strategies for tortricid moth control in apple orchards and vineyards in Italy. *J. Chem. Ecol.* **2016**, *42*, 571–583. [CrossRef]
7. Tewari, S.; Leskey, T.C.; Nielsen, A.L.; Piñero, J.C.; Rodriguez-Saona, C.R. *Use of Pheromones in Insect Pest Management, with Special Attention to Weevil Pheromones*; Elsevier Inc.: Amsterdam, The Netherlands, 2014; ISBN 9780124017092.
8. Lo Verde, G.; Guarino, S.; Barone, S.; Rizzo, R. Can mating disruption be a possible route to control plum fruit moth in mediterranean environments? *Insects* **2020**, *11*, 589. [CrossRef]
9. Ioriatti, C.; Anfora, G.; Tasin, M.; De Cristofaro, A.; Witzgall, P.; Lucchi, A. Chemical ecology and management of *Lobesia botrana* (Lepidoptera: Tortricidae). *J. Econ. Entomol.* **2011**, *104*, 1125–1137. [CrossRef] [PubMed]
10. Maki, E.C.; Millar, J.G.; Rodstein, J.; Hanks, L.M.; Barbour, J.D. Evaluation of mass trapping and mating disruption for managing *Prionus californicus* (Coleoptera: Cerambycidae) in hop production yards. *J. Econ. Entomol.* **2011**, *104*, 933–938. [CrossRef] [PubMed]
11. Rodriguez-Saona, C.R.; Polk, D.; Holdcraft, R.; Koppenhöfer, A.M. Long-term evaluation of field-wide oriental beetle (Col., Scarabaeidae) mating disruption in blueberries using female-mimic pheromone lures. *J. Appl. Entomol.* **2014**, *138*, 120–132. [CrossRef]
12. Barbour, J.D.; Alston, D.G.; Walsh, D.B.; Pace, M.; Hanks, L.M. Mating disruption for managing *Prionus californicus* (Coleoptera: Cerambycidae) in hop and sweet cherry. *J. Econ. Entomol.* **2019**, *112*, 1130–1137. [CrossRef]
13. Vacas, S.; Vanaclocha, P.; Alfaro, C.; Primo, J.; Verdú, M.J.; Urbaneja, A.; Navarro-Llopis, V. Mating disruption for the control of *Aonidiella aurantii* Maskell (Hemiptera: Diaspididae) may contribute to increased effectiveness of natural enemies. *Pest Manag. Sci.* **2012**, *68*, 142–148. [CrossRef]
14. Vacas, S.; Alfaro, C.; Primo, J.; Navarro-Llopis, V. Deployment of mating disruption dispensers before and after first seasonal male flights for the control of *Aonidiella aurantii* in citrus. *J. Pest Sci. 2004* **2015**, *88*, 321–329. [CrossRef]
15. Daane, K.M.; Yokota, G.Y.; Walton, V.M.; Hogg, B.N.; Cooper, M.L.; Bentley, W.J.; Millar, J.G. Development of a mating disruption program for a mealybug, *Planococcus ficus*, in vineyards. *Insects* **2020**, *11*, 635. [CrossRef]
16. Negishi, T.; Uchida, M.; Tamaki, Y.; Mori, K.; Ishiwatari, T.; Asano, S.; Nakagawa, K. Sex pheromone of the comstock mealybug, *Pseudococcus comstocki* KUWANA: Isolation and identification. *Appl. Entomol. Zool.* **1980**, *15*, 328–333. [CrossRef]
17. Bierl-Leonhardt, B.A.; Moreno, D.S.; Schwarz, M.; Fargerlund, J.A.; Plimmer, J.R. Isolation, identification and synthesis of the sex pheromone of the citrus mealybug, *Planococcus citri* (risso). *Tetrahedron Lett.* **1981**, *22*, 389–392. [CrossRef]
18. Hinkens, D.M.; McElfresh, J.S.; Millar, J.G. Identification and synthesis of the sex pheromone of the vine mealybug, *Planococcus ficus*. *Tetrahedron Lett.* **2001**, *42*, 1619–1621. [CrossRef]
19. Arai, T. Attractiveness of sex pheromone of *Pseudococcus cryptus* Hempel (Homoptera: Pseudococcidae) to adult males in a citrus orchard. *Appl. Entomol. Zool.* **2002**, *37*, 69–72. [CrossRef]
20. Zhang, A.; Amalin, D. Sex pheromone of the female pink hibiscus mealybug, *Maconellicoccus hirsutus* (Green) (Homoptera: Pseudococcidae): Biological activity evaluation. *Environ. Entomol.* **2005**, *34*, 264–270. [CrossRef]
21. Millar, J.G.; Midland, S.L.; Mcelfresh, J.S.; Daane, K.M. (2,3,4,4-Tetramethylcyclopentyl)methyl acetate, a sex pheromone from the obscure mealybug: First example of a new structural class of monoterpenes. *J. Chem. Ecol.* **2005**, *31*, 2999–3005. [CrossRef] [PubMed]
22. Figadère, B.A.; McElfresh, J.S.; Borchardt, D.; Daane, K.M.; Bentley, W.; Millar, J.G. trans-α-Necrodyl isobutyrate, the sex pheromone of the grape mealybug, *Pseudococcus maritimus*. *Tetrahedron Lett.* **2007**, *48*, 8434–8437. [CrossRef]
23. Ho, H.Y.; Hung, C.C.; Chuang, T.H.; Wang, W.L. Identification and synthesis of the sex pheromone of the passionvine mealybug, *Planococcus minor* (Maskell). *J. Chem. Ecol.* **2007**, *33*, 1986–1996. [CrossRef]
24. Sugie, H.; Teshiba, M.; Narai, Y.; Tsutsumi, T.; Sawamura, N.; Tabata, J.; Hiradate, S. Identification of a sex pheromone component of the Japanese mealybug, *Planococcus kraunhiae* (Kuwana). *Appl. Entomol. Zool.* **2008**, *43*, 369–375. [CrossRef]

25. Ho, H.Y.; Su, Y.T.; Ko, C.H.; Tsai, M.Y. Identification and synthesis of the sex pheromone of the madeira mealybug, *Phenacoccus madeirensis* green. *J. Chem. Ecol.* **2009**, *35*, 724–732. [CrossRef]
26. Millar, J.G.; Moreira, J.A.; McElfresh, J.S.; Daane, K.M.; Freund, A.S. Sex pheromone of the longtailed mealybug: A new class of monoterpene structure. *Org. Lett.* **2009**, *11*, 2683–2685. [CrossRef]
27. El-Sayed, A.M.; Unelius, C.R.; Twidle, A.; Mitchell, V.; Manning, L.A.; Cole, L.; Suckling, D.M.; Flores, M.F.; Zaviezo, T.; Bergmann, J. Chrysanthemyl 2-acetoxy-3-methylbutanoate: The sex pheromone of the citrophilous mealybug, *Pseudococcus calceolariae*. *Tetrahedron Lett.* **2010**, *51*, 1075–1078. [CrossRef]
28. Unelius, C.R.; El-Sayed, A.M.; Twidle, A.; Bunn, B.; Zaviezo, T.; Flores, M.F.; Bell, V.; Bergmann, J. The absolute Configuration of the sex pheromone of the citrophilous mealybug, *Pseudococcus calceolariae*. *J. Chem. Ecol.* **2011**, *37*, 166–172. [CrossRef]
29. Tabata, J.; Narai, Y.; Sawamura, N.; Hiradate, S.; Sugie, H. A new class of mealybug pheromones: A hemiterpene ester in the sex pheromone of *Crisicoccus matsumotoi*. *Naturwissenschaften* **2012**, *99*, 567–574. [CrossRef]
30. De Alfonso, I.; Hernandez, E.; Velazquez, Y.; Navarro, I.; Primo, J. Identification of the sex pheromone of the mealybug *Dysmicoccus grassii* Leonardi. *J. Agric. Food Chem.* **2012**, *60*, 12959–12964. [CrossRef]
31. Tabata, J.; Ichiki, R.T. A New Lavandulol-related monoterpene in the sex pheromone of the grey pineapple mealybug, *Dysmicoccus neobrevipes*. *J. Chem. Ecol.* **2015**, *41*, 194–201. [CrossRef]
32. Tabata, J.; Ichiki, R.T. Sex pheromone of the cotton mealybug, *Phenacoccus solenopsis*, with an unusual cyclobutane structure. *J. Chem. Ecol.* **2016**, *42*, 1193–1200. [CrossRef]
33. Millar, J.G.; Daane, K.M.; McElfresh, J.S.; Moreira, J.A.; Malakar-Kuenen, R.; Guillén, M.; Bentley, W.J. Development and optimization of methods for using sex pheromone for monitoring the mealybug *Planococcus ficus* (Homoptera: Pseudococcidae) in California vineyards. *J. Econ. Entomol.* **2002**, *95*, 706–714. [CrossRef] [PubMed]
34. Walton, V.M.; Daane, K.M.; Bentley, W.J.; Millar, J.G.; Larsen, T.E.; Malakar-kuenen, R. Pheromone-based mating disruption of *Planococcus ficus* (Hemiptera: Pseudococcidae) in California vineyards. *J. Econ. Entomol.* **2006**, *99*, 1280–1290. [CrossRef] [PubMed]
35. Cocco, A.; Lentini, A.; Serra, G. Mating disruption of *Planococcus ficus* (Hemiptera: Pseudococcidae) in vineyards using reservoir pheromone dispensers. *J. Insect Sci.* **2014**, *14*, 1–8. [CrossRef] [PubMed]
36. Sharon, R.; Zahavi, T.; Sokolsky, T.; Sofer-Arad, C.; Tomer, M.; Kedoshim, R.; Harari, A.R. Mating disruption method against the vine mealybug, *Planococcus ficus*: Effect of sequential treatment on infested vines. *Entomol. Exp. Appl.* **2016**, *161*, 65–69. [CrossRef]
37. Mansour, R.; Grissa-Lebdi, K.; Khemakhem, M.; Chaari, I.; Trabelsi, I.; Sabri, A.; Marti, S. Pheromone-mediated mating disruption of *Planococcus ficus* (Hemiptera: Pseudococcidae) in Tunisian vineyards: Effect on insect population dynamics. *Biologia* **2017**, *72*, 333–341. [CrossRef]
38. Lucchi, A.; Suma, P.; Ladurner, E.; Iodice, A.; Savino, F.; Ricciardi, R.; Cosci, F.; Marchesini, E.; Conte, G.; Benelli, G. Managing the vine mealybug, *Planococcus ficus*, through pheromone-mediated mating disruption. *Environ. Sci. Pollut. Res.* **2019**, *26*, 10708–10718. [CrossRef]
39. Teshiba, M.; Shimizu, N.; Sawamura, N.; Narai, Y.; Sugie, H.; Sasaki, R.; Tabata, J.; Tsutsumi, T. Use of a sex pheromone to disrupt the mating of *Planococcus kraunhiae* (Kuwana) (Hemiptera: Pseudococcidae). *Jpn. J. Appl. Entomol. Zool.* **2009**, *53*, 173–180. [CrossRef]
40. Daane, K.M.; Almeida, R.P.P.; Bell, V.A.; Walker, J.T.S.; Botton, M.; Fallahzadeh, M.; Mani, M.; Miano, J.L.; Sforza, R.; Walton, V.M.; et al. Biology and management of mealybugs in vineyards. In *Arthropod Management in Vineyards: Pests, Approaches, and Future Directions*; Springer: New York, NY, USA, 2012; pp. 271–307. ISBN 9789400740327.
41. Sullivan, N.J.; Butler, R.C.; Salehi, L.; Twidle, A.M.; Baker, G.; Suckling, D.M. Deployment of the sex pheromone of *Pseudococcus calceolariae* (Hemiptera: Pseudococcidae) as a potential new tool for mass trapping in citrus in South Australia. *N. Zeal. Entomol.* **2019**, *42*, 1–12. [CrossRef]
42. Lentini, A.; Mura, A.; Muscas, E.; Nuvoli, M.T.; Cocco, A. Effects of delayed mating on the reproductive biology of the vine mealybug, *Planococcus ficus* (Hemiptera: Pseudococcidae). *Bull. Entomol. Res.* **2018**, *108*, 263–270. [CrossRef]
43. Flores, M.F.; Romero, A.; Oyarzun, M.S.; Bergmann, J.; Zaviezo, T. Monitoring *Pseudococcus calceolariae* (Hemiptera: Pseudococcidae) in fruit crops using pheromone-baited traps. *J. Econ. Entomol.* **2015**, *108*, 2397–2406. [CrossRef]
44. Ricciardi, R.; Lucchi, A.; Benelli, G.; Suckling, D.M. Multiple mating in the citrophilous mealybug *Pseudococcus calceolariae*: Implications for mating disruption. *Insects* **2019**, *10*, 285. [CrossRef] [PubMed]
45. Mafra-Neto, A.; De Lame, F.M.; Fettig, C.J.; Munson, A.S.; Perring, T.M.; Stelinski, L.L.; Stoltman, L.L.; Mafra, L.E.J.; Borges, R.; Vargas, R.I. Manipulation of insect behavior with specialized pheromone and lure application technology (SPLAT®). *ACS Symp. Ser.* **2013**, *1141*, 31–58. [CrossRef]
46. Stelinski, L.L.; Miller, J.R.; Ledebuhr, R.; Siegert, P.; Gut, L.J. Season-long mating disruption of *Grapholita molesta* (Lepidoptera: Tortricidae) by one machine application of pheromone in wax drops (SPLAT-OFM). *J. Pest Sci. 2004* **2007**, *80*, 109–117. [CrossRef]
47. Jenkins, P.E.; Isaacs, R. Mating disruption of *Paralobesia viteana* in vineyards using pheromone deployed in SPLAT-GBM™ wax droplets. *J. Chem. Ecol.* **2008**, *34*, 1089–1095. [CrossRef]
48. Mafra-Neto, A.; Fettig, C.J.; Munson, A.S.; Rodriguez-Saona, C.; Holdcraft, R.; Faleiro, J.R.; El-Shafie, H.; Reinke, M.; Bernardi, C.; Villagran, K.M. Development of specialized pheromone and lure application technologies (SPLAT®) for management of coleopteran pests in agricultural and forest systems. In *Biopesticides: State of the Art and Future Opportunities*; ACS Symposium Series; American Chemical Society: Washington, DC, USA, 2014; Volume 1172, pp. 15–211. ISBN 9780841229990.

49. Arioli, C.J.; Pastori, P.L.; Botton, M.; Silveira Garcia, M.; Borges, R.; Mafra-Neto, A. Assessment of SPLAT formulations to control *Grapholita molesta* (Lepidoptera: Tortricidae) in a Brazilian apple orchard. *Chil. J. Agric. Res.* **2014**, *74*, 184–190. [CrossRef]
50. Teixeira, L.A.F.; Mason, K.; Mafra-Neto, A.; Isaacs, R. Mechanically-applied wax matrix (SPLAT-GBM) for mating disruption of grape berry moth (Lepidoptera: Tortricidae). *Crop Prot.* **2010**, 1514–1520. [CrossRef]
51. Di Rienzo, J.A.; Casanoves, F.; Balzarini, M.G.; Gonzalez, L.; Tablada, M.; Robledo, C.W. *InfoStat Version 2020*; Grupo InfoStat, FCA, Universidad Nacional de Córdoba: Córdoba, Argentina, 2020.
52. R Core Team. *R: A Language and Environment for Statistical Computing*; R Foundation for Statistical Computing: Vienna, Austria, 2021.
53. Carde, R.T.; Minks, A.K. Control of moth pests by mating disruption: Successes and constraints. *Annu. Rev. Entomol.* **1995**, *40*, 559–585. [CrossRef]
54. Gordon, D.; Zahavi, T.; Anshelevich, L.; Harel, M.; Ovadia, S.; Dunkelblum, E.; Harari, A.R. Mating disruption of *Lobesia botrana* (Lepidoptera: Tortricidae): Effect of pheromone formulations and concentrations. *J. Econ. Entomol.* **2005**, *98*, 135–142. [CrossRef]

insects

Article

Influence of Age and Mating Status on Pheromone Production in a Powderpost Beetle *Lyctus africanus* (Coleoptera: Lyctinae)

Titik Kartika [1,*], Nobuhiro Shimizu [2], Setiawan Khoirul Himmi [1], Ikhsan Guswenrivo [1], Didi Tarmadi [1], Sulaeman Yusuf [1] and Tsuyoshi Yoshimura [3]

[1] Research Center for Biomaterials, Indonesian Institute of Sciences, Bogor 16911, Indonesia; khoirul_himmi@biomaterial.lipi.go.id (S.K.H.); ikhsan.guswenrivo@biomaterial.lipi.go.id (I.G.); didi@biomaterial.lipi.go.id (D.T.); sulaeman@biomaterial.lipi.go.id (S.Y.)
[2] Faculty of Bioenvironmental Science, Kyoto University of Advanced Science, Kyoto 621-8555, Japan; shimizu.nobuhiro@kuas.ac.jp
[3] Research Institute for Sustainable Humanosphere, Kyoto University, Kyoto 611-0011, Japan; tsuyoshi@rish.kyoto-u.ac.jp
* Correspondence: titikkartika@biomaterial.lipi.go.id; Tel./Fax: +62-21-8791-4509

Simple Summary: Powderpost beetles, such as *Lyctus africanus*, are a common pest group for dried cured wood. The damage is slow and inconspicuous; thus, the infestation is mostly identified belatedly due to a lack of knowledge of how to locate and monitor it. *L. africanus* produces a pheromone, a chemical compound to attract other beetles. This pheromone has been determined and suggested as a monitoring tool for *L. africanus*. Here, we examined the physiological and behavioral parameters that affect pheromone production. We found that food availability may affect pheromone production in adult *L. africanus* males. In addition, of three components in male *L. africanus* aggregation pheromones, major compounds **2** (3-pentyl dodecanoate) and **3** (3-pentyl tetradecanoate) may be affected by age, not mating status, while compound **1** (2-propyl dodecanoate) was produced steadily and was affected by mating status. This suggests compounds **2** and **3** might have an important function in aggregation behavior, especially in signaling for mating opportunities. We also were able to clarify the minor effect of compound **1** in the aggregation pheromone of *L. africanus*, although not its role. The present information will be helpful in understanding the chemical communication of these insects, which may be important for the development of improved pheromone-based management strategies for controlling *Lyctus* beetles.

Abstract: Powderpost beetles such as *Lyctus africanus* are a common pest group for dried cured wood, causing significant harm to wood and wood products. We examined the life span and effects of aging and mating status on pheromone production in the powderpost beetle *L. africanus* (Coleoptera: Lyctinae). Experiments compared starved and unstarved male groups, and chemical analysis was used to determine factors affecting pheromone production. Regarding lifespan, male beetles provided food survived up to 14 weeks, while starved beetles died before the fifth week. Thus, an adult *L. africanus* male may require food throughout its lifespan, and food availability may affect pheromone production. There was no significant difference in the quantity of two major pheromone compounds, compound **2** (3-pentyl dodecanoate) and **3** (3-pentyl tetradecanoate) between mated and un-mated males. On the other hand, a minor compound, compound **1** (2-propyl dodecanoate) showed increased quantity after mating. The two major compounds were produced in low amounts by young *L. africanus* beetles, increasing until the fifth week, and beginning to decrease at the ninth week. The minor compound was produced steadily without significant change up to 9 weeks. Our results represent a step forward in the knowledge of the chemical communication of this important pest.

Keywords: pheromone; *Lyctus africanus*; powderpost beetle; age; mating

check for **updates**

Citation: Kartika, T.; Shimizu, N.; Himmi, S.K.; Guswenrivo, I.; Tarmadi, D.; Yusuf, S.; Yoshimura, T. Influence of Age and Mating Status on Pheromone Production in a Powderpost Beetle *Lyctus africanus* (Coleoptera: Lyctinae). *Insects* **2021**, *12*, 8. https://dx.doi.org/10.3390/insects12010008

Received: 16 November 2020
Accepted: 22 December 2020
Published: 25 December 2020

Publisher's Note: MDPI stays neutral with regard to jurisdictional claims in published maps and institutional affiliations.

1. Introduction

Powderpost beetles are a common pest group for dried cured wood, especially the sapwood of hardwood. They convert wood into a mass of powdery or pelleted frass during the larval stage [1]. The beetles cause major destruction to wood and its products, such as hardwood flooring, hardwood timbers, plywood, crating, and furniture [2]. The damage is slow and inconspicuous; thus, infestation is generally identified belatedly due to a lack of knowledge on how to locate and monitor them. Powderpost beetles occur worldwide, having indigenous species plus introduced species in each region, but are mainly found in tropical and arid areas [3,4].

Lyctus africanus is a powderpost beetle belonging to the subfamily Lyctinae in the family Bostrichidae [1,5]. The species is endemic to East Asian and East African regions [6,7]. This beetle is recognized as an economically important *Lyctus* species due to the recent explosions of infestation in some areas [8]. The beetle is reported to infest dried roots, seeds, and tubers [6,9] as well as timber and timber products [6,7].

Chemical signaling is a primary strategy in many insect species for communication among individuals. A semiochemical is a chemical signal released from one organism that generates a behavioral or physiological response in other members of the same or different species. Knowing the specific semiochemical of an insect is considered a promising component in integrated pest management (IPM) programs for controlling insects. Different control strategies, such as monitoring, mass trapping, mating disruption, and attract-and-kill and push-pull strategies have been developed using semiochemicals [10]. An aggregation pheromone is one type of semiochemical. Released by an individual, it generates aggregative behavior in members of both sexes or of the same sex as the sender [11,12]. Many non-social Arthropod species, such as beetle, release an aggregation pheromone to attract their conspecific, and some conspicuous similarities of pheromone function are shown in different taxa [11–14]. We recently began the development of a monitoring system by determining the aggregation pheromones in the *L. africanus* species [15]. Three esters (Figure 1) were identified as male-specific compounds: 2-propyl dodecanoate (compound **1**), 3-pentyl dodecanoate (compound **2**), and 3-pentyl tetradecanoate (compound **3**). There was a synergistic effect among the three synthetic esters compounds; however, compound **2** was recognized to play the main role in the aggregation behavior of *L. africanus*, especially in female beetles.

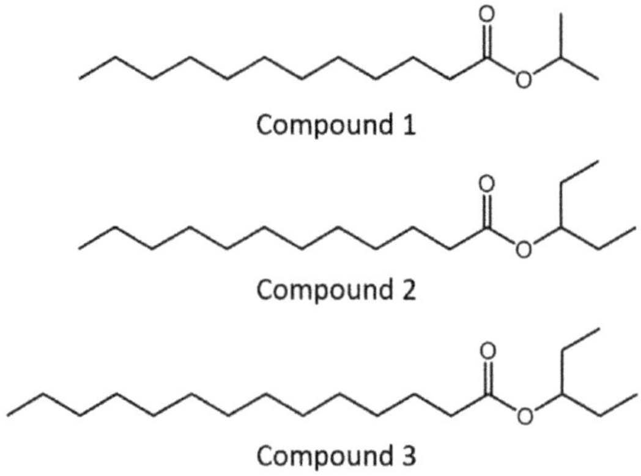

Compound 1

Compound 2

Compound 3

Figure 1. Chemical structure of ester in aggregation pheromone of *Lyctus africanus*.

This study examined the physiological and behavioral parameters that affect pheromone production by *L. africanus* to provide useful information for pest management. Information on lifespan and the effects of aging and mating status on pheromone production in males of *L. africanus* is important to determine the optimum survival conditions for adult beetles. It has been suggested that only individuals with a good ability to survive engage in intensive signaling [16]. Most members of the Bostrichidae family are non-feeding at the adult stage. The *L. africanus* are classified into a group that can taste and feed on wood [17]; hence, optimum lifespan conditions are evaluated in starved and unstarved conditions. Further, observation of the effects of age and mating status of adult male *L. africanus* beetle is essential to understanding the source of variation in pheromone production and the optimum conditions for chemical compound production. Knowing changes in pheromone production of beetles is basic knowledge to develop further research using pheromone as a pest management tool. This is important for further studies on behavioral, experimental, and physiological factors influencing pheromone signaling, and also for developing monitoring strategies for pest management.

2. Materials and Methods

2.1. Insect Source

Adult *L. africanus* beetles were cultured in a mixture of solid wood-based artificial diet [18,19] consisting of dried yeast (24%, Asahi Food and Health Care Co., Ltd., Sumida-ku, Tokyo, Japan), starch (50%, Nacalai Tesque, Kyoto, Japan), and lauan (*Shorea* spp.) wood sawdust (26%). The beetle cultures were kept in glass jars (450 mL) in a dark climatic chamber with a temperature of 26 °C $\pm$ 2 °C and relative humidity of 65% $\pm$ 10%. These beetles have been maintained in the laboratory for at least 10 years (T. Yoshimura, personal communication) in the Deterioration Organisms Laboratory (DOL) at the Research Institute for Sustainable Humanosphere, Kyoto University, Gokasho, Uji, Kyoto, Japan. Newly emerged *L. africanus* beetles were used in this study. Adult female and male beetles were sexed and separated under a stereo microscope (Leica MZ6, Leica Microsystems GmbH, Wetzlar, Germany) by apical hairs present along the hind margin of the abdominal segment [6].

2.2. Life Span of Adult Male L. africanus in Starved and Unstarved Conditions

To test the optimum conditions for adult beetle survival, a lifespan experiment was conducted in starved (cultured without food) and unstarved (cultured with food) conditions. The adults of a related species, *L. brunneus*, have been reported as not feeding on wood [20]. In this study, five pairs of males and females *L. africanus* were put in an individual petri dish (Ø 5 cm, 1 cm in height, BD Biosciences, San Jose, CA, USA) and covered with Ø 5 cm filter paper (Whatman No. 2, GE Healthcare, Buckinghamshire, UK) to facilitate the movement of beetles. A small piece of air-dried wood-based solid food (2 $\times$ 2 $\times$ 1 cm, 3.3 g) was placed in the Ø 5 cm Petri dish for the unstarved condition, and no food was placed in the starved condition dish. All Petri dishes were arranged randomly in the dark climatic chamber and observed weekly for mortality. The observation was terminated when all beetles died. Fifteen replications were made for each condition.

2.3. Effects of Mating Status on Pheromone Production

The experiment was set up to determine whether mating history affect pheromone production. Here, we used unmated and mated male beetles for the chemical analysis of pheromone compounds. To prepare the unmated and mated male beetles, the pupal-stage *L. africanus* beetles were harvested from the artificial diets, then placed individually in a small container (Ø 5 cm, 1 cm in height, BD Biosciences, San Jose, CA, USA) until they became new adults. The adults (1–5 days old) were sexed and separated for further examination. The unmated male beetles were kept separately from female beetles until the chemical analysis. To prepare the mated beetles, males and females of newly emerged adult beetles were transferred into the Ø 5 cm Petri dish covered with a Whatman filter paper No. 2 with 2:3 male-to-female ratio. The greater number of female than male beetles was intended to ensure the opportunity of male beetles to mate. One dish contained 15 adult beetles with eight replications were made for each treatment (mating status).

2.4. Effects of Aging on Pheromone Production

This experiment was set up to determine the effects of time or aging on pheromone production. For one set of experiments, five pairs of newly emerged male and female adult beetles were transferred into a Ø 5 cm Petri dish covered with a Whatman filter paper No. 2. Another small piece of food as described above was placed in the dish to prolong the life span of the beetles. The Petri dishes were positioned randomly in the dark climatic chamber. The males were separated from the females by the above-mentioned method. Observation on pheromone compounds was conducted by GC-MS measurement on 1-, 3-, 5-, and 9-week-old male *L. africanus* beetles using the whole-body extraction method. Eight replications were conducted for each variation.

2.5. Chemical Analysis
2.5.1. Collection of Chemical Compounds

To collect chemical compounds, whole-body extraction with hexane was performed for both mated and unmated male beetles (1–5 days old), using the method described in previous studies to extract pheromones from other beetles and millipede groups [21–23]. Briefly, each beetle was immersed in hexane (10 µL) for 5 min, and 1 µL of aliquot was injected into a gas chromatography–mass spectrometry (GC–MS) instrument. Authentic compounds **1–3** were prepared and synthesized in a similar manner to our previous study [15].

2.5.2. GC–MS Analysis

The GC–MS analysis was conducted with a Network GC System (6890N; Agilent Technologies, Santa Clara, CA, USA) coupled with a mass selective detector (5975 Inert XL; Agilent Technologies, Santa Clara, CA, USA) operated at 70 eV. The column used was an HP-5MS capillary column (Agilent Technologies, Santa Clara, USA, 0.25-mm I.D. × 30 m, 0.25-µm film thickness). The carrier gas was helium with a constant flow rate of 1.00 mL/min. Samples were analyzed in the splitless mode with the temperature programmed to change from 60 °C (initially for 2 min) to 290 °C at a rate of 10 °C/min. The final temperature (290 °C) was then maintained for 5 min. The GC–MS data were recorded using Chemstation (Agilent Technologies, Santa Clara, CA, USA) with reference to an MS database (Agilent NIST05 mass spectral library, Agilent Technologies, Santa Clara, CA, USA).

2.5.3. Quantitative Determination of Three Ester Compounds

A calibration curve was constructed for each compound. The curve was obtained by correlating the GC–MS response data of the crude extract of beetle with each concentration of three standard solutions. A synthetic sample of each ester (2-propyl dodecanoate, compound **1**; 3-pentyldodecanoate, compound **2**; and 3-pentyl tetradecanoate, compound **3**) was diluted with hexane. The following concentrations were prepared: 5, 10, and 25 ng/µL; a 200 ng/µL solution was also prepared for the major compound (**2**). A calibration curve was then constructed.

2.6. Data Analysis

To determine the lifespan of male beetles, survival rates of adult male beetles were recorded weekly defined by fraction surviving. The fraction surviving (Li) was calculated by dividing the number of male beetles surviving at the age i by the initial number of adult male beetles [24]. Then, a Kaplan–Meier analysis followed by a log-rank method with a confidence interval (CI) of 0.05 was used to determine the survival duration of the beetle for 15 weeks of observation. The mean quantities of pheromone detected over the lifespan were subjected to log transformation and one-way analysis of variance (ANOVA), followed by Tukey's HSD as a post hoc test with CI of 0.05. Meanwhile, the means of pheromone responses by mating status effect were subjected to a Mann–Whitney U test with CI of 0.05.

3. Results

3.1. Life Span of Adult Male L. africanus in Starved and Unstarved Conditions

Figure 2 shows lifespan data of male *L. africanus* adults. Throughout their lives, the survival of starved male beetles was lower than that of unstarved beetles. The number of starved male beetles dropped precipitously with time, whereas that of the unstarved beetles decreased gradually until the end of their life period. The longest lifespan of an unstarved male beetle was 14 weeks, while that of starved male beetles was only 5 weeks. The Kaplan–Meier analysis indicated that unstarved beetles survived significantly longer than starved beetles.

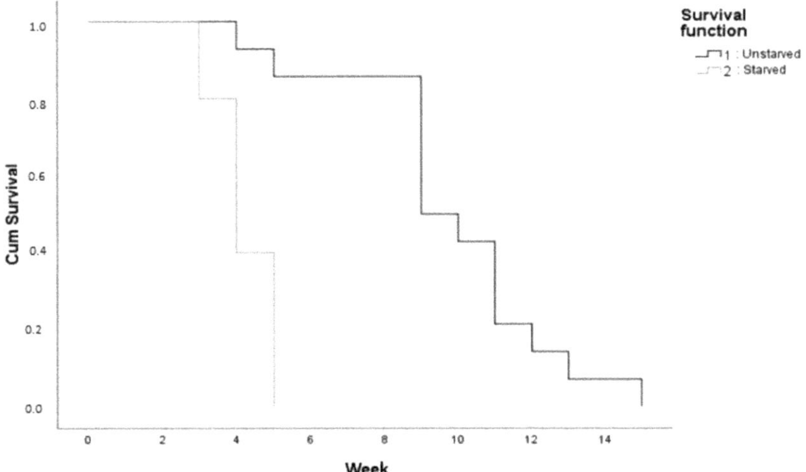

Figure 2. Survival function of starved and unstarved males of *L. africanus*. (Kaplan–Meier analysis, $p = 0.00$).

3.2. Effects of Mating Status on Pheromone Production

Figure 3 shows the quantity of each pheromone compound measured in the mated and unmated males. The figure reveals a tendency toward higher levels of all three ester compounds in the pheromones of mated males compared to unmated ones. The difference between mated and unmated males, however, only reached significance in compound **1**; the quantities of compounds **2** and **3** were not significantly different.

3.3. Effects of Aging on Pheromone Production

As shown in Figure 2, the maximum lifespan of male *L. africanus* was 14 weeks, but very few beetles survived this long. Thus, pheromone measurement was not conducted at 14 weeks.

Table 1 presents the quantitative fluctuation of the pheromone compounds over time. Pheromone compounds, particularly compounds **2** and **3**, were produced in beetles throughout their lives from youth until death. There were no differences in the quantity of pheromone compound **1** by time. However, compounds **2** and **3** were both initially produced at low amounts by young *L. africanus* beetles, then increased until week 5, before subsequently decreasing at week 9.

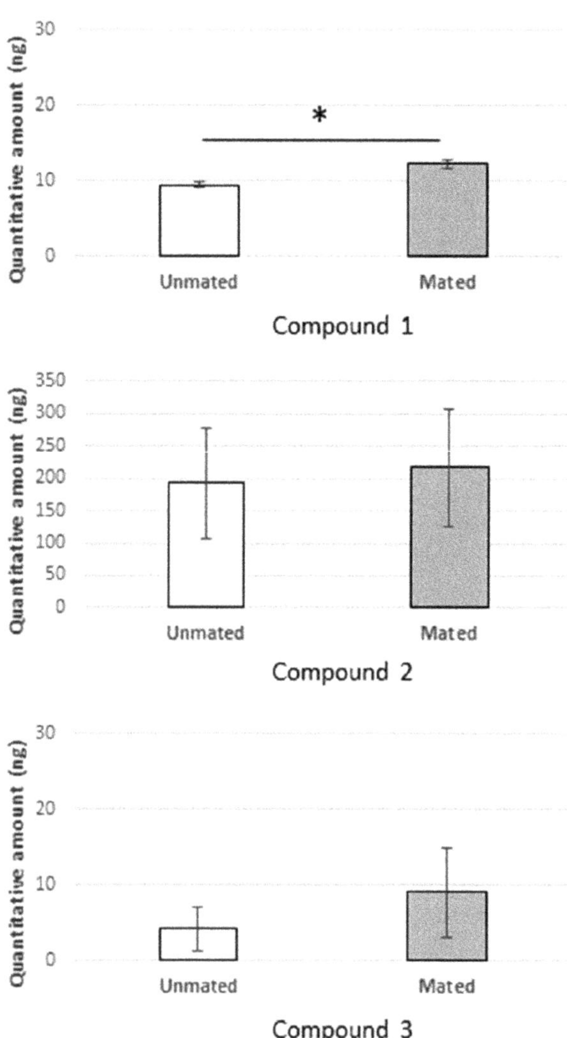

Figure 3. Pheromone production (compound **1**, **2**, and **3**) in the mated and unmated adult male *L. africanus*. Error bar represents the standard deviation, asterisk indicates a significant difference ($p < 0.05$).

Table 1. Mean production of pheromone compounds by *L. africanus* at different ages.

Time (Weeks)	Pheromone Quantity (ng)		
	Compound 1	Compound 2	Compound 3
1	0.00 ± 0.00 [a]	93.14 ± 19.15 [b]	2.22 ± 1.93 [c]
3	8.06 ± 2.31 [a,*]	142.73 ± 45.31 [b]	14.84 ± 6.87 [b,c]
5	21.67 ± 15.04 [a]	1185.15 ± 412.60 [a]	186.73 ± 56.36 [a]
9	5.84 ± 3.60 [a]	376.34 ± 154.50 [a,b]	65.07 ± 20.31 [a,b]

Note: * Same letter within a column indicates no significant difference (Tukey-Kramer HSD test; $p <$ 0.05) following one-way ANOVA.

4. Discussion

This study indicates that the survival of starved male *L. africanus* beetles was lower than that of beetles provided food, which means *L. africanus* might utilize the artificial diet. As mentioned previously, the adults of *L. africanus* are classified into a group that can taste and feed on wood. As reported by Cymorex and Schmidt [17], the gnawing or feeding behavior and feces production of five species of *Lyctus* were quantitatively and qualitatively analyzed using an artificial diet. The *L. africanus* were observed to gnaw, feed, and produce feces on wood materials, such as wood carvings. The feeding behavior can be related to the specialization of gut anatomy, where the development of fine villi in the adult mid-gut section of *L. africanus* has been observed. The *L. africanus* is suggested to have a more functional digestion system than other species, *L. brunneus* or *L. planicollis*. The food may not only act as nutrition and provide a hiding space or means for reducing stress but also as a water source from polysaccharide decomposition in dry wood.

In addition, the male *L. africanus* beetle showed a tasting behavior like males of a closely related species, *L. brunneus* [20]. In general, tasting behavior is performed by females of *Lyctus* beetle in order to select an oviposition site, as the eggs survive best in starchy wood. A 1984 report on other beetle species in the Bostrichidae family stated that beetles with lifespans longer than 1 month require feeding for reproduction [25]. As for *L. africanus*, the lifecycle of beetles is around 2.5 months with a suitable diet for producing a new generation after the introduction of the artificial diet [19]. As mentioned earlier, *L. africanus* adults survived longer in the unstarved condition. Based on these results, the adult *L. africanus* beetles were then maintained with the wood-based diet to prolong their lifespans.

During their life spans, the lyctines are able to mate and oviposit fertile eggs within 24 h of emergence [26]. Mating is known to trigger immediate changes in the physiology and behavior of insects. This study revealed that chemical changes in the male *L. africanus* beetle occurred after mating. The results of the analysis of chemical compounds measured on mated and unmated male *L. africanus* beetles indicate that the quantity of compound **1** was affected by the mating status of male *L. africanus* beetle, while those of compounds **2** and **3** were relatively unaffected. In other words, the production of compounds **2** and **3** of the aggregation pheromone components of the *L. africanus* beetle was unaffected by the mating history. In general, the mating history of male insect beetles had no effect on insect male-produced pheromones [27]. One study reported that the female of an ant species, *Leptothorax gredleri*, was promptly avoided by males after mating, coinciding with immediate changes of the female cuticular hydrocarbon (CHC) profile, known as a pheromone. The modification of the CHC produced by the females was detectable by males, allowing them to distinguish between mated and unmated females [28].

Compound **1** is a minor compound found in the aggregation pheromone of the male *L. africanus* beetle, suggesting it would have a minor effect on aggregation behavior [15]. There was a synergistic effect among the three synthetic compounds when they were blended, indicating the essential role of compound **2** in the aggregation behavior of *L. africanus*. The other single compounds **1** and **3** induced an insignificant effect on the beetle's responses when used alone. Hence, the increased quantity of compound **1** after mating could be affected by some factors in relation to mating activity, such as the stage of sexual maturity and the presence of female beetles. In a study on *Tenebrio molitor* (Tenebrionidae), the presence of other females acted to accelerate sexual maturation as indicated by pheromone emission rate [29]. This mechanism could also occur in *L. africanus*, which generates sexual maturity of the male beetle by the female through a mating process. However, further study is necessary to confirm the function of the compound. Pheromone exchanges frequently occur during mating, as the act of mating appears to initiate and inhibit particular pheromone compounds [30–32].

Measurement of pheromone compounds **1**, **2**, and **3** of the male *L. africanus* beetle indicated a continuous production of the compounds throughout their lives. As illustrates in Table 1, compound **1** production was relatively stable over time. A quantitative fluctuation of the pheromone compounds with time was found in compounds **2** and **3**; however, both were initially produced in low amounts by young *L. africanus* beetles, then gradually increased until week 5, before subsequently decreasing at week 9. Similar patterns have been reported in other Bostrichidae family members, such as *Rhyzopertha dominica* [32]. Dominicalure-1 and Dominicalure-2 as pheromone components of *R. dominica* were released by male *R. dominica* in higher quantities when the insects were relatively young, then declined in significant numbers at 24 weeks, and remained stable thereafter. The decrease of emitted pheromone reduces the insect's lifetime reproductive fitness since the male-produced pheromone primarily attracts females as potential mates. Hence, aging could lower the recruitment rate by the insect. The current study found compounds **2** and **3**, major components of aggregation pheromone in male *L. africanus* beetle, were affected by aging. As previously mentioned, although the unstarved male *L. africanus* beetle could survive until 14 weeks, the production of those major compounds started declining at 9 weeks. In general, aging is known as a process that impacts a broad functional decline in health, increasing vulnerability to diseases and death, and reducing reproductive output. One study reported that composition changes of CHC were found in both male and female *Drosophila melanogaster* consistently with age [33].

Table 1 reveals that compound **2** was by far the dominant aggregation pheromone component in male beetles, as reported in our previous study [15]. This compound was detected in large amounts in every measurement, followed by compound **3** and compound **1**. These results suggest that compounds **2** and **3** might have an important function in the aggregation behavior of *L. africanus*, especially in the signaling behavior for mating opportunities and oviposition. We previously found that compound **1** is a minor compound in the aggregation pheromone of *L. africanus* [15]. This study clarified the minor effect of compound **1** in the aggregation pheromone of *L. africanus*, but we were not able to elucidate its role in the aggregation behavior of *L. africanus*. Further study is required to confirm the roles of each compound and their combination in the aggregation pheromone of *L. africanus*. The present information will help foster a better understanding of the chemical communication of the insect, which may be important for the development of improved pheromone-based management strategies for controlling *Lyctus* beetles.

5. Conclusions

This study indicated that the adult *L. africanus* survived longer in an unstarved condition. The food may act not only as nutrition but also as a hiding space or means for reducing stress. It was suggested that aggregation pheromones produced by male *L. africanus* beetle, which consists of two major compounds **2** and **3**, were affected by age, with maximum production at the 5th week, and not by the mating status. However, the other minor compound **1** was produced steadily without significant change up to 9 weeks, and its production was affected by the mating status of the male *L. africanus* beetle.

Author Contributions: All authors conceived and designed research. Conceptualization, T.K., N.S., and T.Y.; methodology, T.K.; software and analytical tools, N.S.; validation, T.K. and N.S.; formal analysis, T.K., D.T., S.K.H., I.G., and S.Y.; investigation, T.K.; resources, N.S. and T.Y.; data curation, T.K.; writing—original draft preparation, T.K.; writing—review and editing, T.K., N.S., T.Y., D.T., S.K.H., I.G., and S.Y.; visualization, T.K. and N.S.; supervision, T.Y.; project administration, T.Y. All authors have read and agreed to the published version of the manuscript.

Funding: This research received no external funding.

Institutional Review Board Statement: Not applicable.

Informed Consent Statement: Not applicable.

Data Availability Statement: Data set available on request to corresponding author.

Acknowledgments: The authors would like to thank Akio Adachi, Laboratory of Innovative Humano-habitability, Research Institute for Sustainable Humanosphere, Kyoto University, for helping in the preparation of artificial diet. The authors also address their appreciation to Budiman Ismail from Research Center for Biomaterials, Indonesian Institute of Sciences, and Utami Dyah Syafitri from IPB University for helpful guidance in data analysis. The authors also thank JASTIP (Japan-ASEAN Science Technology Innovation Platform) Program, a collaborative project of Kyoto University and Indonesian Institute of Sciences (LIPI) under Work-Package 3: Bioresources and Biodiversity for their technical support.

Conflicts of Interest: The authors declare no conflict of interest.

References

1. Ebeling, W. Wood-Destroying Insects and Fungi. In *Urban Entomology*; Division of Agricultural Sciences, University of California: San Diego, CA, USA, 1978; pp. 167–216.
2. Kliejunas, J.T.; Burdsall, H.H., Jr.; DeNitto, G.A.; Eglitis, A.; Haugen, D.A.; Haverty, M.I.; Micales, J.A.; Tkacz, B.M.; Powell, M.R. *Pest Risk Assessment of the Importation into the United States of Unprocessed Logs and Chips of Eighteen Eucalypt Species from Australia. True Powder Post Beetle*; USDA Forest Service General Technical Report FPL-GTR-137; Forest Products Laboratory, United States Department of Agriculture: Washington, DC, USA, 2003.
3. Liu, L.-Y. New records of Bostrichidae (Insecta: Coleoptera, Bostrichidae, Bostrichinae, Lyctinae, Polycaoninae, Dinoderinae, Apatinae). *Mitt. Munch. Ent. Ges.* **2010**, *100*, 103–117.
4. Liu, L.-Y.; Beaver, R.A. A synopsis of the powderpost beetles of the Himalayas with a key to the genera (Insecta: Coleoptera: Bostrichidae). In *Biodiversität und Naturausstattung im Himalaya VI*; Hartmann, Barclay & Weipert: Erfurt, Germany, 2018; pp. 407–422.
5. Liu, L.-Y.; Schönitzer, K. Phylogenetic analysis of the family Bostrichidae auct. at suprageneric levels (Coleoptera: Bostrichidae). *Mitt. Munch. Ent. Ges.* **2011**, *101*, 99–132.
6. Gerberg, E.J. A revision of the new world species of powder-post beetles belonging to the family Lyctidae. *Tech. Bull.* **1957**, *1157*.
7. Halperin, J.; Geis, K.U. Lyctidae (Coleoptera) of Israel, their damage and its prevention. *Phytoparasitica* **1999**, *27*, 257–262. [CrossRef]
8. Furukawa, N.; Yoshimura, T.; Imamura, Y. Survey of lyctine damages on houses in Japan: Identification of species and infested area. *Wood Preserv.* **2009**, *35*, 260–264. (In Japanese) [CrossRef]
9. Khalsa, H.; Nigam, B.; Agarwal, P. Breeding of powder post beetle, *Lyctus africanus* Lense (Coleoptera) in an artificial medium. *Indian J. Entomol.* **1962**, *24*, 139–142.
10. Beroza, M. Microanalytical Methodology Relating to the Identification of Insect Sex Pheromones and Related Behavior-Control Chemicals. *J. Chromatogr. Sci.* **1975**, *13*, 314–321. [CrossRef]
11. Wertheim, B. *Ecology of Drosophila Aggregation Pheromone: A Multitrophic Approach*; Laboratory of Entomology, Wageningen University: Wageningen, The Netherlands, 2001; p. 200.
12. Wertheim, B.; van Baalen, E.J.A.; Vet, L.E.M. Pheromone-mediated aggregation in non-social arthropods: An evolutionary ecological perspective. *Annu. Rev. Entomol.* **2005**, *50*, 321–346. [CrossRef]
13. Shorey, H.H. Behavioral Responses to Insect Pheromones. *Annu. Rev. Èntomol.* **1973**, *18*, 349–380. [CrossRef]
14. Borden, J.H. Aggregation Pheromones. In *Comprehensive Insect Physiology, Biochemistry and Pharmacology*; Kerkut, G.A., Gilbert, L.I., Eds.; Pergamon Press: Oxford, UK, 1985; pp. 257–285.
15. Kartika, T.; Shimizu, N.; Yoshimura, T. Identification of Esters as Novel Aggregation Pheromone Components Produced by the Male Powder-Post Beetle, *Lyctus africanus* Lesne (Coleoptera: Lyctinae). *PLoS ONE* **2015**, *10*, e0141799. [CrossRef]
16. Papadopoulos, N.T.; Katsoyannos, B.L.; Kouloussis, N.A.; Carey, J.R.; Muller, H.-G.; Zhang, Y. High sexual signaling rates of young individuals predict extended life span in male Mediterranean fruit flies. *Oecologia* **2004**, *138*, 127–134. [CrossRef] [PubMed]
17. Cymorex, V.S.; Schmidt, H. On the gnawing and feeding behavior of powder-post beetles (Lyctidae) with observations on gut anatomy and gut parasites. *Mater. Organ. Beih.* **1976**, *3*, 429–440.
18. Iwata, R.; Nishimoto, K. Studies on the autecology of *Lyctus brunneus* (Stephens): Artificial diets in relationship to beetle supply. *Mokuzai Gakkaishi* **1983**, *29*, 336–343. (In Japanese)
19. Kartika, T.; Yoshimura, T. Evaluation of Wood and Cellulosic Materials as Fillers in Artificial Diets for *Lyctus africanus* Lesne (Coleoptera: Bostrichidae). *Insects* **2015**, *6*, 696–703. [CrossRef]
20. Ito, T. Tasting Behavior of *Lyctus brunneus* Stephens (Coleoptera, Lyctidae). *Appl. Entomol. Zool.* **1983**, *18*, 289–292. [CrossRef]
21. Yinon, U.; Shulov, A. New findings concerning pheromones produced by *Trogoderma granarium* (Everts), (Coleoptera, Dermestidae). *J. Stored Prod. Res.* **1967**, *3*, 251–254. [CrossRef]
22. Korada, R.R.; Griepink, F.C. Aggregation pheromone compounds of the black larder beetle *Dermestes haemorrhoidalis* Kuster (Coleoptera: Dermestidae). *Chemoecology* **2009**, *19*, 177–184. [CrossRef]
23. Shimizu, N.; Kuwahara, Y.; Yakumaru, R.; Tanabe, T. n-Hexyl Laurate and Fourteen Related Fatty Acid Esters: New Secretory Compounds from the Julid Millipede, Anaulaciulus sp. *J. Chem. Ecol.* **2012**, *38*, 23–28. [CrossRef]
24. Bong, L.-J.; Neoh, K.-B.; Jaal, Z.; Lee, C.-Y. Life table of *Paederus fuscipes* (Coleoptera: Staphylinidae). *J. Med. Èntomol.* **2012**, *49*, 451–460. [CrossRef]

25. Burkholder, W. Stored-product insect behavior and pheromone studies: Keys to successful monitoring and trapping. In Proceedings of the 3rd International Working Conference of Stored-Product Entomology, Manhattan, KS, YSA, 23–28 October 1983; pp. 20–33.
26. Gay, F.J. Observation on biology of *Lyctus brunneus* (Steph). *Aust. J. Zool.* **1953**, *1*, 102–110. [CrossRef]
27. Walgenbach, C.A.; Phillips, J.K.; Faustini, D.L.; Burkholder, W.E. Male-produced aggregation pheromone of the maize weevil, *Sitophilus zeamais*, and interspecific attraction between three *Sitophilus* species. *J. Chem. Ecol.* **1983**, *9*, 831–841. [CrossRef] [PubMed]
28. Oppelt, A.; Heinze, J. Mating is associated with immediate changes of the hydrocarbon profile of *Leptothorax gredleri* ant queens. *J. Insect Physiol.* **2009**, *55*, 624–628. [CrossRef] [PubMed]
29. Happ, G.M.; Wheeler, J. Bioassay, Preliminary Purification, and Effect of Age, Crowding, and Mating on the Release of Sex Pheromone by Female Tenebrio molitor1. *Ann. Èntomol. Soc. Am.* **1969**, *62*, 846–851. [CrossRef]
30. Scott, D.; Richmond, R.C.; Carlson, D.A. Pheromones exchanged during mating: A mechanism for mate assessment in Drosophila. *Anim. Behav.* **1988**, *36*, 1164–1173. [CrossRef]
31. Thomas, M.L. Detection of female mating status using chemical signals and cues. *Biol. Rev.* **2010**, *86*, 1–13. [CrossRef] [PubMed]
32. Edde, P.A.; Phillips, T. Longevity and pheromone output in stored-product Bostrichidae. *Bull. Èntomol. Res.* **2006**, *96*, 547–554. [CrossRef] [PubMed]
33. Kuo, T.-H.; Yew, J.Y.; Fedina, T.Y.; Dreisewerd, K.; Dierick, H.A.; Pletcher, S.D. Aging modulates cuticular hydrocarbons and sexual attractiveness in *Drosophila melanogaster*. *J. Exp. Biol.* **2012**, *215*, 814–821. [CrossRef] [PubMed]

 *insects*

Article

Development of Monitoring and Mating Disruption against the Chilean Leafroller *Proeulia auraria* (Lepidoptera: Tortricidae) in Orchards

M. Fernanda Flores [1,2], Jan Bergmann [1], Carolina Ballesteros [3], Diego Arraztio [4] and Tomislav Curkovic [4,*]

1   Instituto de Química, Pontificia Universidad Católica de Valparaíso, Avda. Universidad 330, Curauma, Valparaíso 234000, Chile; fernanda.flores.e@gmail.com or fflores@agroadvance.cl (M.F.F.); jan.bergmann@pucv.cl (J.B.)
2   Agroadvance SpA. Camino Melipilla, Peñaflor, Santiago 26200, Chile
3   Facultad de Agronomía e Ingeniería Forestal, Pontificia Universidad Católica de Chile, Avda. Vicuña Mackenna 4860, Macul, Santiago 7820436, Chile; caballesteros@uc.cl
4   Facultad de Ciencias Agronómicas, Universidad de Chile, Avda. Santa Rosa 11315, La Pintana, Santiago 8820808, Chile; dsarrazt@uchile.cl
*   Correspondence: tcurkovi@uchile.cl

Citation: Flores, M.F.; Bergmann, J.; Ballesteros, C.; Arraztio, D.; Curkovic, T. Development of Monitoring and Mating Disruption against the Chilean Leafroller *Proeulia auraria* (Lepidoptera: Tortricidae) in Orchards. *Insects* **2021**, *12*, 625. https://doi.org/10.3390/insects12070625

Academic Editors: Andrea Lucchi and Giovanni Benelli

Received: 31 March 2021
Accepted: 25 May 2021
Published: 9 July 2021

Publisher's Note: MDPI stays neutral with regard to jurisdictional claims in published maps and institutional affiliations.

Simple Summary: *Proeulia auraria* is a native and growing pest insect in fruit orchards in Chile, which calls for environmentally friendly management methods. Using synthetic pheromone compounds, we conducted field trials to optimize the septa load for monitoring adult moths. Using the optimized blend we studied the phenology of males in vineyards, apples, and blueberries, finding two large flight cycles lasting from September to May. Afterward, based on field trials, we concluded that 250 point sources loaded with a total of 78 g/ha of the pheromone blend, provided high disruption of male-female encounters for mating in all crops tested for at least 5 months. We concluded that mating disruption is feasible for *P. auraria*, needing now the development of a commercial product and of protocols to control this pest.

Abstract: The leafroller *Proeulia auraria* (Clarke) (Lepidoptera: Tortricidae) is a native, polyphagous, and growing pest of several fruit crops in Chile; it also has quarantine importance to several markets, thus tools for management are needed. Using synthetic pheromone compounds, we conducted field trials to optimize the blend for monitoring, and to determine the activity period of rubber septa aged under field conditions. We concluded that septa loaded with 200 µg of E11-14:OAc + 60 µg E11-14:OH allowed for efficient trap captures for up to 10 weeks. Using this blend, we studied the phenology of adult males in vineyards, apple, and blueberry orchards, identifying two long flight cycles per season, lasting from September to May and suggesting 2–3 generations during the season. No or low adult activity was observed during January and between late May and late August. Furthermore, mating disruption (MD) field trials showed that application of 250 pheromone point sources using the dispenser wax matrix SPLAT (Specialized Pheromone and Lure Application Technology, 10.5% pheromone) with a total of 78 g/ha of the blend described above resulted in trap shutdown immediately after application, and mating disruption >99% in all orchards for at least 5 months. We concluded that MD is feasible for *P. auraria*, needing now the development of a commercial product and the strategy (and protocols) necessary to control this pest in conventional and organic orchards in Chile. As far as we know, this is the first report on MD development against a South American tortricid pest.

Keywords: pheromone; field trials; moth phenology; vineyards; apple orchards; blueberry orchards; SPLAT wax matrix; remaining pheromone in point sources

1. Introduction

The Chilean fruit or pear leafroller, *Proeulia auraria* (Clarke) (Lepidoptera: Tortricidae: Tortricinae), is distributed in Chile, from Atacama (coordinates −27.37, −70.33) to Los Lagos (−41.47, −72.94) [1–3]. The species is also mentioned in the literature from Argentina but restricted to Chile based on the Centre for Agricultural and Bioscience International [4,5], therefore considered a native moth. *Proeulia auraria* is the economically most important species of the Genus *Proeulia*, endemic to Chile [6–8]. It is a polyphagous insect that has moved from native hosts to exotic plants [9]; the hosts reported for either *P. auraria* or *Proeulia* sp, include citrus, figs, grapes, kiwifruits, loquats, pome fruits, pomegranates, stone fruits, and walnuts; several native plants (e.g., *Ugni molinae* Turcz), weeds (e.g., *Galega officinalis* L.), and ornamental trees (e.g., *Platanus orientalis* L.) [3,9–12]. During the last four decades, *P. auraria* has become a key pest in some vineyards and orchards in Chile [3,10]. This is probably due to changes in management tactics for other pests (e.g., withdrawal of insecticide applications because of increasing use of mating disruption against *Cydia pomonella* L. and *Lobesia botrana* (Denis & Schiffermüller)), the decreasing amount of natural enemies due to insecticide use, and/or replacement of the original vegetation including *P. auraria* native hosts [9,10]. The larvae cause damage by folding foliage and eating flowers, reducing the photosynthetic potential, the canopy development, and the fruit set. Larvae also feed on the fruit epidermis, and cause surface tunneling, which makes the fruits unmarketable, and facilitates the development of saprophagous insects (e.g., *Drosophila melanogaster* Meigen) or fungal diseases (e.g., *Botrytis cinerea* Pers.) [9]. However, the most important economic impact is due to its quarantine status, frequently causing rejections of Chilean fresh fruits for export when immature stages (mainly larvae) are found during fruit inspections after harvest [10,11]. In fact, several reports [13,14] mentioned *P. auraria* within the list of new pests representing a risk of introduction to different new areas importing plants and fresh fruits from South America. For this reason, many growers currently rely on different insecticides (e.g., *Bacillus thuringiensis*, organophosphates, etc.) as the most common pre-harvest tactic used against this pest [15,16].

Monitoring *P. auraria* males for timing insecticide applications has been conducted in Chile using pheromone-baited traps [16]. The first report of *P. auraria* pheromone established a 7:3-ratio of (*E*)-11-tetradecenyl acetate (E11-14:OAc) and (*E*)-11-tetradecenol (E11-14:OH) as the main components [17], whereas Roelofs and Brown [18] reported only the alcohol as the sex pheromone. During several years, *P. auraria* monitoring in Chile was conducted using pheromone lures for the tufted apple bud moth (TBM), *Platynota ideausalis* Walker (Tortricidae) [10] containing a 1:1 ratio of E11-14:OAc and E11-14:OH. However, a significantly more attractive 4-component sex pheromone for *P. auraria* populations from central Chile was reported by our group [19]: tetradecyl acetate (14:OAc), E11-14:OAc, (*Z*)-11-tetradecenyl acetate (Z11-14:OAc), and E11-14:OH, in a relative ratio of 11:100:1:37. In fact, recently this blend was successfully tested in Southern areas of Chile [20]. Additionally, preliminary data obtained from small-scale field trials suggested that *P. auraria* males might be susceptible to mating disruption [21].

Based on that information, the objectives of this study were to develop and optimize monitoring and to further explore the potential of mating disruption (MD) for *P. auraria*. Particularly, we conducted field trials to determine the optimum blend and the longevity of septa for efficient monitoring, and the initial load and emitter density necessary to maximize disruption.

2. Materials and Methods

2.1. Orchards

Field trials were conducted in commercial fields in the Region of O'Higgins, central Chile: a conventional 5-years-old blueberry orchard (cvs. O'Neill and Brigitta planted at 3 × 0.8 m) near San Francisco de Mostazal (coordinates −33.98, −70.68); two organic (>20 years old) vineyards (cvs. Cabernet Sauvignon, Carmenere, and Syrah at 2–2.5 × 1–1.3 m) near Requinoa (−34.28, −70.81) and Nancagua (−34.63, −71.22); and an organic 10-years-old apple orchard (cvs. Granny Smith and Gala at 2.7 × 2 m) near San Fernando (-34.59, -70.98). All four orchards have historically had infestations by *P. auraria*, thus, spray programs using *Bacillus thuringiensis* (Bt) and/or Spinosad were being applied in all of them. However, neither Bt nor Spinosad were applied in the MD plots during our study.

2.2. Chemicals, Traps, and Pheromone Matrix

E11-14:OAc (purity 99%, containing ca. 1% of Z11-14:OAc), Z11-14:OAc (99%), and E11-14:OH (>99%) were purchased from Bedoukian Research Inc (Danbury, CT, USA) and were either used as received (hereafter "NP" or "non-purified") or were purified by column chromatography on silica gel impregnated with silver nitrate to remove the respective geometric isomer and other impurities (hereafter "PU" or "purified"). The isomeric purity of the compounds after purification was >99%, as determined by gas chromatography. Tetradecyl acetate (14:OAc) was obtained according to the procedure described previously [19] and had a purity of >99%. All compounds were dissolved in hexane (Suprasolv®, Merck, Darmstadt, Germany), and appropriate amounts were loaded on white rubber septa (Sigma Aldrich) to prepare the treatments described below. Septa for control treatments were loaded with hexane alone. After evaporation of the solvent, septa were placed inside Pherocon VI Delta traps (Trécé Inc., Adair, OK, USA) which were hung at canopy level (1.5–1.8 m) separated 30 m from each other for composition, proportion, and septa aging trials, or separated >50 m (for phenology and mating disruption studies). Unlike the vineyard and the blueberry orchard, the plots used in the MD trial in the apple orchard were uprooted just after harvest (late February) and the respective traps destroyed, preventing us to complete that study. Traps were placed at the south-west quadrant inside the tree canopy, considering always a distance of ca. 10 m from the plot edges. For mating disruption trials, an emulsified microcrystalline wax matrix called SPLAT® (an acronym for "Specialized Pheromone and Lure Application Technology", [22]) was mixed with *P. auraria* pheromone components (10.5% *w/w* of a blend consisting of non-purified E11-14:OAc and E11-14:OH at a 1:0.3 ratio), loaded in manual silicone sealant guns, and stored at $-20\ °\mathrm{C}$ until application in the fields. The mixture was prepared by ISCA Technologies (Riverside, CA, USA). SPLAT not loaded with pheromone was used as a control treatment in the mating disruption trials, as also recently reported by other authors [23,24].

2.3. Optimizing Pheromone Load in Septa

The first experiment studied the composition of lures. Treatments for this field trial ("composition trial") were: (1) E11-14:OAc (100 µg), E11-14:OH (30 µg)—both NP -, and 14:OAc (14 µg); (2) E11-14:OAc (100 µg) and E11-14:OH (30 µg)—both NP -; (3) E11-14:OAc (100 µg) and E11-14:OH (30 µg)—both PU -; (4) E11-14:OAc (100 µg), E11-14:OH (30 µg), Z11-14:OAc (1 µg)—all PU -, and 14:OAc (14 µg); (5) control (only hexane). This experiment was carried out in the Requínoa vineyard starting 12 November 2014. The second experiment evaluated the effect of the two main pheromonal compounds ratio. Treatments for this field trial ("proportion trial") were blends of E11-14:OAc and E11-14:OH (both NP) at ratios (µg/septum): (1) 100:100; (2) 100:60; (3) 100:30; (4) 100:10; (5) hexane only. This experiment was carried out in the Nancagua vineyard starting 18 February 2015. In both trials, the traps were deployed in a randomized complete block design with four replicates. Blocks, and traps within blocks, were at least 30 m apart from each other. Traps were checked for captures 7 days after setup in the field. Results are expressed as the average cumulative catches/trap ($n = 4$) $\pm$ SE. Data were analyzed by straight 2-way fixed-effect ANOVA and the Tukey HSD test for mean discrimination ($p \leq 0.05$) [25]. Statistical analyses were done with MINITAB version 2019.

To evaluate lures longevity, rubber septa ($n = 30$/treatment) were loaded with mixtures of E11-14:OAc and E11-14:OH (both NP) as follows: (1) 50 µg + 15 µg, (2) 200 µg + 60 µg, (3) 800 µg + 240 µg, (4) hexane only (control). Afterward, twelve septa (3 per treatment) were placed in a 6 × 4.5 cm tulle bag ($n = 10$). Thus, on 17 December 2014, all bags were placed on vines in the vineyard (Nancagua) at ca. 1.5 m height and ca. 10 m apart, and were covered by a piece of corrugated cardboard (as a roof) to avoid direct sun exposure. During up to 10 weeks, one bag per week was removed and maintained in a sealed container at $-20\ ^\circ$C until use. On 18 February 2015, septa were placed inside Delta traps (1 septum per trap) which were deployed in the field in a completely randomized block design (n = 3), and male catches were counted 5 days later. Trap distribution and setup were as described previously. Data were analyzed by General Linear Model-ANOVA with a factorial arrangement and means were separated by the Tukey HSD test ($p \leq 0.05$) [25].

2.4. Remaining Pheromone in Septa after Aging under Field Conditions

To determine the amount of remaining pheromone after aging septa under field conditions, thirty rubber septa were loaded with 200 µg of E11-14:OAc and 60 µg of E11-14:OH (both non-purified), and 5 septa were placed together in tulle bags (as described above) in the Nancagua vineyard on 12 October 2016. One bag was immediately taken to the freezer and the remaining 5 bags were kept under field conditions as described above. Every 30 days another bag was collected and stored in a sealed container at $-20\ ^\circ$C until 10 March 2017. The remaining pheromone was extracted from individual septa with 5 mL hexane for 30 min assisted by ultrasound. Extracts were analyzed by gas chromatography-mass spectrometry (GC-MS), using a Shimadzu GCMS-QP2010 Ultra equipped with a fused silica RTX-5 capillary column (30 m $\times$ 0.25 mm id, 0.25 µm film, Restek) and a temperature oven program from 50 $^\circ$C for 5 min to 270 $^\circ$C at 8 $^\circ$C min^{-1}. Quantification was based on the area under the respective peak, compared to an external calibration curve obtained using authentic standards.

2.5. Proeulia Auraria Adult Male Phenology

Monitoring of *P. auraria* males was conducted during two consecutive seasons (2015/16 and 2016/17) in one vineyard (Requinoa) and the two orchards described above. Septa were loaded with 200 µg of E11-14:OAc and 60 µg of E11-14:OH (both NP) and replaced every 10 weeks. Delta traps (n = 3/orchard or vineyard) with those septa were installed in 2015: monitoring started at the beginning of September (vineyards and apples) or October (blueberries) and ended in May 2017. The numbers of trapped males were counted once a week, except between mid-June and late August, when traps were revised once or twice a month.

2.6. Proeulia Auraria Mating Disruption Trials

The SPLAT was applied in 30 mL plastic cups (to prevent the material from running off) in amounts equivalent to 78 g/ha (MD treatment; SPLAT 10.5% *w/w* of E11-14:OAc and E11-14:OH both NP, at 1:0.3 ratio) and 0 g/ha (control treatment; SPLAT pheromone-free), with 250 point sources/ha (based on previous results from small scale field trials [21]). At the beginning of September 2016, cups were homogeneously distributed in the plant canopy (1.5–1.8 m height) in the respective plots (ca. 4 ha each); plots were spaced at least 50 m apart in the Requinoa vineyard and the orchards. Two pheromone-baited Delta traps (as described above, see *Proeulia auraria adult male phenology*) were placed in July 2016 in each plot, and captured males were counted weekly until March 2017. Disruption indexes (DI) for seasonal male captures/treatment were calculated for each fruit species, using the equation DI = 100 * (C $-$ D)/C (where C = captures in control plots, D = captures in respective MD plots).

Fruit damage was evaluated at harvest by contrasting two randomly chosen fruit clusters per tree, in 150 (blueberries), 421 (apples), and 559 (grapes) plants/plot, in the control and the MD treatment, respectively. The results were analyzed using the test for the equality of proportions ($p \leq 0.05$) [25].

2.7. Remaining Pheromone in SPLAT Matrix after Ageing under Field Conditions

To determine the amount of remaining pheromone in the SPLAT matrix after aging under field conditions, 30 plastic cups containing the pheromone-loaded matrix as described above were placed in the Requinoa vineyard on 15 October 2016. After 30, 60, 90, 120, 150, and 210 days, 5 cups were removed at a time and submitted to analysis. The remaining pheromone was extracted as described above. After filtration through grade 1 filter paper, extracts were analyzed and quantified as described above (see *Remaining pheromone in septa after aging under field conditions*).

3. Results

3.1. Optimizing Pheromone Load in Septa

3.1.1. Pheromone Composition Trial

The field trial designed to evaluate optimum pheromone composition showed significantly larger *P. auraria* male catches ($F = 24.21$, $df = 4$, $p < 0.001$, Figure 1) when either the 2- or the 3-component blends, prepared with NP compounds, were used. Treatments using blends prepared with purified compounds were statistically less attractive. All blends were significantly more attractive than the control.

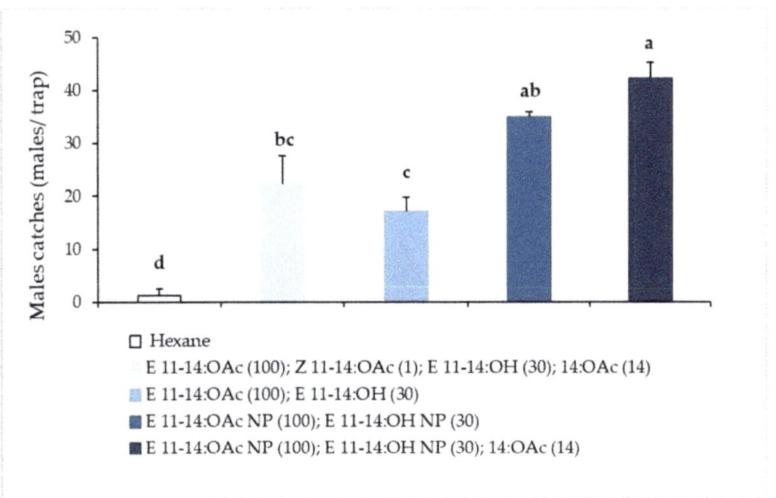

Figure 1. Male *P. auraria* cumulative catches (avg. ± SE) in traps ($n = 4$) loaded with different pheromone compositions or hexane (control), Requinoa, central Chile, 12–19 November 2014. Different small letters above a column mean significant differences between treatments. NP = commercial compounds not purified; if not indicated, the compounds were purified. Figures in parenthesis represent µg/septa.

3.1.2. Pheromone Proportion Trial

Results from a field trial designed to identify the most attractive ratio of E11-14:OAc and E11-14:OH, showed that the largest *P. auraria* captures ($F = 215.78$, $df = 3$, $p < 0.001$, Figure 2) were obtained with 100 µg of E11-14:OAc and either 10 or 30 µg of E11-14:OH (1:0.1 or 1:0.3 ratios). Treatments with lower ratios (1:0.6 or 1:1) provided significantly lower captures. No catches occurred with the control.

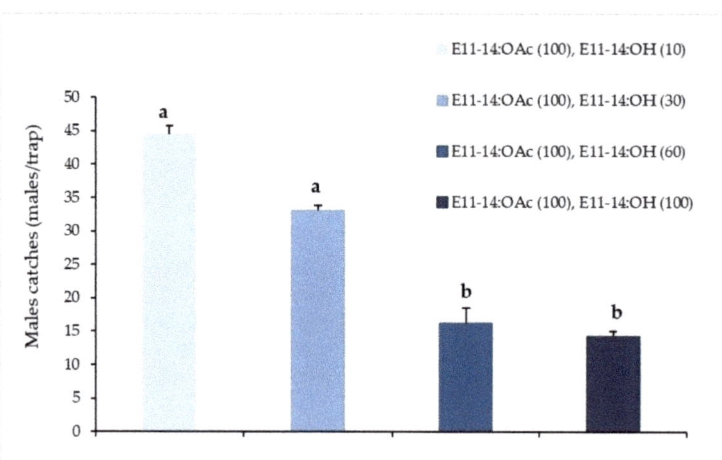

Figure 2. Male *P. auraria* cumulative catches (avg. ± SE) in traps (*n* = 4) loaded with different ratios of E11-14:OAc and E11-14:OH. Nancagua, central Chile, 18–25 February 2015. Different small letters above a column mean significant differences between treatments. Figures in parenthesis represent µg/septa.

3.1.3. Septa Longevity

The field trial designed to evaluate the longevity of septa loaded with three different amounts of E11-14:OAc and E11-14:OH (1:0.3 ratio), aged for up to 10 weeks, showed significant interaction (pheromone load x aging time; $F = 16.57$, $df = 18$, $p < 0.001$), so both factors did not act independently. Thus, we perform the contrast with the Tukey test for every week x pheromone load (Table 1, SE = 1.28). Results shown in most cases (7 out of 10) 50 µg of E11-14:OAc + 15 µg E11-14:OH significantly lower captures, despite some week-to-week variability. On the other hand, 200 µg of E11-14:OAc + 60 µg E11-14:OH was equal to 800 µg of E11-14:OAc + 240 µg E11-14:OH also in seven cases, in two cases the former treatment was greater, and only in one the latter treatment was greater. Based on that, we selected the 200 µg of E11-14:OAc + 60 µg E11-14:OH pheromone load for further studies, considering a septa replacement time of 10 weeks. No males were captured in control traps.

Table 1. Mean cumulative *P. auraria* catches in traps (*n* = 3) within aging weeks for the three tested pheromone loads, Nancagua, central Chile, 18–23 February 2015.

Aging Weeks	E11-14:OAc + E11-14:OH (µg)		
	50 + 15	200 + 60	800 + 240
1	13.7 b	24.7 a	26.3 a *
2	26.3 a	20.7 a	20.7 a
3	21.7 a	24.0 a	11.0 b
4	22.7 a	27.7 a	24.0 a
5	8.0 b	27.7 a	23.0 a
6	17.7 b	19.7 b	30.0 a
7	16.0 b	28.3 a	23.7 a
8	13.3 b	21.0 a	22.3 a
9	16.7 b	27.7 a	17.0 b
10	14.7 b	26.0 a	21.0 ab

* Different lowercase letters in a row indicate significant differences (*p* < 0.05) between pheromone doses within an aging time (weeks).

3.2. Remaining Pheromone in Septa after Ageing under Field Conditions

The amount of pheromone remaining in septa originally loaded with 200 µg E11-14:OAc + 60 µg E11-14:OH, and aged up to 5 months under field conditions, decreased over time, reaching close to 38% of the initial amount (the acetate and the alcohol) after 150 days, with a release rate equivalent to 12%/month. Analysis of the individual components showed that the initial proportion of 1:0.3 (at day 0) had increased to 1:0.1 five months later (Figure 3).

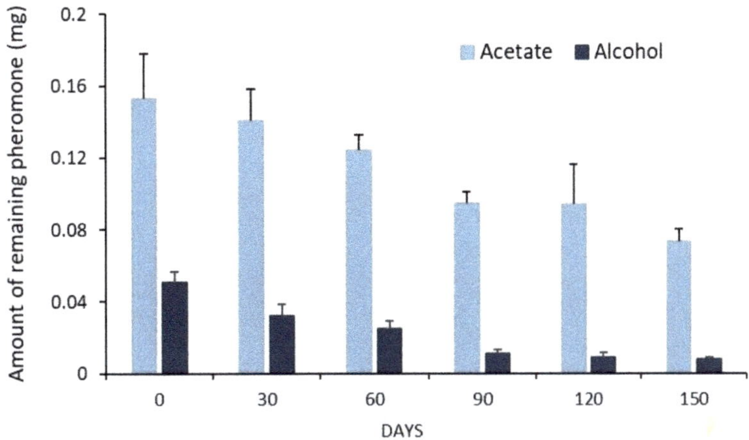

Figure 3. Amount of pheromone remaining (µg ± SE) in septa (original load of 200 µg E11-14:OAc and 60 µg E11-14:OH) aged up to 150 days under field conditions from October 2016 through March 2017.

3.3. Proeulia Auraria Adult Male Phenology

Regarding *P. auraria* phenology, adult male catches in the vineyard and the orchards showed similar patterns in both seasons (2015–2016 and 2016–2017), with two very long flight cycles, the first one from late September until early January, and the second one from early February through early/mid-May, in three localities from the O'Higgins region (Figure 4). Very low or no male adult catches occurred during January, and from early/mid-May until late August or early September.

3.4. Proeulia auraria Mating Disruption Trials

During the MD field trials conducted in all orchards, *P. auraria* catches and disruption showed a similar trend (Figure 5A–C). Captures started at least a couple of weeks before the application of SPLAT (mid-September) in both plots (MD and control), except in apples, where no catches in the MD plot were observed during that period, probably due to a more delayed rise in male activity in the area (as shown in Figure 5B). One week after the application of pheromone-loaded SPLAT, the captures fell to zero (trap shutdown) in all disruption plots, whereas captures continued in the control plots. During the following months, male captures continued to be almost nonexistent in MD plots (0 in blueberries and apples, and only 8 in grapes between September–March/April), whereas relatively high numbers of males were caught in the respective control plots during the season (apples: 292, blueberries: 343; and vines: 838 males). Thus, disruption over the season reached 100% in blueberries and apples, and 99.1% in grapes. The marginal reduction of disruption in grapes occurs essentially in the last week of April.

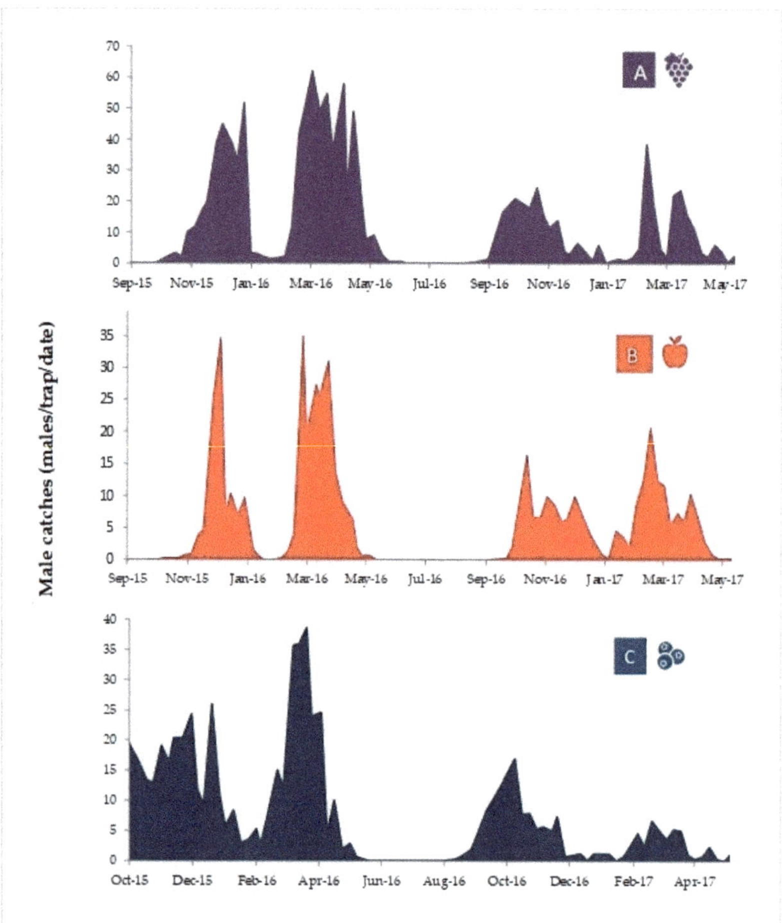

Figure 4. Captures of *P. auraria* (total males/3 delta traps/date) in a vineyard (**A**: Requinoa), an apple orchard (**B**: San Fernando), and a blueberry orchard (**C**: San Francisco de Mostazal), all located in the O′Higgins region, central Chile, during two consecutive seasons (September (**A**,**B**)/October (**C**) 2015 through May 2017), using septa loaded with 200 µg of E11-14:OAc and 60 µg of E11-14:OH.

At harvest, no fruit damage was observed in blueberries. In the apple orchard, the percentage of damaged fruit was not significantly different between the MD plot (3.39%) and the control plot (3.95%) ($p = 0.6$). In grapes, the MD plot had 5.99% of damaged fruits, while in the control plot, 3.48% of fruits were damaged. These values are significantly different ($p = 0.005$).

3.5. Remaining Pheromone in SPLAT Matrix after Aging under Field Conditions

The amount of pheromone remaining in SPLAT originally loaded with 3 g/cup of 10.5% *w/w* E11-14:OAc and E11-14:OH at 1:0.3 ratio, and aged up to 7 months under field conditions, decreased over time for both components, reaching close to 29% of the initial loading after 150 days and keeping close to the original ratio during that period of time. On the other hand, the evaluation at 210 days showed 82% reduction from the original load and a significant deviation from the original ratio (Figure 6).

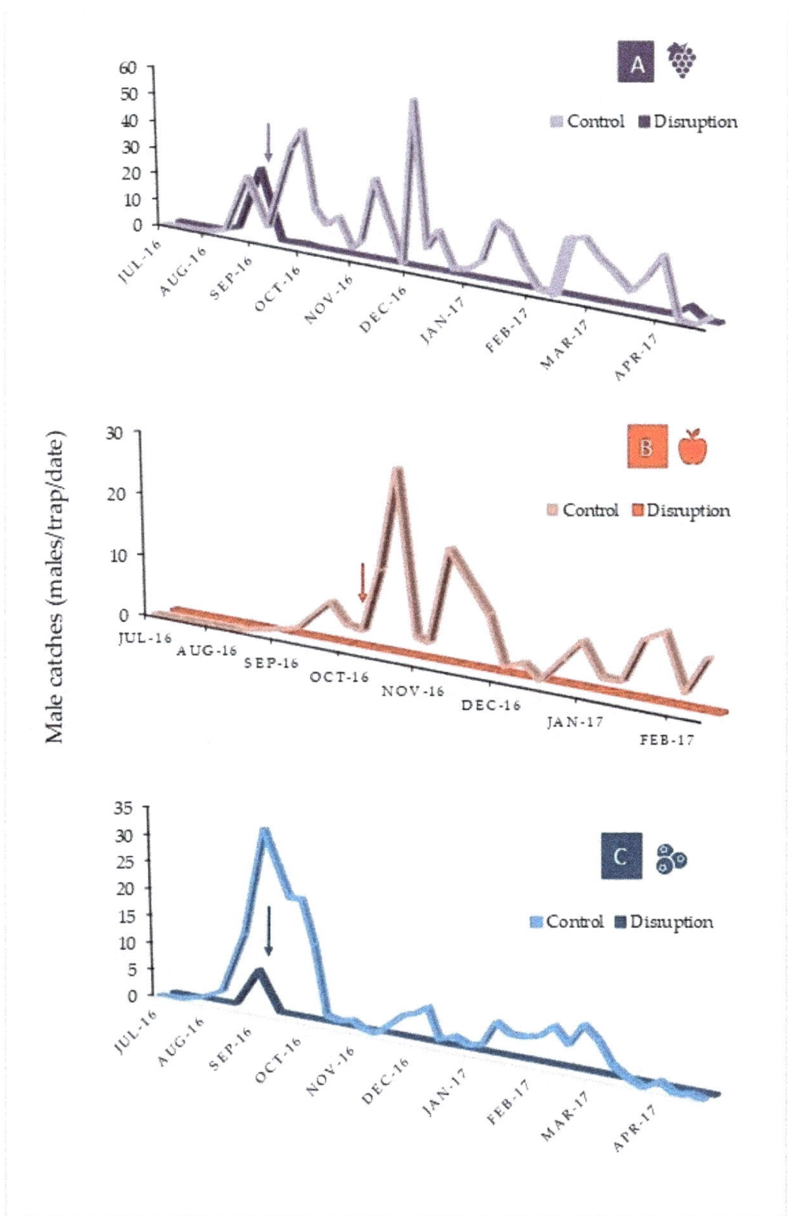

Figure 5. *Proeulia auraria* male captures (avg. males/2 traps*plot/date) in disruption plots (darker color line) and control plots (lighter color line) in a vineyard (**A**: Requinoa), an apple orchard (**B**: San Fernando), and a blueberry orchard (**C**: San Francisco de Mostazal), all located in the O´Higgins region, central Chile, July 2016 through April 2017. The arrow indicates the respective date of SPLAT application.

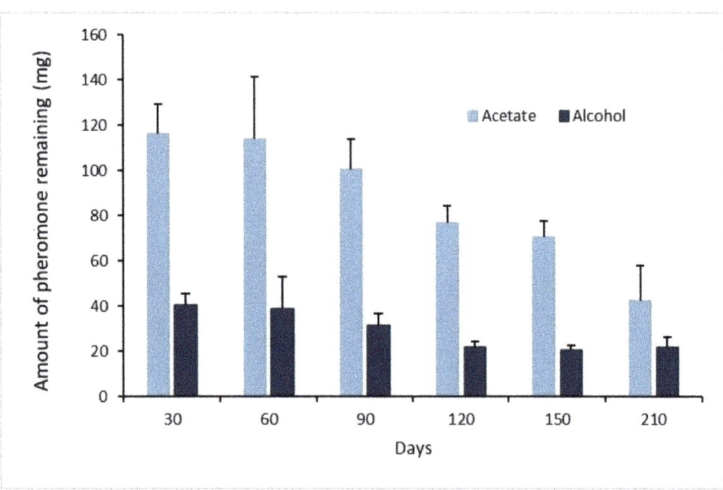

Figure 6. Average amount of pheromone remaining (µg ± SE) per point source (SPLAT matrix loaded as described above, *n* = 5) aged up to 210 days under field conditions.

4. Discussion

The results from our studies designed to optimize the pheromone composition load in septa for *P. auraria* male attraction to traps are generally consistent with our previous reports about optimum composition, where the 4-component blend (in a 100:37:4:11 ratio) attracted more males than several incomplete combinations [19]. Furthermore, the synergistic effect of small amounts of the geometric isomer (Z11-14:OAc) of the main compound (E11-14:OAc) was also confirmed in our present experiments. This is of particular interest for applications in the field, because commercially available E11-14:OAc usually contains small amounts of the geometric isomer. As long as this small impurity does not exceed 1%, a time-consuming purification process is not only not necessary, but even would be detrimental, as the synergist would be removed. It has been discussed in the literature whether or not the whole blend is needed for moth monitoring or mating disruption, or if a partial blend is sufficient [26]. In fact, most monitoring or control strategies using pheromones are conducted only with the main pheromone component, despite the fact that many more chemicals have been identified in most tortricid pheromone blends [27], but there are other examples [28–30] for successful MD studies using binary blends against lepidopteran pest species. On the other hand, there are also some reports on the effects of impurities (including those produced by isomerization from the main component within the septum) modifying attractiveness as well [31]. In the case of *P. auraria*, we selected an incomplete 2-component blend (E11-14:OAc and E11-14:OH), considering its efficiency, ease of use (no further purification needed), and lower cost.

The results from our trial comparing different proportions of E11-14:OAc and E11-14:OH are not fully comparable to previous studies since the 1:0.1 and the 1:0.6 ratios were not tested before. On the other hand, the 1:1 ratio was previously tested [19], but this blend was never the most attractive treatment in field trials, despite the fact it is the ratio used in a lure commercialized in Chile for *P. auraria* monitoring [10]. Our results mostly agree with previous work where a ratio of 1:0.37 [19] was usually the most attractive treatment among several multi-component blends tested. Thus, considering its efficiency and previous data, the 1:0.3 ratio was selected for the next studies. Regarding the results obtained in our study to determine optimum septa replacement timing, we concluded septa loaded with 200 µg of E11-14:OAc + 60 µg E11-14:OH replaced every 10 weeks was more convenient for monitoring *P. auraria*. The duration of lures targeting tortricids for monitoring is very variable (from a few weeks to several months), depending mostly on the dispenser type and pheromone loads [32]. Some studies have concluded, similar to our results [33], a 10-week period for septa replacement when loaded with similar compounds for trapping tortricids in vineyards. The remaining pheromone in septa over time showed a behavior approaching the first kinetic dynamic described for this type of chemicals evaporating from similar dispensers [34], which would explain the relatively long activity observed in our trials. Considering the aforementioned results, a blend of 200 µg of E11-14:OAc and 60 µg E11-14:OH (1:0.3 ratio) was selected, with replacement of the septa every 10 weeks, for *P. auraria* monitoring.

The monitoring of the flight activity of *Proeulia auraria* based on male captures in pheromone-baited traps suggests at least two generations during the season in central Chile. The second-generation complete development the following season, since winter populations (June–August) are composed mostly of second, third, and fourth larval instars. These larvae overwinter (with low activity) in refuges [10], and the occurrence of overlapping stages [9] led to a large period of adult emergence and flight during next spring, as was observed in our results. This pattern differs from other tortricid species affecting the same crops in Chile, for example, *C. pomonella* and *L. botrana*, which spend the winter as a uniform population consisting of a single instar/stage [35,36]. Thus, in the case of *P. auraria*, it is necessary to consider a complementary treatment against immatures along (or before) the installation of mating disruption targeting adult males.

Our results show that male captures in traps tend to be greater in the vineyard, medium in the apples, and lower in the blueberry orchard, being lesser in the second consecutive study season in all localities. Previous studies [1,2] conducted in the Metropolitan region (100–150 km to the north from our tested area), mostly in pear orchards using the TBM lure (the tufted apple bud moth, *P. ideausalis*), showed similar results to those reported here, but with shorter flights and events occurring a couple of weeks earlier. Those reports also suggested the occurrence of two generations per season for *P. auraria*. It should be mentioned that up to three generations for *P. auraria* have been estimated for perennial orchards such as oranges, or in seasons with warmer autumns in central Chile [1,10]. In general, the first generation of adults and its offspring are present in orchards during sprouting/bloom, whereas the second generation originates most immatures which are present between fruit set and harvest, and these juveniles (mainly larvae) are the reason for detections occurring during post-harvest inspections for fruit species.

Previous preliminary studies on mating disruption (MD) carried out in small plots per treatment (0.1 ha/plot, [21]) suggested that *P. auraria* was disruptible with the pheromone deployed in the SPLAT matrix. Those results oriented us to define the pheromone load per point source (78 g pheromone/ha) and the source density (250/ha) reported herein. In the present study, we conducted trials at a larger scale (4 ha/plot), aiming at the validation of these parameters. The values of both parameters -pheromone dose and SPLAT point source density, per ha- are within the ranges reported for other tortricids using the same matrix for MD purposes [37,38]. The trap shutdown in all MD plots in our trials was also observed in other studies conducted to develop mating disruption against different tortricids [39,40]. The very high levels of disruption obtained in our experiments (99–100%) suggest high efficiency and are above the values reported (77–89%) in several similar reports [41–44]. The marginal reduction of disruption in grapes occurred mostly in the last week of April, almost seven months after pheromone application, which is longer than the usual protection required for MD formulations (e.g., 180 days for *L. botrana*) used against tortricids [36]. Considering the varieties used in our trials and the respective localities, where the latest harvest time is usually in March, the MD tested formulation would cover the whole season. This reduction in disruption in grapes occurred when the remaining amount of pheromone in the SPLAT matrix was reduced by 82% and the original components ratio was significantly changed from 1:0.3 to 1:0.5, which we demonstrated previously to be less efficient to maximize attraction [19]. Thus, it seems to be important to develop a dispenser matrix able to keep the proportion of compounds as close to the original as long as possible (i.e., covering the whole period of male flights) for efficient disruption. The possible mechanism for mating disruption in *P. auraria* has not been identified. The literature indicates that the mechanisms in tortricids are either non-competitive (e.g., sensory adaptation) or competitive based on dispenser attraction (e.g., false-trail following) [44], being the latter cited for many tortricid species, including the target of that study (*Cydia* (*Grapholita*) *molesta* (Busck)) [22]. However, researchers in the same group, in a subsequent study [45], concluded that a competitive mechanism acted on *C. molesta* when using monitoring lures (with a low pheromone load per septum) for mating disruption, but a non-competitive one operated when mating disruption dispensers (with the equivalent to 1500× the release rate from monitoring lures) were used, demonstrating a switch of the disruption mechanism depending on the atmospheric pheromone concentration.

Regarding the level of fruit damage at harvest, in apples, no differences occurred between the control plot (sprayed with Bt and/or Spinosad) and the MD plot (no other control methods applied), whereas in grapes the damage was significantly greater in the MD plot. Similar results have been reported previously, where high levels of mating disruption on males (as observed in our field trials) do not necessarily match with the absence of damage in crops because fertilized females can migrate from nearby areas [46], particularly if the MD plot is within an orchard with high pest pressure (as we observed at the Requinoa vineyard). Thus, mating disruption treatments should consider covering the whole orchard in order to properly evaluate crop damage when using this technique.

5. Conclusions

In conclusion, our results set the basis for the development of more sustainable management of this important pest in Chile. We established an optimized blend of pheromone compounds for monitoring based on commercially available material and also established that the pheromone lures (septa) remain attractive for at least 10 weeks. The information generated about male flight dynamics will help to develop a phenological model for *P. auraria* for better timing of insecticide application or installation of mating disruption dispensers. As far as we know, our findings on mating disruption against *P. auraria* are the first report for a South American tortricid affecting vineyards/orchards and demonstrate that it is possible to confuse males of this increasing pest using this technique. However, there is still the need to produce a commercial product and to develop protocols for the application in vineyards and orchards in Chile. The challenge of future work is: to test the technology on larger scales (whole vineyards/orchards) and to evaluate the respective fruit damage level; to determine the effects on other synchronic and sympatric pests and natural enemies; to evaluate the economic feasibility for a commercial product development [47] given the relatively small size of the Chilean market, and to adapt the technology to the particular cycle of *P. auraria*. However, once developed, this will be a valuable tool for local, particularly organic growers not having many efficient options to manage this pest or those who look for new and more environmentally friendly tactics against *P. auraria*.

Author Contributions: Conceptualization, T.C., M.F.F.; methodology, T.C., M.F.F., J.B.; formal analysis, T.C., M.F.F., J.B., C.B., D.A.; resources, T.C., M.F.F., J.B.; writing—original draft preparation, T.C., J.B., M.F.F., D.A.; writing—review and editing, T.C., M.F.F., J.B., C.B., D.A.; supervision, T.C., M.F.F., C.B., D.A.; project administration, M.F.F., T.C., C.B.; funding acquisition, M.F.F., T.C. All authors have read and agreed to the published version of the manuscript.

Funding: Chilean Dept. of Agriculture through the grant (FIA PYT 0014-2014) founded by the Agricultural Research Found (FIA) to FF and TC.

Institutional Review Board Statement: Not applicable.

Informed Consent Statement: Not applicable.

Data Availability Statement: The data presented in this study are available on request from the corresponding author.

Acknowledgments: The authors would like to thank the following persons: Mayerly Prieto, Agustina Valverde, and Yuri Cuevas (Universidad de Chile) for field and laboratory help, and Jaime Tapia and Valery Rodriguez (PUCV) for chemical analysis.

Conflicts of Interest: The authors declare no conflict of interest. The funders had no role in the design of the study; in the collection, analyses, or interpretation of data; in the writing of the manuscript, or in the decision to.

References

1. Campos, L.; Faccin, M.; Echeverria, N.; Sazo, L. Distribución y ciclo evolutivo del tortrícido enrollador de la vid *Proeulia auraria* (Clarke). *Agric. Tec.* **1981**, *41*, 249–256.
2. Álvarez, P.; González, R.H. Biología de la polilla enrolladora del peral *Proeulia auraria* (Clarke). *Rev. Frutic.* **1982**, *3*, 75–80.
3. Klein Koch, C.; Waterhouse, D.F. *The Distribution and Importance of Arthropods Associated with Agriculture and Forestry in Chile*; Australian Centre for International Agricultural Research (ACIAR): Canberra, Australia, 2000; p. 231.
4. Trematerra, P.; Brown, J.W. Argentine *Argyrotaenia* (Lepidoptera: Tortricidae): Synopsis and descriptions of two new species. *Zootaxa* **2004**, *574*, 1–12. [CrossRef]
5. CABI (Centre for Agriculture and Bioscience International). Proeulia auraria (Chilean Fruit Tree Leaf Folder). Available online: https://www.cabi.org/isc/datasheet/44569 (accessed on 13 March 2021).
6. Razowski, J.; Pelz, V. *Tortricidae* from Chile (Lepidoptera: Tortricidae). *SHILAP Rev. Lepid.* **2010**, *38*, 5–55.
7. Cepeda, D.; Curkovic, T. A new species of *Proeulia* Clarke (Lepidoptera: Tortricidae) from Central Chile. *Rev. Chil. Entomol.* **2020**, *46*, 493–499.
8. Cepeda, D.; González, R. Nueva especie del género *Proeulia* Clarke, con registros adicionales de distribución geográfica para cinco especies (Lepidoptera: Tortricidae). *Rev. Chil. Entomol.* **2017**, *40*, 5–12.
9. Artigas, J.N. Tortricidae. In *Entomología Económica. Insectos de Interés Agrícola, Forestal, Médico y Veterinario (Nativos, Introducidos y Susceptibles de ser Introducidos)*; Universidad de Concepción: Concepción, Chile, 1994; pp. 751–779. ISBN 9562271013.
10. González, R.H. *Las Polillas de la Fruta en Chile (Lepidoptera: Tortricidae; Pyralidae)*; Serie Ciencias Agronómicas; Universidad de Chile: Santiago, Chile, 2003; pp. 107–138.
11. Curkovic, T.; Ballesteros, C.; Carpio, C. Manejo integrado de plagas del granado. In *Bases Para el Cultivo del Granado en Chile*; Henríquez, J.L., Franck, N., Eds.; Serie Ciencias Agronómicas; Universidad de Chile: Santiago, Chile, 2015; pp. 159–231.
12. Chacón-Fuentes, M.A.; Lizama, M.G.; Parra, L.J.; Seguel, I.E.; Quíroz, A.E. Insect diversity, community composition and damage index on wild and cultivated murtilla. *Cienc. Investig. Agrar.* **2016**, *43*, 57–67. [CrossRef]
13. Biosecurity, N.Z. Biosecurity New Zealand, Standard 155.02.06: Importation of Nursery Stock. Available online: https://www.ippc.int/static/media/files/publications/en/2013/06/05/1125929849079_Biosecurity_New_Zealand.pdf (accessed on 13 March 2021).
14. Suffert, M.; Wilstermann, A.; Petter, F.; Schrader, G.; Grousset, F. Identification of new pests likely to be introduced into Europe with the fruit trade. *EPPO Bull.* **2018**, *48*, 144–154. [CrossRef]
15. González, R.; Curkovic, T. Manejo de plagas y degradación de residuos de pesticidas en kiwi. *Rev. Frutic.* **1994**, *15*, 5–20.
16. Ripa, R.; Larral, P. Manejo de plagas en paltos y cítricos. *Colecc. Libros Inst. Investig. Agropecu.* **2008**, *23*, 399.
17. Faccin, M. Feromona Sexual del Enrollador de la vid *Proeulia auraria* (Clarke) (Lepidoptera: Tortricidae). Bachelor Thesis, Universidad de Chile, Santiago, Chile, 1979.
18. Roelofs, W.L.; Brown, R.L. Pheromones and evolutionary relationships of Tortricidae. *Annu. Rev. Ecol. Syst.* **1982**, *13*, 395–422. [CrossRef]
19. Reyes-Garcia, L.; Cuevas, Y.; Ballesteros, C.; Curkovic, T.; Löfstedt, C.; Bergmann, J. A 4-component sex pheromone of the Chilean fruit leaf roller *Proeulia auraria* (Lepidoptera: Tortricidae). *Cienc. Investig. Agrar.* **2014**, *41*, 9–10. [CrossRef]
20. Kahler, E. Evaluación de Cebos Basados en Feromona Sexual para el Monitoreo de *Proeulia auraria* y *Proeulia triquetra* en Cultivos de Arándanos. Bachelor Thesis, Universidad Austral de Chile, Valdivia, Chile, 2018.
21. Valverde-Rodríguez, A. Formulación SPLAT para el control de *Proeulia auraria* (Lepidoptera: Tortricidae) a través del método de confusión sexual en frutales. *Rev. Invest. Agr.* **2019**, *1*, 67–75. [CrossRef]
22. Stelinski, L.L.; Miller, J.R.; Ledebuhr, R.; Siegert, P.; Gut, L.J. Season-long mating disruption of *Grapholita molesta* (Lepidoptera: Tortricidae) by one machine application of pheromone in wax drops (SPLAT-OFM). *J. Pest Sci.* **2007**, *80*, 109–117. [CrossRef]
23. Babu, A.; Rodríguez-Saona, C.; Mafra-Neto, A.; Sial, A.A. Efficacy of Attract-and-Kill Formulations Using the Adjuvant Acttra SWD TD for the Management of Spotted-Wing Drosophila in Blueberries, 2020. *Arthropod Manag. Tests* **2021**, *46*, 1–2. [CrossRef]
24. Wallingford, A.K.; Connelly, H.L.; Dore Brind'Amor, G.; Boucher, M.T.; Mafra-Neto, A.; Loeb, G.M. Field evaluation of an oviposition deterrent for management of spotted-wing drosophila, *Drosophila suzukii*, and potential nontarget effects. *J. Econ. Entomol.* **2016**, *109*, 1779–1784. [CrossRef] [PubMed]
25. Zar, J.H. *Biostatistical Analysis*, 4th ed.; Prentice Hall: Upper Saddle River, NJ, USA, 1999; p. 929.
26. Carde, R.T.; Minks, A.K. Control of moth pests by mating disruption: Successes and constraints. *Annu. Rev. Entomol.* **1995**, *40*, 559–585. [CrossRef]
27. Ioriatti, C.; Anfora, G.; Tasin, M.; De Cristofaro, A.; Witzgall, P.; Lucchi, A. Chemical ecology and management of *Lobesia botrana* (Lepidoptera: Tortricidae). *J. Econ. Entomol.* **2011**, *104*, 1125–1137. [CrossRef]

28. Evenden, M.L.; Judd, G.J.R.; Borden, J.H. Pheromone-mediated mating disruption of *Choristoneura rosaceana*: Is the most attractive blend really the most effective? *Entomol. Exp. Appl.* **1999**, *90*, 37–47. [CrossRef]

29. Lapointe, S.L.; Stelinski, L.L.; Evens, T.J.; Niedz, R.P.; Hall, D.G.; Mafra-Neto, A. Sensory imbalance as mechanism of orientation disruption in the leafminer *Phyllocnistis citrella*: Elucidation by multivariate geometric de-signs and response surface models. *J. Chem. Ecol.* **2009**, *35*, 896–903. [CrossRef]

30. Higbee, B.S.; Burks, C.S.; Cardé, R.T. Mating disruption of the navel orangeworm (Lepidoptera: Pyralidae) using widely spaced aerosol dispensers: Is the pheromone blend the most efficacious disruptant? *J. Econ. Entomol.* **2017**, *110*, 2056–2061. [CrossRef] [PubMed]

31. Coracini, M.D.A.; Bengtsson, M.; Reckziegel, A.; Löfqvist, J.; Francke, W.; Vilela, E.F.; Eiras, A.E.; Kovaleski, A.; Witzgall, P. Identification of a Four-Component Sex Pheromone Blend in *Bonagota cranaodes* (Lepidoptera: Tortricidae). *J. Econ. Entomol.* **2001**, *94*, 911–914. [CrossRef] [PubMed]

32. Knight, A.L. A comparison of gray halo-butyl elastomer and red rubber septa to monitor codling moth (Lepidoptera: Tortricidae) in sex pheromone-treated orchards. *J. Entomol. Soc. B. C.* **2002**, *99*, 123–132.

33. Biever, K.D.; Hostetter, D.L. Phenology and pheromone trap monitoring of the grape berry moth, *Endopiza viteana* Clemens (Lepidoptera: Tortricidae) in Missouri. *J. Entomol. Sci.* **1989**, *24*, 472–481. [CrossRef]

34. Butler, U.I.; McDonough, L.M. Insect Sex Pheromones: Evaporation Rates of Alcohols and Acetates from Natural Rubber Septa 1. *J. Chem. Ecol.* **1981**, *7*, 627–633. [CrossRef] [PubMed]

35. Fuentes-Contreras, E.; Reyes, M.; Barros, W.; Sauphanor, B. Evaluation of Azinphos-Methyl Resistance and Activity of Detoxifying Enzymes in Codling Moth (Lepidoptera: Tortricidae) from Central Chile. *J. Econ. Entomol.* **2007**, *100*, 551–556. [CrossRef] [PubMed]

36. Lobesia botrana o Polilla del Racimo de la Vid, SAG. Available online: https://www.sag.gob.cl/ambitos-de-accion/lobesia-botrana-o-polilla-del-racimo-de-la-vid (accessed on 19 March 2021).

37. Steffan, S.A.; Chasen, E.M.; Deutsch, A.E.; Mafra-Neto, A. Multi-species mating disruption in cranberries (Ericales: *Ericaceae*): Early evidence using a flowable emulsion. *J. Insect Sci.* **2017**, *17*, 1–6. [CrossRef]

38. Svensson, G.P.; Wang, H.L.; Jirle, E.V.; Rosenberg, O.; Liblikas, I.; Chong, J.M.; Löfstedt, C.; Anderbrant, O. Challenges of pheromone-based mating disruption of *Cydia strobilella* and *Dioryctria abietella* in spruce seed orchards. *J. Pest Sci.* **2018**, *91*, 639–650. [CrossRef] [PubMed]

39. Kovanci, O.B.; Schal, C.; Walgenbach, J.F.; Kennedy, G.G. Comparison of Mating Disruption with Pesticides for Management of Oriental Fruit Moth (Lepidoptera: Tortricidae) in North Carolina Apple Orchards. *J. Econ. Entomol.* **2005**, *98*, 1248–1258. [CrossRef]

40. Light, D.M. Control and Monitoring of Codling Moth (Lepidoptera: Tortricidae) in Walnut Orchards Treated with Novel High-Load, Low-Density "Meso" Dispensers of Sex Pheromone and Pear Ester. *Environ. Entomol.* **2016**, *45*, 700–707. [CrossRef]

41. Vacas, S.; Alfaro, C.; Primo, J.; Navarro-Llopis, V. Studies on the development of a mating disruption system to control the tomato leafminer, *Tuta absoluta* Povolny (Lepidoptera: Gelechiidae). *Pest Manag. Sci.* **2011**, *67*, 1473–1480. [CrossRef] [PubMed]

42. Arioli, C.J.; Pastori, P.L.; Botton, M.; Garcia, M.S.; Borges, R.; Mafra-Neto, A. Assessment of SPLAT formulations to control *Grapholita molesta* (Lepidoptera: Tortricidae) in a Brazilian apple orchard. *Chil. J. Agric. Res.* **2014**, *74*, 184–190. [CrossRef]

43. Suckling, D.M.; Baker, G.; Salehi, L.; Woods, B.; Dickens, J.C. Is the Combination of Insecticide and Mating Disruption Synergistic or Additive in Lightbrown Apple Moth, *Epiphyas postvittana*? *PLoS ONE* **2016**, *11*, e0160710. [CrossRef]

44. Miller, J.R.; Gut, L.J. Mating disruption for the 21st century: Matching technology with mechanism. *Environ. Entomol.* **2015**, *44*, 427–453. [CrossRef] [PubMed]

45. Reinke, M.D.; Siegert, P.Y.; McGhee, P.S.; Gut, L.J.; Miller, J.R. Pheromone release rate determines whether sexual communication of Oriental fruit moth is disrupted competitively or non-competitively. *Entomol. Exp. Appl.* **2014**, *150*, 1–6. [CrossRef]

46. Curkovic, T.; Sazo, L.; Araya, J.E.; Agurto, L.; Polanco, J. Effect of mating disruption pheromones on *Cydia pomonella* and other arthropods in pome orchards in Central and Southern Chile. *Bol. Sanid. Veg.* **2004**, *31*, 309–318.

47. Brunner, J.; Welter, S.; Calkins, C.; Hilton, R.; Beers, E.; Dunley, J.; Unruh, T.; Knight, A.; Van Steenwyk, R.; Van Buskirk, P. Mating disruption of codling moth: A perspective from the Western United States. *IOBC-WPRS Bull.* **2002**, *25*, 11–20.

insects

Article

Mating Disruption for Managing the Honeydew Moth, *Cryptoblabes gnidiella* (Millière), in Mediterranean Vineyards

Renato Ricciardi [1,†] , Filippo Di Giovanni [1,†] , Francesca Cosci [1] , Edith Ladurner [2] , Francesco Savino [2] , Andrea Iodice [2] , Giovanni Benelli [1,*] and Andrea Lucchi [1]

[1] Department of Agriculture, Food and Environment, University of Pisa, via del Borghetto 80, 56124 Pisa, Italy; renato_ricciardi@hotmail.it (R.R.); aphelocheirus@gmail.com (F.D.G.); francesca.cosci1@virgilio.it (F.C.); andrea.lucchi@unipi.it (A.L.)
[2] CBC (Europe) srl, Biogard Division, via Zanica, 25, 24050 Grassobbio, Italy; eladurner@cbceurope.it (E.L.); fsavino@cbceurope.it (F.S.); aiodice@cbceurope.it (A.I.)
* Correspondence: giovanni.benelli@unipi.it; Tel.: +39-050-2216141
† These authors contributed equally.

Citation: Ricciardi, R.; Di Giovanni, F.; Cosci, F.; Ladurner, E.; Savino, F.; Iodice, A.; Benelli, G.; Lucchi, A. Mating Disruption for Managing the Honeydew Moth, *Cryptoblabes gnidiella* (Millière), in Mediterranean Vineyards. *Insects* **2021**, *12*, 390. https://doi.org/10.3390/insects12050390

Academic Editor: Thomas W. Phillips

Received: 26 March 2021
Accepted: 22 April 2021
Published: 28 April 2021

Publisher's Note: MDPI stays neutral with regard to jurisdictional claims in published maps and institutional affiliations.

Simple Summary: *Cryptoblabes gnidiella* has recently become one of the most feared pests in the Mediterranean grape-growing areas. Its expanding impact requires the development of effective strategies for its management. Since insecticide strategy has shown several weaknesses, we developed a pheromone-based mating disruption (MD) approach as a possible sustainable control technique for this pest. Between 2016 and 2019, field trials were carried out in two study sites in central and southern Italy, using experimental pheromone dispensers. The number of adult captures in pheromone-baited traps and the percentage of infestation recorded on ripening grapes were compared among plots treated with MD dispensers, insecticide-treated (no MD) plots, and untreated plots. Results highlighted that the application of MD may contribute to lowering the damage significantly. However, further studies aimed at clarifying the still little-known aspects of the biology and population dynamics of the honeydew moth are needed.

Abstract: The demand for a reduced use of pesticides in agriculture requires the development of specific strategies for managing arthropod pests. Among eco-friendly pest control tools, pheromone-based mating disruption (MD) is promising for controlling several key insect pests of economic importance, including many lepidopteran species. In our study, we evaluated an MD approach for managing the honeydew moth (HM), *Cryptoblabes gnidiella*, an emerging threat for the grapevine in the Mediterranean basin. The trials were carried out in two study sites, located in Tuscany (central Italy, years 2017–2019) and Apulia (southern Italy, years 2016 and 2018–2019), and by applying MD dispensers only in April, in April and July, and only in July. To evaluate the effects of MD, infested bunches (%), damaged area (%) per bunch, and number of living larvae per bunch were compared among plots covered with MD dispensers, insecticide-treated plots (Apulia only), and untreated control plots. Male flights were monitored using pheromone-baited sticky traps. Except for the sampling carried out in Tuscany in 2018, where HM infestation level was very low, a significant difference was recorded between MD and control plots, both in terms of HM damage caused to ripening grapes and/or number of living larvae per bunch. Overall, our study highlighted that MD, irrespective of the application timing, significantly reduced HM damage; the levels of control achieved here were similar to those obtained with the application of insecticides (no MD). However, MD used as stand-alone strategy was not able to provide complete pest control, which may instead be pursued by growers with an IPM approach.

Keywords: chemical ecology; grapevine; Integrated Pest Management; sex pheromone

1. Introduction

In recent years, the honeydew moth (HM), *Cryptoblabes gnidiella* (Millière, 1867) (Lepidoptera, Pyralidae, Phycitinae), has become one of the most harmful moths in Mediterranean vineyards, raising concerns about the damage caused to ripening grapes [1–3]. Native to the Mediterranean basin, this moth is highly polyphagous, feeding on about 60 different plant species, and it has been recorded in several African and Asian countries, as well as in India, New Zealand, Hawaii, and North and South America [1,4].

Although widespread throughout the Mediterranean area, the HM has never been considered a key pest of vineyards, since it usually occurred at low density [1,3], and its impact on yield has been regarded as negligible compared to that of major vineyard pests, such as the European grapevine moth, *Lobesia botrana* (Denis & Schiffermüller) (Lepidoptera: Tortricidae), or the vine mealybug, *Planococcus ficus* (Signoret) (Hemiptera: Pseudococcidae) [5–8]. Most likely because of the increase in average temperatures, the economic impact of the HM in vineyards throughout the Mediterranean basin has been growing rapidly over the last decade [3]. In the warmer coastal vineyards of central and southern Italy, as well as in Provence (southern France), the HM has already been reported to cause severe losses in late-ripening grapes, e.g., Aglianico, Montepulciano, Sangiovese, or Grenache [3].

To date, the control of this pest has been achieved mainly through the chemical management of other key pests such as *L. botrana* [1]. For example, the use of *Bacillus thuringiensis* subsp. *kurstaki* has proven to be effective in controlling HM larval populations [1,9]. However, research focusing on the life cycle of this pest [1–4] has shown that—although the first flights are recorded in April–May—no larvae or grape damage is found before the end of July, in correspondence with the phenological phase of "majority of berries touching/beginning of ripening" (79–81 BBCH scale). Consequently, treatments against the second generation of *L. botrana* may not ensure the control of HM populations. In addition, the growing impact of this moth and the general demand for an increasingly reduced use of pesticides in agriculture require the development of specific and eco-friendly strategies for the containment of this pest.

In this context, a shift towards the development of an Integrated Pest Management (IPM) strategy for the sustainable management of the HM in vineyards is an incoming challenge for the wine industry. Among eco-friendly pest control strategies, pheromone-based mating disruption (MD) turned out to be a promising tool for the control of several insect pests [10–12]. Moth sexual communication is based on the release of a sex pheromone by the female, which is eventually detected by males through appropriate neurosensory structures [13,14]. The release of synthetic pheromone plumes interferes with the mate finding process [15–17]; thus, this technique affects the chance of reproduction of the target species, with a consequent impact on its population dynamics [18]. As mentioned above, MD has proved to be a valid control tool to manage the populations of several moths of economic importance [15,19–24] and has been successfully applied in the control of the European grapevine moth and the vine mealybug in European vineyards [5,6,8,10,25–27].

The aim of our study was to develop and validate a novel MD approach for managing the infestations of the HM in Italian vineyards by comparing the percentage of infested bunches, the percentage of damaged area per bunch, the number of living HM larvae per bunch, and HM trap catches among MD plots, plots treated with insecticides (hereinafter, grower's standard), and untreated control plots. The flight pattern of *C. gnidiella* in the Mediterranean area shows four peaks [1,28–31]: the first two are in May–June and July, respectively, with relatively low captures; the third and fourth peaks are in August–October, partially overlapping and with high captures. However, although trap catches are recorded, larvae infesting bunches are hardly ever seen in May–June and appear gradually in bunches from the second half of July on [3]. For this reason, three different timings of dispenser application were tested: (*i*) dispensers applied once in April (i.e., before the beginning of the first flight); (*ii*) dispensers applied once in July (at the first appearance of larvae on bunches); and (*iii*) dispensers applied twice, once in April and once in July, with the second application aiming at reinforcing the pheromone cloud in the field before the fourth flight peak.

2. Materials and Methods

2.1. Experimental Design

Trials were carried out between 2016 and 2019, in two typical wine-growing areas: one located in Apulia (Southern Italy), in the municipality of Minervino Murge (province of Barletta-Andria-Trani), characterized by the late-ripening wine grape variety Aglianico (harvesting period: middle of October), and the other one located in Tuscany (Central Italy), in the municipality of Capalbio (province of Grosseto), on the wine grape variety Cabernet Sauvignon (harvesting period: end of September). Vineyard details are given in Table 1.

Table 1. Sites and crop details of the vineyards where Isonet CGX111 dispensers were tested to manage *Cryptoblabes gnidiella* populations.

Site	Variety	Harvesting Period	Row Spacing (m)	Space within Rows (m)
Tuscany (Capalbio) 42°25′51.78″ N 11°25′4.82″ E	Cabernet Sauvignon	End of September	1.3	0.6
Apulia (Minervino Murge) 41°8′56.72″ N 16°2′16.09″ E	Aglianico	Middle of October	2.3	0.8

Plastic hand-applied Isonet CGX111 dispensers (Shin-Etsu Chemical Co. Ltd., Naoetsu, Japan) were tested at the application rate of 500 units/ha. The dispensers were set in a regular grid, spaced out about 4.5 m from each other. Isonet CGX111 is a reservoir pheromone dispenser consisting of two parallel brown-red polymer tubes, one filled with the HM synthetic pheromone blend (Z)-11-hexadecenal >27% and (Z)-13-octadecenal >27% and the other containing an aluminum wire that enables their placement on supports.

In Apulia, the experiment was carried out by dividing the study area into plots of 6–8 ha. The first study was conducted in 2016. Four sampling plots were established: an untreated control plot; an MD plot where Isonet CGX111 dispensers were applied only once in April (11 April); an MD plot where the dispensers were applied twice, once in April and once in July (18 July); and a plot managed according to local practices for HM control, hereinafter called the grower's standard (treated 4 times during berry ripening with Delfin at 0.75 kg/ha, a.i. *B. thuringiensis* subsp. *kurstaki* strain SA 11). In 2018, the study was carried out by testing an MD plot with Isonet CGX111 dispensers applied in April (9 April) vs. an untreated control plot and the grower's standard plot (treated 5 times with Delfin at 0.75 kg/ha and once with Laser 0.25 L/ha, a.i. spinosad, during fruit development/berry ripening). The trial was repeated in 2019, adding two MD plots: one with two applications, the first in April (15 April) and the second in July (22 July), both at a rate of 500 dispensers per ha, and another one where the dispensers were applied only once in July (500 per ha).

In Tuscany, the experiment was conducted by dividing the study area into plots of about 6 ha. In 2017 and 2018, the sampling was carried out testing an MD plot with Isonet CGX111 dispensers applied in April (14 April and 18 April, respectively) vs. an untreated control. In 2019, the survey was repeated, adding an MD plot where 500 dispensers per ha were applied in April (11 April) and then again in July (23 July) and another MD plot where 500 dispensers per ha were applied only in July. Experimental plots details per year are given in Table 2.

A randomized block design is not applicable to large plots required for studies on MD products (see European and Mediterranean Plant Protection Organization 2019 guidelines), and therefore each treatment (i.e., Isonet CGX111 at 500 dispensers per ha applied once in April, once in July and/or twice, grower's standard, and untreated control) was applied to one large plot of 6–8 ha. For each treatment, 10 sampling units were selected, distributed in a grid pattern, except for the survey of 2016 in Apulia, where 6 sampling units were selected. In each of these units, 50 bunches (for a total of 300 bunches per treatment in 2016 and 500 bunches per treatment in the following years) were randomly selected and checked for HM infestation.

Table 2. Treatments tested in the different years in Minervino Murge and Capalbio vineyards. A: Isonet CGX111 dispensers applied only in April; AJ: Isonet CGX111 dispensers applied twice, once in April and once in July; J: Isonet CGX111 dispensers applied only in July; C: untreated control; GS: grower's standard (Delfin at 0.75 kg/ha, a.i. *Bacillus thuringiensis kurstaki*—SA 11 and Laser 0.25 L/ha, a.i. spinosad); '-': treatment not included in the study.

Year	Site		Treatements			
2016	Minervino Murge	A	AJ	-	C	GS
2017	Capalbio	A	-	-	C	-
2018	Minervino Murge	A	-	-	C	GS
	Capalbio	A	-	-	C	-
2019	Minervino Murge	A	AJ	J	C	GS
	Capalbio	A	AJ	J	C	-

2.2. Adult Captures and Evaluation of Infestation Levels

Adult flights were monitored each year in each plot using Biogard Delta Traps (BDT sticky traps, CBC (Europe) S.r.l., Grassobbio (BG), Italy), baited with HM sex pheromone lures (NovaPher, Settimo Milanese (MI), Italy). Sticky plates and sex pheromone lures were replaced monthly; adults were counted regularly by direct observation in the field and periodically removed.

Infestation levels were evaluated on selected bunches in September, just before harvest. Three parameters were considered: percentage of HM-infested bunches, number of living larvae of *C. gnidiella* per bunch, and percentage of damaged area per bunch. The percentage of damaged area per bunch was evaluated visually as the portion of the bunch with wilted and/or damaged berries, rousers on green parts, feces, and silk.

2.3. Pheromone Release Rate Over the Season

In order to evaluate the gradual release of pheromone, Isonet CGX111 dispensers ($n = 5$ for sampling) during the field experiment in Apulia in 2016 were periodically collected and analyzed with GC-MS, following the method recently described by Lucchi et al. [10]. Based on internal (SEC) standard GC-MS analysis, an Agilent 6890 N (Santa Clara, CA, USA) gas chromatograph equipped with a 5973N mass spectrometer (MS) was used. MS was set as follows: EI mode, 70 eV, mass-to-charge ratio (m/z) scan between 35 and 400. HP capillary column (30 m × ID 0.25 mm × 0.25 μm film thickness, J&W Scientific, Folsom, CA, USA) with helium gas flow (1.0 mL/min) was employed for separation. The GC temperature program was 50 °C for 5 min, then increasing at 20 °C/min to 300 °C. The injector temperature was 150 °C.

2.4. Statistical Analysis

Estimated parameters were not normally distributed in all the plots of our samplings, and the variance was not homogeneous (Shapiro–Wilk test, goodness of fit $p < 0.05$; Levene's test, goodness of fit $p < 0.05$). Data transformation to ln (x + 1) was not able to normalize the distribution or homogenize the variance. Therefore, differences between plots were tested using nonparametric statistics, i.e., the Wilcoxon test (for comparisons between two plots) or the Kruskal–Wallis test (for multiple comparisons among different plots), the latter followed by the Nemenyi post hoc test; a p-value of 0.05 was used as the threshold to assess significant differences. Statistical analysis was performed using R software (www.R-project.org).

3. Results

3.1. Adult Flights

In both trials, adult flights showed a typical trend, with a first and second peak in May and July, respectively, followed by a third peak in August and a higher peak in September–October (Figures 1 and 2). In Tuscany, the maximum number of male catches per trap per week occurred in the control plot during the fourth flight, with peaks of 25 adults in 2017, 62 in 2018, and 39 in 2019; MD-treated plots accounted for about 0.09% of total catches in 2017, 0.02% in 2018, and 0.10% in 2019. In Apulia, the maximum number of catches per trap per week in the untreated control plot occurred in 2016 and 2018 during the fourth flight, with peaks of 132 and 34 adults, respectively, while a maximum of 299 adults was found in 2019 in the grower's standard plot; MD-treated plots accounted for about 0.07% of total catches in 2016, 0.22% in 2018, and 0.12% in 2019.

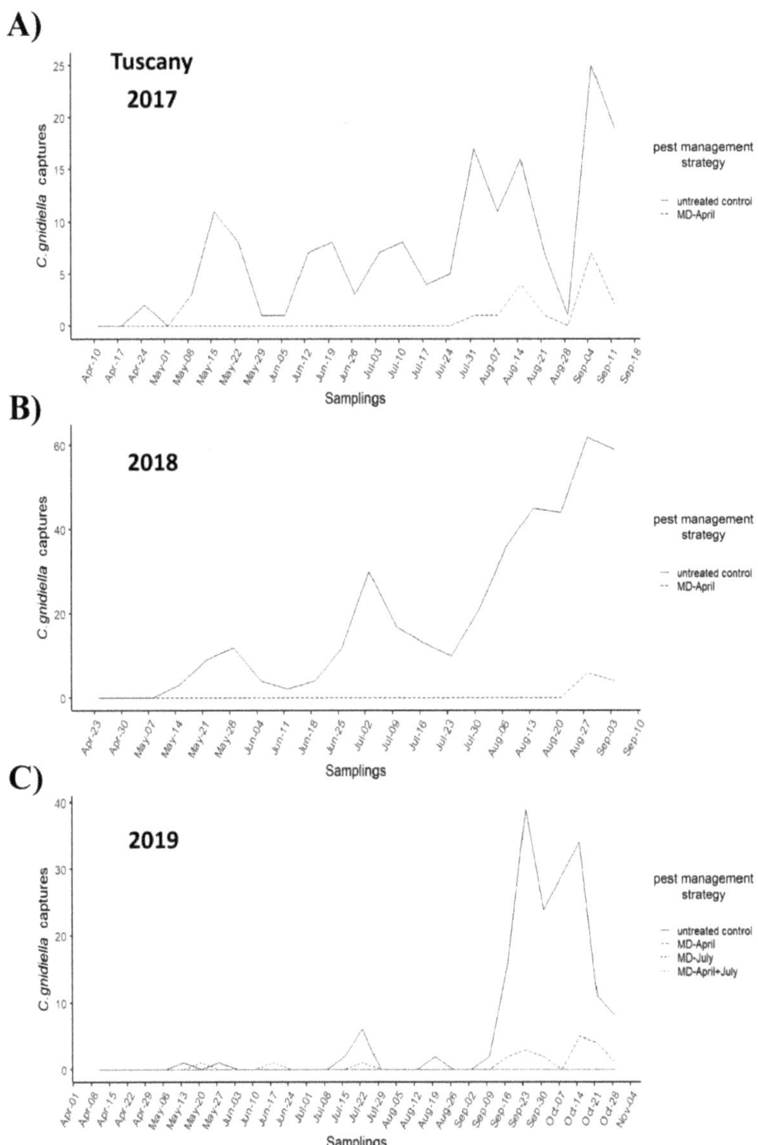

Figure 1. *Cryptoblabes gnidiella* flight trends in mating disruption (MD)-managed and untreated control plots in Tuscany in 2017 (**A**), 2018 (**B**), and 2019 (**C**). In all years, there is a marked contrast in the number of catches between plots under MD (dashed lines) and untreated control plots (solid line), especially from August onwards.

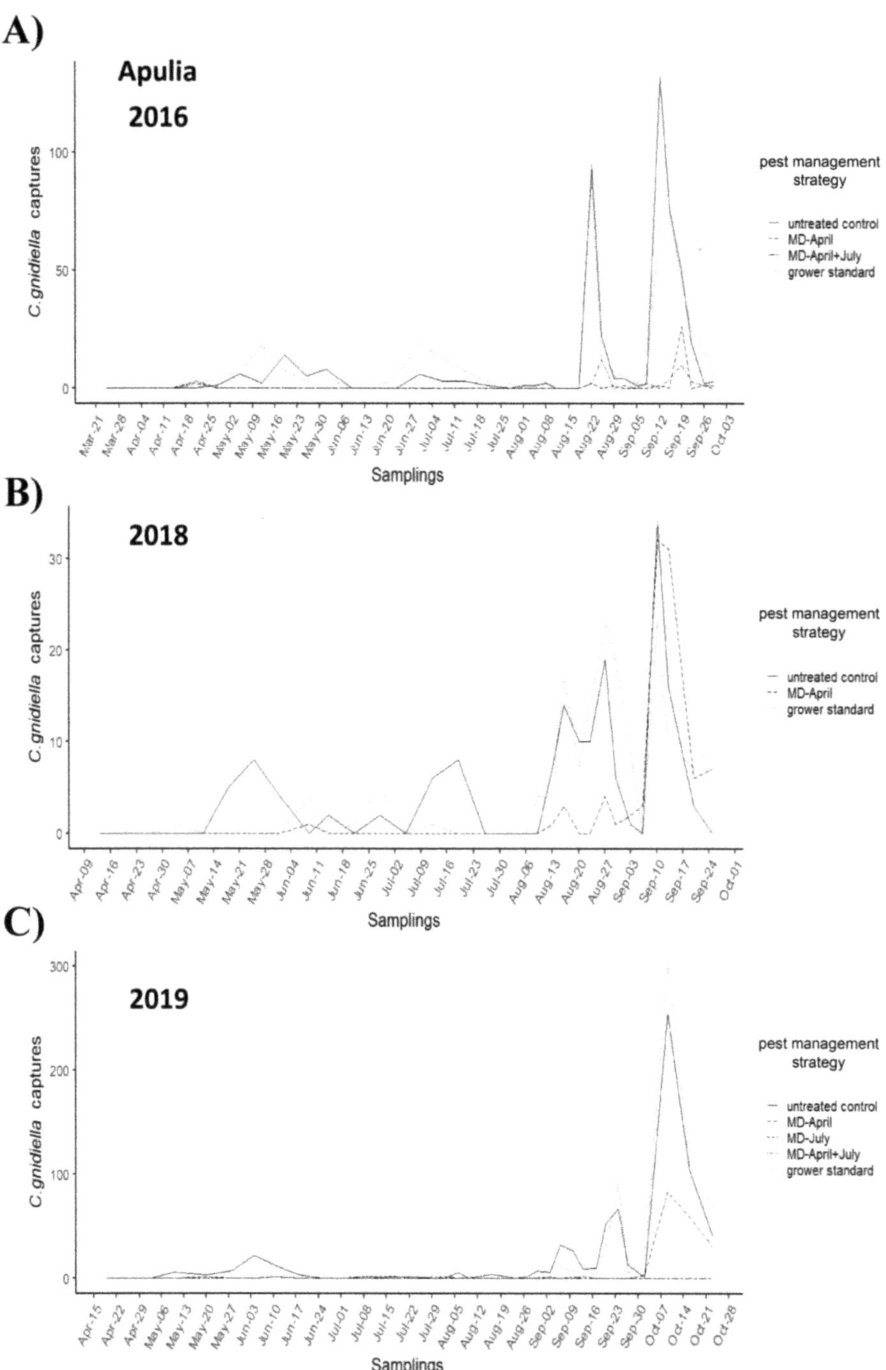

Figure 2. *Cryptoblabes gnidiella* flight trends in mating disruption (MD)-managed and untreated control plots in Apulia in 2016 (**A**), 2018 (**B**), and 2019 (**C**). In all years, there is a marked contrast in the number of catches between plots under MD (dashed lines), untreated control plots (solid line), and grower standard (dotted lines) especially from August onwards.

3.2. Pheromone Release Rate over the Season

Figure 3 shows the pheromone release of the Isonet CGX111 dispensers placed in the field experiment in Apulia during the 2016 grape-growing season. GC-MS results underline how the pheromone content of the dispensers decreased progressively throughout the season. During the third HM flight in August, the percentage of residual pheromone in MD dispensers applied in April was approx. 10%, and it decreased to levels below 5% during the preharvest period. Concerning MD dispensers applied in July, the residual pheromone content was about 75–80% in August, and about 25% during the preharvest period.

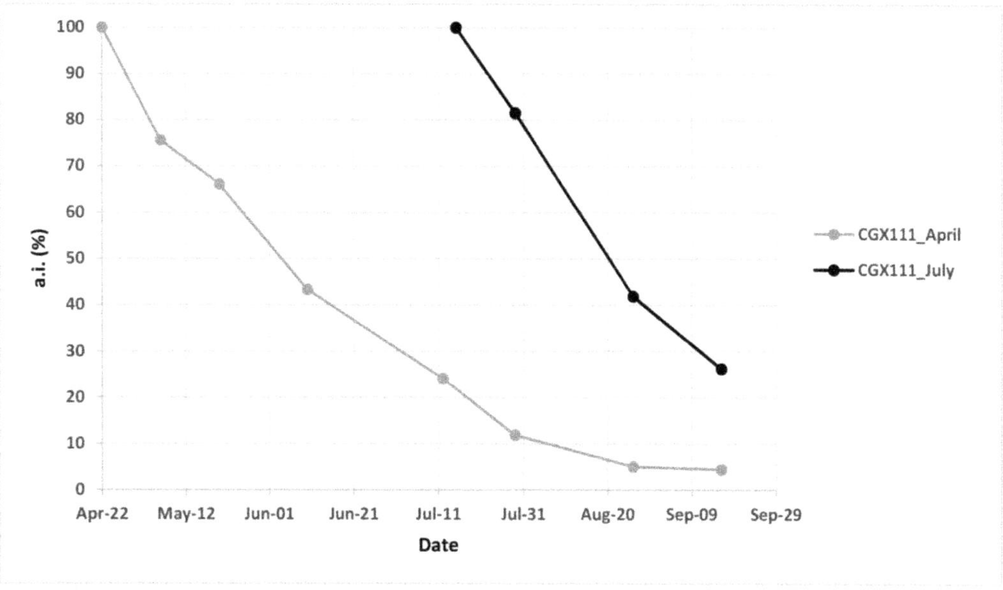

Figure 3. *Cryptoblabes gnidiella* pheromone load in the mating disruption dispensers applied in April or July, tested in Apulian vineyards during the 2016 growing season. a.i. = active ingredient.

3.3. Field Experiments in Central Italy

In 2017 (Figures 4–6), significant differences were observed between the untreated control and the plot with dispensers applied in April in terms of percentage of infested bunches ($\chi^2 = 7.86$, $d.f. = 1$, $p = 0.006$) and percentage of damaged area per bunch ($\chi^2 = 7.83$, $d.f. = 1$, $p = 0.006$), while the difference was not significant for number of living larvae per bunch ($\chi^2 = 2.11$, $d.f. = 1$, $p = 0.167$). In 2018 (Figures 4–6), no significant differences between treatments were found for any of the considered variables (percentage of infested bunches, $\chi^2 = 3.22$, $d.f. = 1$, $p = 0.079$; percentage of damaged area per bunch, $\chi^2 = 1.76$, $d.f. = 1$, $p = 0.197$; number of living larvae per bunch, $\chi^2 = 0.58$, $d.f. = 1$, $p = 0.467$). The sampling in 2019 (Figures 4–6) showed significant differences among treatments in terms of percentage of infested bunches ($\chi^2 = 14.472$, $d.f. = 3$, $p = 0.002$) and percentage of damaged area per bunch ($\chi^2 = 13.071$, $d.f. = 3$, $p = 0.004$), while no significant difference emerged in terms of number of living larvae per bunch ($\chi^2 = 7.060$, $d.f. = 3$, $p = 0.070$). Both for percentage of infested bunches and percentage of damaged area per bunch, a significant difference was recorded between the untreated control and the plot with double application of dispensers in April and July ($q = 3.976$, $p = 0.025$ and $q = 3.719$, $p = 0.042$, respectively).

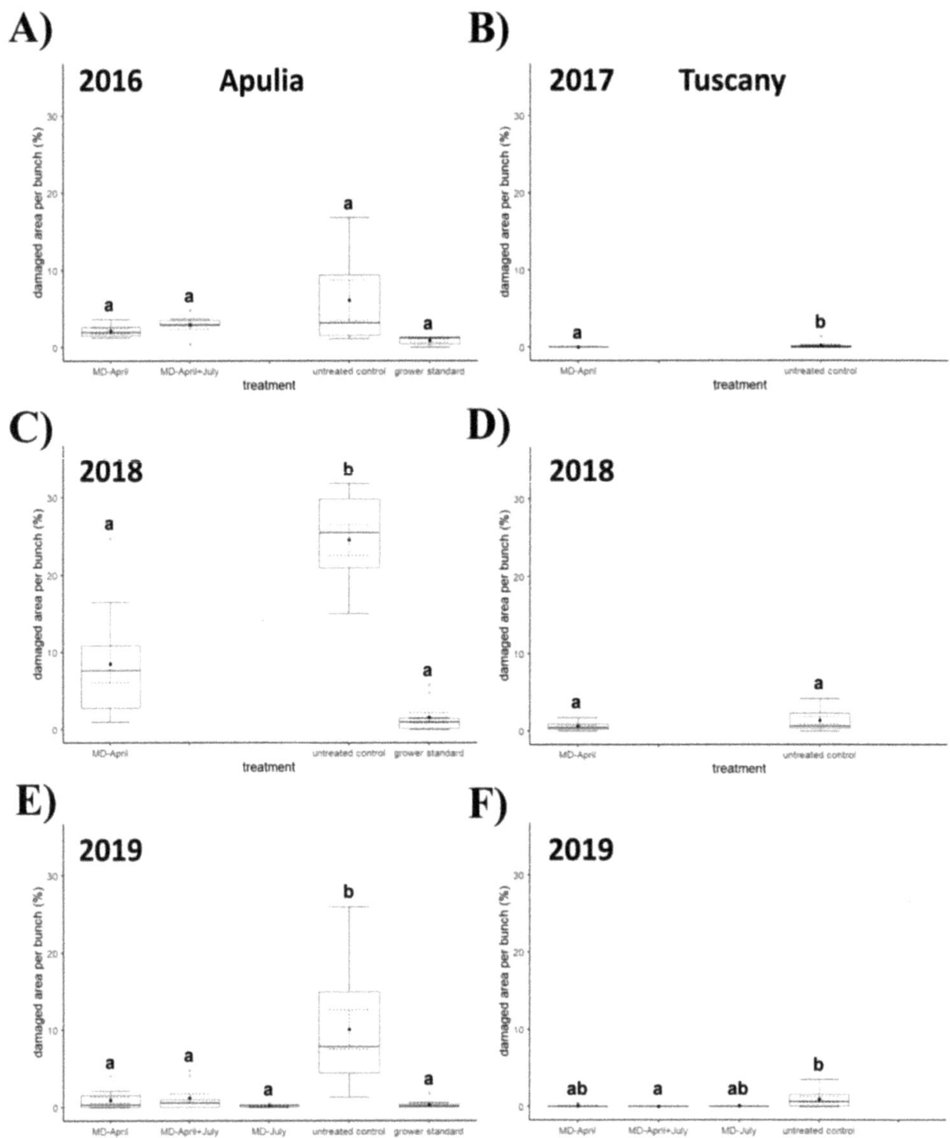

Figure 4. Impact of mating disruption (MD) using Isonet CGX111 dispensers on *Cryptoblabes gnidiella* damaged area (%) per bunch among treatments and untreated control plots in Apulia (**A,C,E**) and Tuscany (**B,D,F**) during three-year samplings. Boxplots indicate the median (solid line) and the range of dispersion (lower and upper quartiles and outliers) of the parameter; black dot and dashed line indicate mean value and standard error, respectively. Different letters above boxplots indicate significant differences among treatments (Wilcoxon test for comparisons between two plots or Kruskal–Wallis test for multiple comparisons, the latter followed by Nemenyi post hoc test; $p < 0.05$).

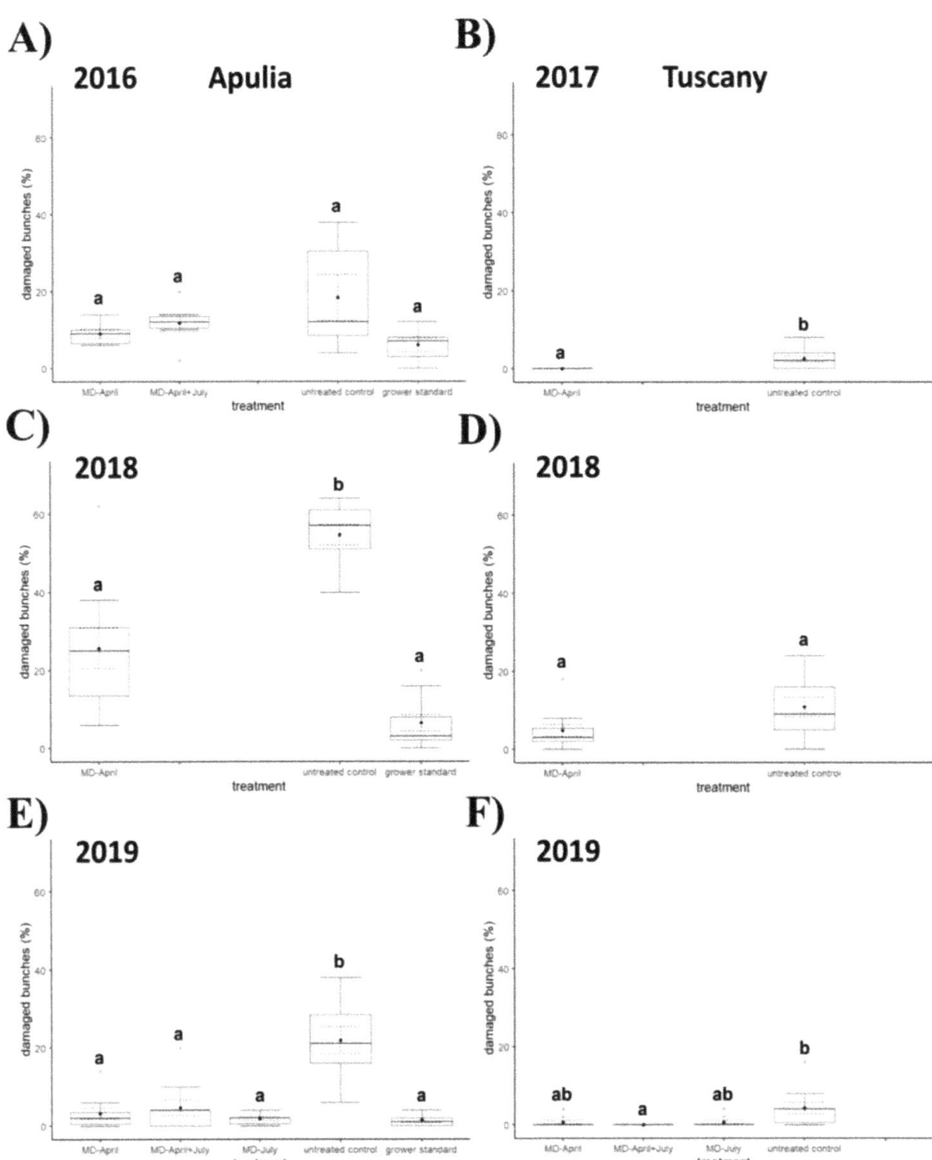

Figure 5. Impact of mating disruption (MD) using Isonet CGX111 dispensers on *Cryptoblabes gnidiella* infested bunches (%) among treatments and untreated control plots in Apulia (**A,C,E**) and Tuscany (**B,D,F**) during three-year samplings. Boxplots indicate the median (solid line) and the range of dispersion (lower and upper quartiles and outliers) of the parameter; black dot and dashed line indicate mean value and standard error, respectively. Different letters above boxplots indicate significant differences among treatments (Wilcoxon test for comparisons between two plots or Kruskal–Wallis test for multiple comparisons, the latter followed by Nemenyi post hoc test; $p < 0.05$).

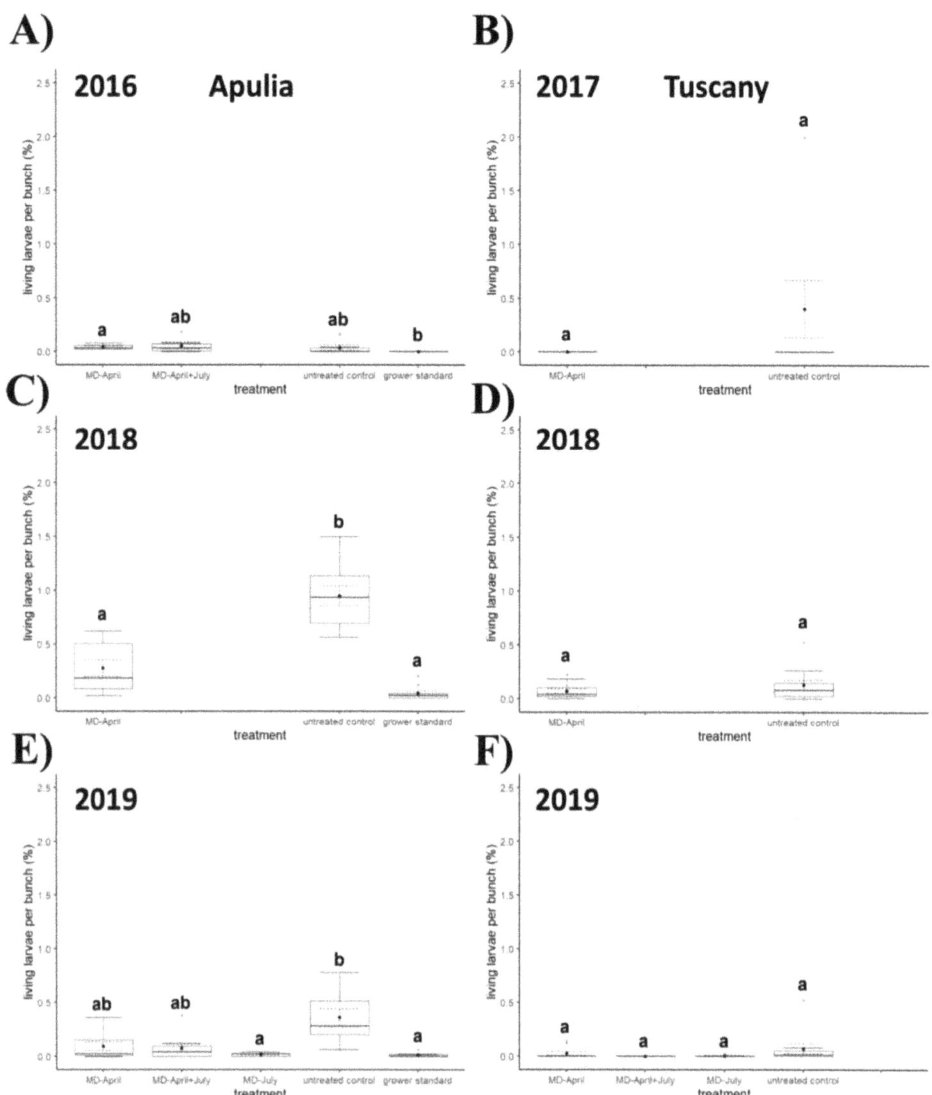

Figure 6. Impact of mating disruption (MD) using Isonet CGX111 dispensers on number of living larvae of *Cryptoblabes gnidiella* per bunch among treatments and untreated control plots in Apulia (**A,C,E**) and Tuscany (**B,D,F**) during three-year samplings. Boxplots indicate the median (solid line) and the range of dispersion (lower and upper quartiles and outliers) of the parameter; black dot and dashed line indicate mean value and standard error, respectively. Different letters above boxplots indicate significant differences among treatments (Wilcoxon test for comparisons between two plots or Kruskal–Wallis test for multiple comparisons, the latter followed by Nemenyi post hoc test; $p < 0.05$).

3.4. Field Experiments in Southern Italy

The Kruskal–Wallis test showed a significant difference (Figures 4–6) among treatments in 2016 in terms of percentage of damaged area per bunch ($\chi^2 = 8.130$, *d.f.* = 3, $p = 0.043$) and number of living larvae per bunch ($\chi^2 = 9.205$, *d.f.* = 3, $p = 0.027$), but the Nemenyi post hoc test confirmed a significant difference only in terms of number of living larvae between the grower's standard and the area with MD dispensers applied once in

April ($q = 3.695$, $p = 0.044$). No significant differences emerged in percentage of infested bunches between treatments ($\chi^2 = 5.284$, *d.f.* $= 3$, $p = 0.152$).

In 2018 (Figures 4–6), all the variables showed significant differences among treatments (percentage of infested bunches, $\chi^2 = 21.690$, *d.f.* $= 2$, $p < 0.0001$; percentage of damaged area per bunch, $\chi^2 = 21.194$, *d.f.* $= 2$, $p < 0.0001$; number of living larvae per bunch, $\chi^2 = 22.041$, *d.f.* $= 2$, $p < 0.0001$); significant differences were found for percentage of infested bunches, percentage of damaged area per bunch, and number of living larvae between the untreated control and the area with MD treatment in April ($q = 3.341$, $p = 0.048$; $q = 3.520$, $p = 0.034$; and $q = 3.987$, $p = 0.013$ respectively), and between the untreated control and the grower's standard ($q = 6.574$, $p < 0.0001$; $q = 6.502$, $p < 0.0001$; and $q = 6.574$, $p < 0.0001$ respectively).

In 2019 (Figures 4–6) all three variables again showed significant differences among treatments (percentage of infested bunches, $\chi^2 = 24.089$, *d.f.* $= 4$, $p < 0.0001$; percentage of damaged area per bunch, $\chi^2 = 22.780$, *d.f.* $= 4$, $p = 0.0001$; number of living larvae per bunch, $\chi^2 = 23.089$, *d.f.* $= 4$, $p = 0.0001$). In post hoc tests, significant differences were noted between the untreated control and the area with MD application in April about the percentage of infested bunches ($q = 4.892$, $p = 0.005$) and the percentage of damaged area per bunch ($q = 4.382$, $p = 0.017$); between the untreated control and the area with MD application in July, in the percentage of infested bunches ($q = 5.434$, $p = 0.001$), percentage of damaged area per bunch ($q = 5.716$, $p = 0.0005$), and number of living larvae per bunch ($q = 5.206$, $p = 0.002$); between the untreated control and the area with the application of MD dispensers in April and July about the percentage of infested bunches ($q = 4.371$, $p = 0.017$) and percentage of damaged area per bunch ($q = 4.328$, $p = 0.019$); and between the untreated control and the grower's standard in the percentage of infested bunches ($q = 6.020$, $p = 0.0002$), percentage of damaged area per bunch ($q = 5.694$, $p = 0.0005$), and number of living larvae per bunch ($q = 6.128$, $p = 0.0001$).

4. Discussion

This study shows that the application of Isonet CGX111 MD dispensers contributed to reducing grape damage due to HM larval feeding (i.e., percentage of infested bunches and percentage of damaged bunch area) in most of the trials. MD has been widely recognized as an effective strategy for managing several important agricultural pests such as *Cydia pomonella* (L.) [5,24], *Grapholita molesta* (Busck) [23], *Tuta absoluta* (Meyrick) [22], *Eupoecilia ambiguella* (Hübner), and *L. botrana* [6,26,27,32], as well as mealybugs such as *Planococcus ficus* (Signoret) and scale insects such as *Aonidiella aurantii* (Maskell) [8,33]. Concerning *C. gnidiella*, pheromone-baited traps have been used to track flights of this moth in vineyards, orchards, and uncultivated areas [3], but little has been done aiming at the application of MD for the control of this moth [2,34,35].

In our experiments, differences in bunch damage between MD treated and untreated control plots failed to show significance in Apulia in 2016 and in Tuscany in 2018, but it must be pointed out that a high variability among sampling units within the untreated control plot was recorded in these trials. The irregular distribution of the damage is a consequence of the irregular distribution of larvae, which is likely related to the vigor of the plants and the concomitant presence of other pests [1–4]. As far as the number of living larvae per bunch is concerned, it was generally higher in control plots than in MD plots in Apulia, while differences were never significant in Tuscany; the number of living larvae found was too low in all study years to be considered conclusive.

In all trials, a considerable reduction was observed in the number of trap catches in MD-treated plots compared to control plots. When the dispenser application was performed also or only in July, an almost complete trap catch shut-down was achieved. However, although pheromone-baited traps can provide useful information about the efficacy of MD products [36], crop damage reduction was less pronounced than trap catch reduction. These findings confirm what is stated in the official EPPO guidelines for efficacy evaluation of mating disruption products. HM catches in the monitoring traps should not be considered as the only indicator for MD efficacy assessment [36–38].

No substantial differences emerged in terms of percentage of infested bunches among MD-treated plots. In all the trials, no significant differences in bunch damage emerged between the plots treated with the application of MD dispensers in April, the application of MD dispensers in July, or the double application of MD dispensers in April and July. However, as indicated by the pheromone release curves and the trap catches, the application of MD dispensers in April was not sufficient to provide an adequate pheromone release during the preharvest period. The level of residual pheromone in August–October may not be sufficient to provide adequate HM control for late-ripening varieties. Furthermore, as already mentioned above, when MD dispensers were applied in July, a considerable reduction in adult catches during the fourth flight was observed in comparison to the untreated control, the grower's standard, and the plot where MD was applied in April. It may therefore be assumed that by placing the MD dispensers in July, MD efficacy could be extended up to harvest time also on very late ripening varieties.

In Apulia, the efficacy of MD in bunch damage reduction was always comparable to that of the grower's standard treatments. The level of HM damage reduction achieved both with MD alone and with insecticide applications only (no MD) was not complete and was considered not acceptable by growers. However, strategies consisting in the combined use of MD and a reduced number of insecticidal applications (e.g., *B. thuringiensis*) are being successfully used for the control of many lepidopteran pests, and they may constitute a valuable tool for HM control as well.

It is known that insecticide treatments focused on controlling the third-generation eggs and young larvae of *L. botrana* in late July/beginning of August may have an effect against *C. gnidiella*. However, as the life cycles of the two pests are not synchronous, insecticide treatments against the HM must be extended to the end of August or, in the case of late-ripening varieties, all through September.

As the HM is commonly present alongside *L. botrana*, a further strategy could be represented by the application of double dispensers for the control of both pests, followed by insecticide treatments as needed. Although the life cycles of the two species are not strictly synchronous, the gradual release of pheromone provided by the dispensers can cover the period of activity of both species. In a first trial conducted in Israel with MD against *L. botrana* and *C. gnidiella*, no significant differences were observed between plots treated against one vs. two pests. Moreover, in plots treated with pheromones against both pests, no pesticides were applied [2]. However, in some cases where MD was applied against more than a single target pest, mixed results were obtained [2]. In addition, it should be pointed out that the efficacy of MD in the control of both *L. botrana* and the HM has not been verified yet at high population densities of the latter [2].

Further studies aimed at clarifying the still poorly explored aspects of HM biology are needed. It is unclear whether the high number of catches from August onwards are adults, which have completed the life cycle (egg to adult) in the vineyard or are adults migrating from the surrounding environment. Moreover, the life cycle of the HM may be influenced by several factors such as temperature and relative humidity [39,40], causing wide fluctuations in HM populations from year to year. Shedding light on the issues concerning the ecology of this moth in the field and the factors influencing the growth of populations in the vineyard are crucial for calibrating the use of MD and its further development.

5. Conclusions

This study shows that MD may represent an effective and sustainable method for the control of the HM in vineyards. However, at least with the formulations used in our trials, MD was not able to provide complete pest control as a stand-alone strategy, and it may need to be supported by additional insecticide treatments to reduce the population of this pest to a level, which could be considered acceptable by growers. Moreover, the HM is a highly polyphagous species, which likely spends part of its life cycle on wild plants and then appears in the vineyards at the beginning of grape ripening, most probably attracted by volatile compounds [1]. One of the caveats for determining a pest's susceptibility to

Insects **2021**, *12*, 390

pheromone MD is the prevalence of migrating females, as MD cannot provide protection against mated females coming from outside the area treated with the disruptant [15]. For this reason, further studies on the HM phenology are necessary, especially in the months of May and June. In addition, when high populations occur, MD should be regarded as a control method within an IPM strategy [15,17,41]. An increase in knowledge about HM occurrence on wild host plants and abiotic factors influencing its population dynamics is crucial to enhance the application of MD against this emerging threat to Mediterranean vineyards.

Author Contributions: Conceptualization, R.R., F.D.G., A.I., G.B., and A.L.; investigation: R.R., F.D.G., F.C., E.L., F.S., A.I., G.B., and A.L.; methodology: E.L., F.S., A.I., G.B., and A.L.; data curation: R.R., F.D.G., and G.B.; software: R.R. and G.B.; supervision: G.B. and A.L.; writing—original draft: R.R., F.D.G., E.L., G.B., and A.L.; writing—review and editing: R.R., F.D.G., F.C., E.L., F.S., A.I., G.B., and A.L. All authors have read and agreed to the published version of the manuscript.

Funding: This research received no external funding.

Institutional Review Board Statement: Not applicable.

Data Availability Statement: Data are contained within the article.

Acknowledgments: The authors are grateful to V. Zeni (University of Pisa, Italy) for her kind assistance during data analysis.

Conflicts of Interest: Edith Ladurner, Francesco Savino, and Andrea Iodice work for CBC (Europe) S.r.l., Biogard Division (Grassobbio, Italy), a company that sells biocontrol solutions, including mating disruption products. The study is not biased by their position. The company had no role in the design of the study, analyses or interpretation of data, writing of the manuscript, or decision to publish the results.

References

1. Bagnoli, B.; Lucchi, A. Bionomics of *Cryptoblabes gnidiella* (Millière) (Pyralidae Phycitinae) in Tuscan Vineyards. *IOBC-WPRS Bull.* **2001**, *24*, 79–84.
2. Harari, A.R.; Zahavi, T.; Gordon, D.; Anshelevich, L.; Harel, M.; Ovadia, S.; Dunkelblum, E. Pest Management Programmes in Vineyards Using Male Mating Disruption. *Pest Manag. Sci.* **2007**, *63*, 769–775. [CrossRef] [PubMed]
3. Lucchi, A.; Ricciardi, R.; Benelli, G.; Bagnoli, B. What Do We Really Know on the Harmfulness of *Cryptoblabes gnidiella* (Millière) to Grapevine? From Ecology to Pest Management. *Phytoparasitica* **2019**, *47*, 1–15. [CrossRef]
4. CABI *Cryptoblabes gnidiella* (Citrus Pyralid). Available online: https://www.cabi.org/isc/datasheet/16381 (accessed on 22 January 2021).
5. Ioriatti, C.; Lucchi, A. Semiochemical Strategies for Tortricid Moth Control in Apple Orchards and Vineyards in Italy. *J. Chem. Ecol.* **2016**, *42*, 571–583. [CrossRef] [PubMed]
6. Lucchi, A.; Sambado, P.; Juan Royo, A.B.; Bagnoli, B.; Conte, G.; Benelli, G. Disrupting Mating of *Lobesia botrana* Using Sex Pheromone Aerosol Devices. *Environ. Sci. Pollut. Res. Int.* **2018**, *25*, 22196–22204. [CrossRef]
7. Thiéry, D.; Louâpre, P.; Muneret, L.; Rusch, A.; Sentenac, G.; Vogelweith, F.; Iltis, C.; Moreau, J. Biological Protection against Grape Berry Moths. A Review. *Agron. Sustain. Dev.* **2018**, *38*, 15. [CrossRef]
8. Lucchi, A.; Suma, P.; Ladurner, E.; Iodice, A.; Savino, F.; Ricciardi, R.; Cosci, F.; Marchesini, E.; Conte, G.; Benelli, G. Managing the Vine Mealybug, *Planococcus ficus*, through Pheromone-Mediated Mating Disruption. *Environ. Sci. Pollut. Res.* **2019**, *26*, 10708–10718. [CrossRef]
9. Wysoki, M.; Izhar, Y.; Gurevitz, E.; Swirski, E.; Greenberg, S. Control of the Honeydew Moth, *Cryptoblabes gnidiella* Mill. (Lepidoptera: Phycitidae), with *Bacillus thuringiensis* Berliner in Avocado Plantations. *Phytoparasitica* **1975**, *3*, 103–111. [CrossRef]
10. Lucchi, A.; Ladurner, E.; Iodice, A.; Savino, F.; Ricciardi, R.; Cosci, F.; Conte, G.; Benelli, G. Eco-Friendly Pheromone Dispensers—a Green Route to Manage the European Grapevine Moth? *Environ. Sci. Pollut. Res.* **2018**, *25*, 9426–9442. [CrossRef] [PubMed]
11. Witzgall, P.; Kirsch, P.; Cork, A. Sex Pheromones and Their Impact on Pest Management. *J. Chem. Ecol.* **2010**, *36*, 80–100. [CrossRef]
12. Rodriguez-Saona, C.R.; Stelinski, L.L. Behavior-Modifying Strategies in IPM: Theory and Practice. In *Integrated Pest Management: Innovation-Development Process*; Peshin, R., Dhawan, A.K., Eds.; Springer: Dordrecht, The Netherlands, 2009; Volume 1, pp. 263–315. ISBN 978-1-4020-8992-3.
13. Cardé, R.T.; Haynes, K.F. Structure of the pheromone communication channel in moths. In *Advances in Insect Chemical Ecology*; Millar, J.G., Cardé, R.T., Eds.; Cambridge University Press: Cambridge, UK, 2004; pp. 283–332. ISBN 978-0-521-79275-2.
14. Cardé, R.T.; Willis, M.A. Navigational Strategies Used by Insects to Find Distant, Wind-Borne Sources of Odor. *J. Chem. Ecol.* **2008**, *34*, 854–866. [CrossRef]
15. Cardé, R.T.; Minks, A.K. Control of Moth Pests by Mating Disruption: Successes and Constraints. *Annu. Rev. Entomol.* **1995**, *40*, 559–585. [CrossRef]

16. Suckling, D.M. Issues Affecting the Use of Pheromones and Other Semiochemicals in Orchards. *Crop. Prot.* **2000**, *19*, 677–683. [CrossRef]
17. Lo Verde, G.; Guarino, S.; Barone, S.; Rizzo, R. Can Mating Disruption Be a Possible Route to Control Plum Fruit Moth in Mediterranean Environments? *Insects* **2020**, *11*, 589. [CrossRef]
18. Mori, B.A.; Evenden, M.L. When Mating Disruption Does Not Disrupt Mating: Fitness Consequences of Delayed Mating in Moths. *Entomol. Exp. Appl.* **2013**, *146*, 50–65. [CrossRef]
19. Onufrieva, K.S.; Hickman, A.D.; Leonard, D.S.; Tobin, P.C. Ground Application of Mating Disruption against the Gypsy Moth (Lepidoptera: Erebidae). *J. Appl. Entomol.* **2019**, *143*, 1154–1160. [CrossRef]
20. Frank, D.L.; Starcher, S.; Chandran, R.S. Comparison of Mating Disruption and Insecticide Application for Control of Peachtree Borer and Lesser Peachtree Borer (Lepidoptera: Sesiidae) in Peach. *Insects* **2020**, *11*, 658. [CrossRef] [PubMed]
21. Trematerra, P.; Colacci, M.; Athanassiou, C.G.; Kavallieratos, N.G.; Rumbos, C.I.; Boukouvala, M.C.; Nikolaidou, A.J.; Kontodimas, D.C.; Benavent-Fernández, E.; Gálvez-Settier, S. Evaluation of Mating Disruption For the Control of *Thaumetopoea pityocampa* (Lepidoptera: Thaumetopoeidae) in Suburban Recreational Areas in Italy and Greece. *J. Econ. Entomol.* **2019**, *112*, 2229–2235. [CrossRef]
22. Jallow, M.F.A.; Dahab, A.A.; Albaho, M.S.; Devi, V.Y.; Jacob, J.; Al-Saeed, O. Efficacy of Mating Disruption Compared with Chemical Insecticides for Controlling *Tuta absoluta* (Lepidoptera: Gelechiidae) in Kuwait. *Appl. Entomol. Zool.* **2020**, *55*, 213–221. [CrossRef]
23. Kong, W.; Wang, Y.; Guo, Y.; Chai, X.; Niu, G.; Li, Y.; Ma, R.; Li, J. Effects of Disruption of *Grapholita molesta* (Lepidoptera: Tortricidae) Using Sex Pheromone on Moth Pests and Insect Communities in Orchards. *Appl. Entomol. Zool.* **2020**, *55*, 367–377. [CrossRef]
24. Angeli, G.; Anfora, G.; Baldessari, M.; Germinara, G.S.; Rama, F.; De Cristofaro, A.; Ioriatti, C. Mating Disruption of Codling Moth *Cydia pomonella* with High Densities of Ecodian Sex Pheromone Dispensers. *J. Appl. Entomol.* **2007**, *131*, 311–318. [CrossRef]
25. Cocco, A.; Muscas, E.; Mura, A.; Iodice, A.; Savino, F.; Lentini, A. Influence of Mating Disruption on the Reproductive Biology of the Vine Mealybug, *Planococcus ficus* (Hemiptera: Pseudococcidae), under Field Conditions. *Pest Manag. Sci.* **2018**, *74*, 2806–2816. [CrossRef]
26. Anfora, G.; Baldessari, M.; De Cristofaro, A.; Germinara, G.S.; Ioriatti, C.; Reggiori, F.; Vitagliano, S.; Angeli, G. Control of *Lobesia botrana* (Lepidoptera: Tortricidae) by Biodegradable Ecodian Sex Pheromone Dispensers. *J. Econ. Entomol.* **2008**, *101*, 444–450. [CrossRef]
27. Hummel, H.E. A Brief Review on *Lobesia botrana* Mating Disruption by Mechanically Distributing and Releasing Sex Pheromones from Biodegradable Mesofiber Dispensers. *Biochem. Mol. Biol. J.* **2017**, *3*. [CrossRef]
28. Hashem, A.G.; Tadros, A.W.; Abo-Sheasha, M.A. (Ministry of A. Monitoring the Honeydew Moth, *Cryptoblabes gnidiella* Mill in Citrus, Mango and Grapevine Orchards, Lepidoptera: Pyralidae. *Ann. Agric. Sci. Ain Shams Univ. Egypt* **1997**, *42*, 335–343.
29. Yehuda, S.B.; Wysoki, M.; Rosen, D. Phenology of the Honeydew Moth, *Cryptoblabes gnidiella* (Millière) (Lepidoptera: Pyralidae), on Avocado in Israel. *Isr. J. Entomol.* **1991**, *45*, 149–160.
30. Abdel-Moaty, R.M.; Hashim, S.M.; Tadros, A.W. Monitoring the Honeydew Moth, *Cryptoblabes gnidiella* Millière (Lepidoptera: Pyralidae) in Pomegranate Orchards in the Northwestern Region of Egypt. *J. Plant. Prot. Pathol.* **2017**, *8*, 505–509. [CrossRef]
31. Demirel, N. Seasonal Flight Patterns of the Honeydew Moth, *Cryptoblabes gnidiella* Millière (Lepidoptera: Pyralidae) in Pomegranate Orchards as Observed Using Pheromone Traps. *Entomol. Appl. Sci. Lett.* **2016**, *3*, 1–5.
32. Arn, H.; Rauscher, S.; Schmid, A.; Jaccard, C.; Bierl-Leonhardt, B.A. Field Experiments to Develop Control of the Grape Moth, *Eupoecilia ambiguella*, by Communication Disruption. In *Management of Insect Pests with Semiochemicals: Concepts and Practice*; Mitchell, E.R., Ed.; Springer: Boston, MA, USA, 1981; pp. 327–338. ISBN 978-1-4613-3216-9.
33. Franco, J.C.; Lucchi, A.; Mendel, Z.; Suma, P.; Vacas, S.; Mansour, R.; Navarro-Llopis, V. Scientific and Technological Developments in Mating Disruption of Scale Insects. *Entomol. Gen.* **2021**. accepted. [CrossRef]
34. Sellanes, C.; González, A. The Potential of Sex Pheromones Analogs for the Control of *Cryptoblabes gnidiella* (Lepidoptera: Pyralidae), an Exotic Pest in South America. *IOBC WPRS Bull.* **2014**, *99*, 55–60.
35. Acín, P. Management of the honeydew moth by mating disruption in vineyard. In Proceedings of the Joint Meeting of the IOBC-WPRS Working Groups "Pheromones and Other Semiochemicals in Integrated Production" & "Integrated Protection of Fruit Crops", Lisbon, Portugal, 20–25 January 2019. *IOBC WPRS Bull.* **2019**, *146*, 28–31.
36. European and Mediterranean Plant Protection Organization Principles of Efficacy Evaluation for Mating Disruption Pheromones. *EPPO Bull.* **2019**, *49*, 426–430. [CrossRef]
37. Ioriatti, C.; Bagnoli, B.; Lucchi, A.; Veronelli, V. Vine Moths Control by Mating Disruption in Italy: Results and Future Prospects. *Redia* **2004**, *87*, 117–128.
38. Miller, J.R.; Gut, L.J. Mating Disruption for the 21st Century: Matching Technology with Mechanism. *Environ. Entomol.* **2015**, *44*, 427–453. [CrossRef] [PubMed]
39. Elnagar, H.M. Population Dynamic of Honeydew Moth, *Cryptoblabes gnidiella* Miller in Vineyards Orchards. *Egypt. Acad. J. Biol. Sci. Entomol.* **2018**, *11*, 73–78. [CrossRef]

40. Barker, B.; Coop, L. *Honeydew Moth Cryptoblabes gnidiella (Lepidoptera: Pyralidae). Phenology/Degree-Day and Climate Suitability Model Analysis*; Prepared for USDA APHIS PPQ. Version 1.0, 12/09/2019; Department of Horticulture and Integrated Plant Protection Center, Oregon State University: Corvallis, OR, USA, 2019.
41. Louis, F.; Schirra, K.-J. Mating Disruption of *Lobesia botrana* (Lepidoptera: Tortricidae) in Vineyards with Very High Population Densities. *IOBC-WPRS Bull.* **2001**, *24*, 75–79.

Article

Individual and Additive Effects of Insecticide and Mating Disruption in Integrated Management of Navel Orangeworm in Almonds

Bradley S. Higbee [1],* and Charles S. Burks [2]

1   Trécé Inc., Adair, OK 74330, USA
2   USDA, Agricultural Research Service, San Joaquin Valley Agricultural Sciences Center, 9611 South Riverbend Avenue, Parlier, CA 93648, USA; charles.burks@usda.gov
*   Correspondence: buglimo@gmail.com; Tel.: +1-661-301-3225

Simple Summary: Mating disruption is an increasingly important part of pest management for the navel orangeworm *Amyelois transitella*. Industry groups have long supported mating disruption research and development with the divergent objectives of both minimizing damage from this key pest and reducing insecticide used on these crops. It is therefore important to know whether the benefits of mating disruption and insecticide are additive or, alternatively, if using both together provides no additional benefit over either alone. Ten years of data from research trials in a large commercial almond orchard found that the benefits of mating disruption are generally additive with lower damage if both are used together than either alone. Substantial year-to-year variability in navel orangeworm damage was also evident, even with stringent management. These findings indicate that the combination of mating disruption and insecticide can reduce the impact of navel orangeworm damage on the almond industry. Further improvements in monitoring and predictions of navel orangeworm abundance and damage are necessary for mating disruption to effectively contribute to the industry goal of reduction of insecticide use by 25%.

Abstract: Damage from *Amyelois transitella*, a key pest of almonds in California, is managed by destruction of overwintering hosts, timely harvest, and insecticides. Mating disruption has been an increasingly frequent addition to these management tools. Efficacy of mating disruption for control of navel orangeworm damage has been demonstrated in experiments that included control plots not treated with either mating disruption or insecticide. However, the navel orangeworm flies much farther than many orchard pests, so large plots of an expensive crop are required for such research. A large almond orchard was subdivided into replicate blocks of 96 to 224 ha and used to compare harvest damage from navel orangeworm in almonds treated with both mating disruption and insecticide, or with either alone. Regression of navel orangeworm damage in researcher-collected harvest samples from the interior and center of management blocks on damage in huller samples found good correlation for both and supported previous assumptions that huller samples underreport navel orangeworm damage. Blocks treated with both mating disruption and insecticide had lower damage than those treated with either alone in 9 of the 10 years examined. Use of insecticide had a stronger impact than doubling the dispenser rate from 2.5 to 5 per ha, and long-term comparisons of relative navel orangeworm damage to earlier- and later-harvested varieties revealed greater variation than previously demonstrated. These findings are an economically important confirmation of trade-offs in economic management of this critical pest. Additional monitoring tools and research tactics will be necessary to fulfill the potential of mating disruption to reduce insecticide use for navel orangeworm.

Keywords: mating disruption; navel orangeworm; *Amyelois transitella*; almond; integrated pest management

1. Introduction

Mating disruption [1–5] is an increasingly important tool in integrated pest management (IPM) [6,7] and area-wide control of insect pests. It is used primarily against lepidopteran pests, although there are examples of mating disruption for control of Hemiptera [8,9], Coleoptera [10–12], and Hymenoptera [13]. Historically, synthetic pheromones and dispenser systems have been expensive [14,15]. Mating disruption use is most widespread in protection of high-value commodities such as horticultural crops, or in programs for management of invasive pests on public lands or across entire jurisdictions where management tactics are determined by policy objectives rather than cost-return criteria [5]. In some cases, mating disruption is used to reduce insecticide input and achieve the IPM goal of controlling pests with the least non-target impact, and in other cases, it is used with insecticides to achieve another IPM goal of maintaining economic sustainability. The degree of efficacy of mating disruption and the precise mechanisms by which it works varies with the target pest and the dispensing system, so the degree to which mating disruption is used to reduce insecticides vs. the degree to which it is used to reduce environmental impact varies with particular situations.

Mating disruption has become an increasingly prominent part of pest management for the navel orangeworm *Amyelois transitella* (Walker) (Lepidoptera: Pyralidae) [16]. The navel orangeworm is the principal pest of almonds and pistachios, and is an important pest of walnuts [16]. The area planted in each of these crops has increased substantially in the last 20 years [16]. Biological features of the navel orangeworm important to its pest status include its wide host range, its multivoltine life history, and a strong dispersal capacity [17–20]. The navel orangeworm directly attacks fruit, making it economically destructive. This polyphagous pest depends on host vulnerability for larval entry through lesions from disease, attack by another insect pest, or increased exposure of fruit with maturity [16]. Its robust dispersal capacity allows the navel orangeworm to move between orchards of different host crops, and its multivoltine nature and wide host range allows the navel orangeworm to amplify in one crop then go to another as earlier-maturing crops (e.g., almonds) are harvested and later-maturing crops (e.g., pistachios and walnuts) become vulnerable [16]. The adult does not feed, and the larva remains in its host through pupation until adult eclosion [16]. The navel orangeworm overwinters in unharvested fruit remaining on trees or dropped to the ground [16].

Integrated management of the navel orangeworm has long been based on the biology of the host and the phenology of the crop [16]. The most abundant and economically valuable almond variety in California is Nonpareil, which has a thinner shell and matures earlier than other varieties [21]. Nonpareil, like most almond varieties, requires cross-pollination with a different variety for optimal fruit set and so it is grown in orchards with multiple varieties [22]. Most of the pollinizer varieties planted alongside Nonpareil have a harder shell and a later harvest date, so they are less vulnerable to navel orangeworm but can be challenged by greater abundance [21,22]. Almonds in California generally flower in February and leaf out immediately afterward. For Nonpareil, hull split and vulnerability in healthy fruit generally begin in mid-June to early July, and harvest maturity is attained around six weeks later. The pollinizer varieties are typically three to four weeks behind in hull split and harvest maturity relative to Nonpareil. Almonds are harvested by shaking the nuts onto the ground. They are allowed to dry for several days, gathered into windrows on the ground, and taken to a huller for processing. Navel orangeworm from the overwintering generation are reproductively active in April and May and depend on almonds left from the previous year for a host. These overwintered navel orangeworm are referred to as first flight [23–28]. Their progenies form a second flight, which emerge in June around the time of Nonpareil hull split. The progeny of this second flight can develop faster because of better host quality, and the resulting third flight typically arrives in August around the time of Nonpareil harvest. A fourth flight typically arrives in September, around the time of harvest of the pollinizer varieties [16].

The pest management strategy based on this pest biology and host phenology relies most fundamentally on cultural practices of sanitation (rigorous removal and destruction of fruit left after harvest) and timely harvest [16,19]. If these tactics are not sufficient to keep navel orangeworm damage below an acceptable threshold, then insecticide treatments are also used [29–33]. Mating disruption for navel orangeworm has most often been used in addition to insecticide rather than as a replacement for it [34,35]. However, there is potential for reduced insecticide use.

The insecticides currently used most often for navel orangeworm include methoxyfenozide, chlorantraniliprole, various pyrethroids, and spinetoram [16]. All are targeted to eggs and/or neonates. Although the pyrethroids have had historically more activity than the others against adults, resistance has been documented (BH, unpublished data, and Niu et al. [36,37]). The time period when these products are most effective is the period immediately after the initiation of Nonpareil hull split [38]. The second most important time is a second preharvest application sometime between two weeks after the hull split application and the last possible application point before the preharvest interval [38]. These are both targeted against the second and third flights. In some cases, an application is made in April or May targeting first flight [38]. Applications also occasionally target the third flight in the period between the Nonpareil and pollinizer harvests, but often this is not done because of the complexity of coordinating the restricted access interval and other activities necessary during the harvest period. Use of more selective insecticides like methoxyfenozide or chlorantraniliprole is encouraged earlier in the season because these have a narrower spectrum of activity and are less likely to kill natural enemies that prevent defoliation by web-spinning mites. Decisions about the number of insecticide applications tend to be based on previous history and current crop prices. Monitoring assists in timing of insecticide applications, but predicting navel orangeworm damage based on in-season monitoring remains an ongoing challenge [39]. A further challenge to insecticide control results from the requirement that insecticide residue coverage prevents the larva entering the host where it is therefore sheltered from further exposure.

Currently, the most well-established formulation for mating disruption for navel orangeworm uses aerosol dispensers [40–44]. Peer-reviewed studies have also demonstrated efficacy for a hand-applied meso-dispenser formulation based on polyvinylchloride emitters [35]. Experimental formulations based on a more complete and attractive pheromone blend suppress males in pheromone traps more effectively than a single-component formulation, but all commercial formulations still use the single-component blend because of economic and regulatory considerations [43]. Mating disruption mechanisms are broadly categorized as competitive (the male interacts with the dispenser) or non-competitive (the male is made unresponsive to females without interacting directly with dispensers). The mechanism seems to be a hypothesized hybrid which initially involves attraction to the dispenser but then makes males unresponsive to females without continued interaction with the dispenser [2,15,45]. Like a purely non-competitive mechanism, the hybrid mechanism is less density dependent than competitive mechanisms [2,15,45]. Mating disruption for navel orangeworm provides the greater economic return with greater pressure within a range from moderate to high baseline damage [35].

Here, we present the damage data from ongoing mating disruption trials at a commercial almond site between 2006 and 2015 near the town of Lost Hills, CA. Methods that have been used to improve cost-effectiveness of aerosol mating disruption include limiting the part of the field season during which it is used, limiting the amount of pheromone loaded in each dispenser, and limiting the number of aerosol dispensers per ha. Previous studies analyzed the data from this and another site between 2009 and 2015 to examine the association of various monitoring methods with subsequent navel orangeworm damage, and to examine the relationship between variety composition and damage in these varieties [22,39]. In this paper, the Lost Hills data are analyzed using the randomized complete block design with which this site was arranged to compare navel orangeworm damage between plots treated with mating disruption alone, insecticide alone, or both. This long-term data set is used to examine effects of year-to-year variation on outcomes of management strategies for the navel orangeworm, and also year-to-year variation on relative impact of navel orangeworm on two major almond varieties, Nonpareil and Monterey.

2. Materials and Methods

2.1. Site and Plot Arrangement

Trials were performed on 971 hectares of almond trees from 2006 to 2015 at the Lost Hills Ranch, planted in 1990 and 1993, owned and operated by Wonderful Orchards (formerly Paramount Farming). From 2008 to 2012, these were part of a USDA Agricultural Research Service area-wide integrated pest management project to improve navel orangeworm management [22,46]. General features of this site are illustrated by a plot map from 2011 (Figure 1). The basic management units at this site were partial or complete 54 ha (160 ace) quarter-sections [47]. Collections of these quarter-sections were referred to as ranches. In this figure, almond ranches (delineated by heavier green lines) include 3450, 3440, and 3460. The pistachio ranch 4390 was not included in the current analysis. East-west tiers of quarter-sections served as replicate blocks. For example, in Ranch 3450, the third and fourth tiers from the top (respectively, purple and pink) served as replicate blocks for a mating disruption treatment with and without insecticide. All blocks were subject to an intensive sanitation regime that combined machine and hand removal of nuts from trees and flail mowing of residue following harvest to eliminate mummies from managed areas. This resulted in 5–10 mummies/tree on the ground and less than 0.3 mummies/tree in the canopy [19]. All blocks had Nonpareil almonds. The most common pollinizer variety was Monterey. Other pollinizer varieties included Butte, Carmel, Fritz, Mission, Price, Ruby, and Wood Colony.

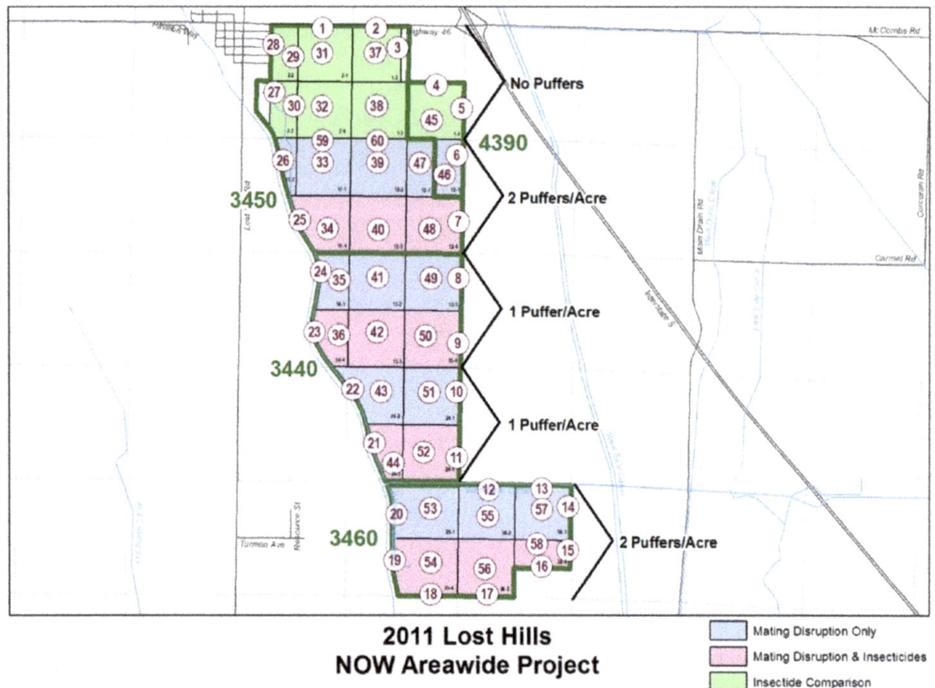

Figure 1. Representative map for treatment areas (blocks) in a commercial almond orchard in 2011. The two blocks with sites 4, 5, 6, 45, and 46 (upper right portion of map) are pistachio orchards and not included in this study.

Four different experiments were performed at this site from 2006 to 2015: (1) An experiment in 2006 and 2007 comparing mating disruption at label rate used alone, insecticide used alone, and both together; (2) an experiment from 2008 to 2011 examining aerosol mating disruption dispensers per hectare; (3) an experiment from 2012 to 2014 examining different concentrations of pheromone emitted from dispensers at a fixed spatial density; and (4) an experiment in 2015 examining the date that mating disruption started. The latter three experiments were elaborations of the initial experiment; thus, they were amenable to a common analysis.

2.2. Evaluation of Navel Orangeworm Damage

Damage to almonds by navel orangeworm was evaluated by two methods, which we refer to here as windrow and huller. For research evaluation, windrow samples of approximately 500 nuts were gathered when each variety was harvested from the ground or windrow at each of the numbered monitoring sites in Figure 1. Sampled nuts were evaluated individually in the laboratory. Nuts were opened and scored for whether the kernel was damaged (navel orangeworm damage), and for whether there were navel orangeworm larvae associated with the hull in almonds with undamaged kernels (proportion of navel orangeworm infestation was the proportion of nuts with either kernel damage or infestation without kernel damage). The numbered sites corresponded to monitoring stations and most quarter-sections had a monitoring site on the edge of the orchard and one on the center. The monitoring sites on the edge were important for trapping and monitoring, but only the center sites were used for evaluation of effects of mating disruption and insecticide treatments.

Huller evaluation used methods similar to industry-standard practice. Windrowed almonds are picked up from windrows, placed in hopper trailers, and taken to a huller for initial processing. At the huller, after field debris has been removed and the kernels are removed from the hulls and shells, damage to kernels is evaluated from a 20 kg sample from each trailer as a basis of quality control and payment. In standard industry procedure, kernels are determined to be edible or non-edible, and non-edible kernels are noted by a broad range of damage categories. In the current study, a further assessment was made to determine if non-edible status was the result of navel orangeworm damage.

2.3. Individual and Additive Effects of Mating Disruption and Insecticide (2006 and 2007)

In 2006 and 2007, the treatments were: (1) aerosol mating disruption with 5 dispensers per ha (the first label rate [15,41,43]); (2) applications of the insecticide methoxyfenozide targeting the first and second navel orangeworm flights (Table S1); and (3) a combination of both of these treatments.

2.4. Mating Disruption Dispensers per Hectare (2008–2011)

Subsequent experiments examined more and less intensive forms of mating disruption with and without an insecticide treatment. From 2008 to 2011, aerosol mating disruption at two different dispenser densities, with or without insecticide, was compared to insecticide treatment alone. Treatments were thus: (1) insecticide treatment without mating disruption; (2) 2.5 mating disruption dispensers per ha without insecticide; (3) 2.5 mating disruption dispensers per ha with insecticide; (4) 5 mating disruption dispensers per ha without insecticide; and (5) 5 mating disruption dispensers per ha with insecticide. Mating disruption treatments used Suterra Checkmate Puffer NOW aerosol dispensers, each of which contained 3.8 g of the active ingredient (a.i.) (Z11,Z13)-hexadecadienal and releasing 0.38 mg every 15 min from 17:00 to 05:00 local time for a total of 18.24 mg per dispenser per night [15,41,43]. The two replicates of the no-mating disruption insecticide treatment were placed adjacent to each other and at either the north or south end of the site to minimize the effect of the mating disruption treatments on these no-mating disruption treatment blocks. Insecticide treatments for navel orangeworm consisted of two applications per year, approximately as described in the previous section (Table S1).

2.5. Mating Disruption Active Ingredient Per Hectare (2012–2014)

An experiment from 2012 to 2014 examined aerosol mating disruption with or without insecticide in a manner similar to the previous experiment. However, all mating disruption blocks were treated using 5 dispenser per ha, using either the standard rate or half of the standard rate. Amount of a.i. per ha was varied by the amount of a.i. in the aerosol cannister (3.8 or 1.9 mg), and therefore 0.38 or 0.19 mg a.i. per emission and 91 or 45 mg a.i. per ha per night. Treatments were thus: (1) insecticide treatment without mating disruption; (2) 5 mating disruption dispensers per ha, each containing 1.9 mg a.i., without insecticide; (3) 5 mating disruption dispensers per ha, each containing 1.9 mg a.i. with insecticide; (4) 5 mating disruption dispensers per ha, each containing 3.8 mg a.i., without insecticide; and (5) 5 mating disruption dispensers per ha, each containing 3.8 mg a.i., with insecticide. In 2012, methoxyfenozide was applied in spring and at hull split, similar to the previous years. In 2013 and 2014, three applications were made against navel orangeworm, with the pyrethroid, bifenthrin, applied post-hullsplit, and prior to the Nonpareil harvest (Table S2).

2.6. Time of Start of Mating Disruption (2015)

The variable for aerosol mating disruption for 2105 was the time that mating disruption began: either early season (17 March, 336 NOW degree-days from 1 January) or normal deployment (13 April, 577 NOW degree-days from 1 January) of mating disruption in combination with conventional treatment. Treatments were thus: (1) insecticide treatment without mating disruption; (2) the standard mating disruption timing without insecticide; (3) the standard mating disruption timing with insecticide; (4) the early mating disruption timing without insecticide; and (5) the early mating disruption timing with insecticide. The insecticide regime in 2015 was similar to 2013 and 2014 (Table S2). Mating disruption trials at this site were discontinued after a single year of this experiment.

2.7. Data Analysis

Data were processed and plotted using R 4.0 [48]. Correlation, linear regression, and nonparametric analysis was performed in R, and generalized linear mixed model (GLMM) analysis was conducted using the SAS system [49]. An initial analysis compared damage effects between edge windrow samples, interior windrow samples, and huller samples. In this case, the experimental units were the 65 ha "sections" rather than the tiers that formed replicate blocks. This was done because the sections were the smallest unit for which there were independent data for the huller samples. The edge and interior windrow samples were aggregated (there were not edge samples for all sections, since some were bounded on all sides by other sections). Damage over 10 years was compared between three types of treatment (mating disruption, insecticide, or both) for the three types of samples using the non-parametric Kruskal–Wallis ANOVA followed by the Dunn post hoc test, with the holm procedure for means separation. In addition, ordinary least squares regression was used to compare percent *A. transitella* damage in interior and edge windrow samples with *A. transitella* damage in huller samples.

A subsequent analysis compared damage across all 10 years, based on the fact that the general treatment structure was used in 2008 to 2015 was an overlay on the three-way comparison in 2006 and 2007. Based on the initial analysis, the interior windrow samples were used for this analysis. Damage and total nuts examined were pooled across the tier that formed replicate blocks (Figure 1), and analyzed as using a GLMM (PROC GLIMMIX) with a binomial error distribution, and Kenward–Roger degrees of freedom [50]. The treatment (insecticide, mating disruption, or both) was a fixed factor, and the year and tier (replicate block) were random factors. The binomial samples were based on a mean sample size of 5336 (range 1072 to 12,753).

In addition to analyzing the entire 10-year data set, experimental variations were analyzed separately. The same fixed and random independent variables were used for data from 2006 and 2007. For the experiments from 2008–2011, 2012, 2014, and 2015, data from the insecticide-only plots were set aside and the experiments were analyzed as a 2 × 2 factorial design with one factor representing two different intensities of mating disruption, and the other factor representing presence or absence of insecticide treatments.

Damage in the varieties Nonpareil and Monterey was compared only in the plots treated with insecticides and not mating disruption, in order to minimize pest management treatment as a confounding factor in this comparison over the 10 years. Comparisons were made using the windrow interior samples. In addition, navel orangeworm degree-days from January 1 were calculated using the UC IPM degree-day calculator [51] and data from the Lost Hills (Kern County) California Irrigation Management Information System (CIMIS) site [52], and degree-day accumulation on 15 June was compared between years in which Nonpareil had more damage than Monterey and years in which the converse was true.

3. Results

3.1. Comparison of Huller and Windrow Samples

Damage in all almond varieties over the 10-year study differed based on treatment type and the type of sample used to evaluate the treatment (Table 1). In all cases, damage over the 10-year period was numerically higher in mating disruption plots than in plots treated only with insecticide, but this difference was not significant ($p > 0.05$) in huller samples and in interior windrow samples collected by researchers. Regression revealed a significant association of the internal and the edge windrow samples with the huller sample (Table 2). These comparisons are based on all blocks with both edge and internal collection sites. Based on these observations, the windrow internal samples were used for subsequent analysis because the raw data were more uniform than the huller reports, and more suitable for statistical analysis (i.e., direct quantification of damage and total sample size was preserved).

Table 1. Median damage (percent) of all almond varieties by treatment type over 10 years.

Treatment Type	Huller	Windrow (Interior)	Windrow (Edge)
Mating disruption only	0.61a	0.91a	2.22a
Insecticide only	0.56a	0.73a	1.58ab
Both	0.3b	0.37b	1.01b

Medians in the same column followed by different letters are significantly different (Kruskal–Wallis ANOVA, experiment-wise $p < 0.05$).

Table 2. Linear regression of percent damage of windrow samples on huller samples.

	Windrow (Interior)	Windrow (Edge)
Intercept	−0.01	0.69 ***
Slope	1.42 ***	1.75 ***
Adjusted r^2	0.69	0.59

*** $p < 0.001$.

3.2. Comparison of Overall Damage over the 10-Year Study

Interior windrow samples from all blocks (aggregated into replicate tiers) were used for a more comprehensive comparison of damage over the period of the study. Over the 10 years, navel orangeworm damage was different among treatments ($F_{2,91.8} = 14.06$, $p < 0.0001$). Plots treated with both insecticide and mating disruption had significantly less damage than those treated with either insecticide alone or mating disruption alone, while there was no significant different among the latter two treatments (Table 3). A graph of damage based on interior windrow samples revealed that it was numerically lower in the plots treated with both mating disruption and insecticide compared to those treated with only mating disruption or only insecticide in 9 of the 10 years examined (Figure 2). There was substantial variation in navel orangeworm damage with the average across all treatments ranging from 0.36% in 2010 to 5.4% in 2015.

Table 3. Percent navel orangeworm damage (mean ± SE) from interior windrow samples, by insecticide and mating disruption treatment and across all varieties, 2006–2015.

Treatment	Percent NOW Damage
Insecticide only	1.9 ± 0.47a
Mating disruption only	1.8 ± 0.49a
Both insecticide and mating disruption	1.0 ± 0.18b

Means followed by different letters are significantly different (generalized linear mixed model (GLMM) with binomial distribution, $p < 0.05$).

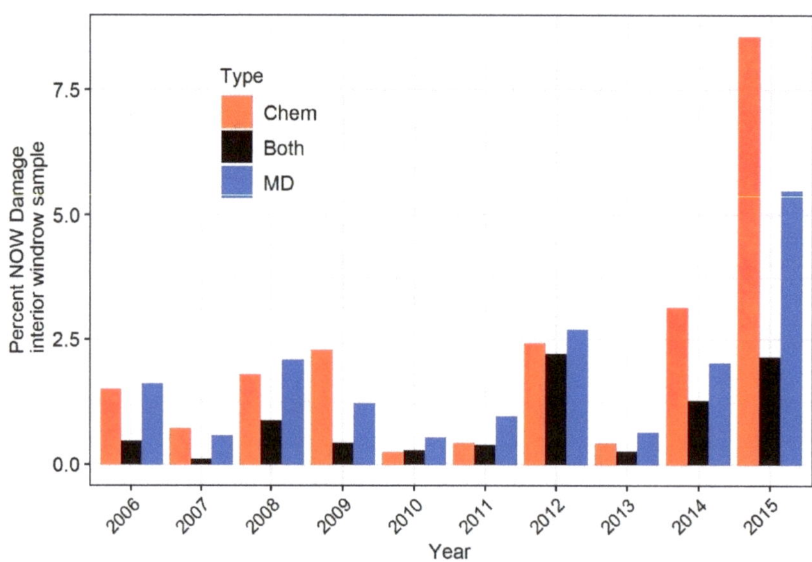

Figure 2. Mean navel orangeworm damage for all almond varieties for 2006 to 2015, as determined from interior windrow samples. The insecticide (Chem) indicates plots that received insecticide but not mating disruption; the mating disruption (MD) treatment category includes all mating disruption treatments when used without insecticides, and the Both treatment category includes all mating disruption treatments when used with insecticide.

3.3. Comparison of Damage in Specific Experiments

Analysis of data from the first two years, in which there were only three treatments, revealed similar trends to the 10-year data set. There were significant differences among the treatments ($F_{2,7.05} = 6.59$, $p = 0.02$), and the mating disruption and insecticide treatments were not different while the combined treatment had significantly less damage (Table 4).

Table 4. Percent navel orangeworm damage (mean ± SE) from interior windrow samples, by insecticide, and mating disruption treatment and across all varieties, 2006 and 2007.

Treatment	n (Replicate Block by Year)	Percent NOW Damage
Insecticide only	9	1.0 ± 0.24a
Mating disruption only	4	1.1 ± 0.32a
Both insecticide and mating disruption	3	0.4 ± 0.13b

Means followed by different letters are significantly different (GLMM with binomial distribution, $p < 0.05$).

$\rightarrow$ For the experiment from 2008 to 2011, there were numerical differences among all levels of the factorial comparison of 2.5 or 5 mating disruption dispensers with or without insecticide (Table 5). The GLMM analysis of fixed effects revealed significant effects due to insecticide ($F_{1,13.75} = 11.34$, $p = 0.0047$), not quite significant effects due to dispenser density ($F_{1,13.76} = 3.33$, $p = 0.0896$), and no significant interaction ($F_{1,24.59} = 0.42$, $p = 0.52$).

Table 5. Percent navel orangeworm infestation (mean $\pm$ SE, n = 8) from windrow samples from Nonpareil and pooled pollinizer varieties by insecticide and mating disruption (MD) treatment, 2008–2011.

Mating Disruption Dispensers per ha	Without Insecticide	With Insecticide
2.5	1.67 ± 0.64	0.64 ± 0.20
5	0.93 ± 0.22	0.33 ± 0.09

The row-wise differences (insecticide effect) are significant ($p < 0.05$), the column-wise differences (dispensers per ha) are not quite significant ($0.1 > p > 0.05$), and the interaction is not significant (($p > 0.1$) (GLMM with negative binomial distribution).

$\rightarrow$ For the experiment from 2012 to 2014, percent navel orangeworm damage in mating disruption plots was, respectively, 1.6 ± 0.57 and 1.1 ± 0.39 for half and full label concentration mating disruption from 5 dispensers per ha in the absence of insecticide treatments. In the presence of insecticide, these figures were, respectively, 0.9 ± 0.61 and 0.7 ± 0.38 percent damage. There were no significant effects from either insecticide or mating disruption release rate, and the interaction was not significant ($p > 0.1$).

The analysis also found no significant effects for the 1-year experiment at the time of the start of mating disruption, conducted in 2015. Percent damage in mating disruption plots was, respectively, 10.3 ± 9.4 and 3.0 ± 0.93 for the early and standard mating disruption start times in the absence of insecticide treatments. In the presence of insecticide, these figures were, respectively, 2.6 ± 0.44 and 1.9 ± 1.31 percent damage.

3.4. Relationship Between Damage in Early-Harvested Nonpareil and a Later Harvested Pollinizer Variety

Over a 10-year period, the mean navel orangeworm damage in the insecticide-treated plots was numerically higher in Nonpareil almonds (1.4 ± 0.38) than in Monterey almonds (1.0 ± 0.32). This difference was not statistically significant (Welch unequal variance t-test, t = 0.86, df = 17.402, $p = 0.40$). A year-by-year graph of damage shows that the relative damage between the varieties was highly variable, and in 4 of the 10 years, damage was higher in Monterey than in Nonpareil (Figure 3). The mean navel orangeworm degree-day ($^\circ$C) accumulation on June 15 was 638 ± 57 for the 4 years when Monterey damage was higher than Nonpareil damage, and 726 ± 50 for the 6 years when Nonpareil damage was higher. This difference was not significant (Welch unequal variance t-test, t = -1.17, df = 6.915, $p = 0.28$).

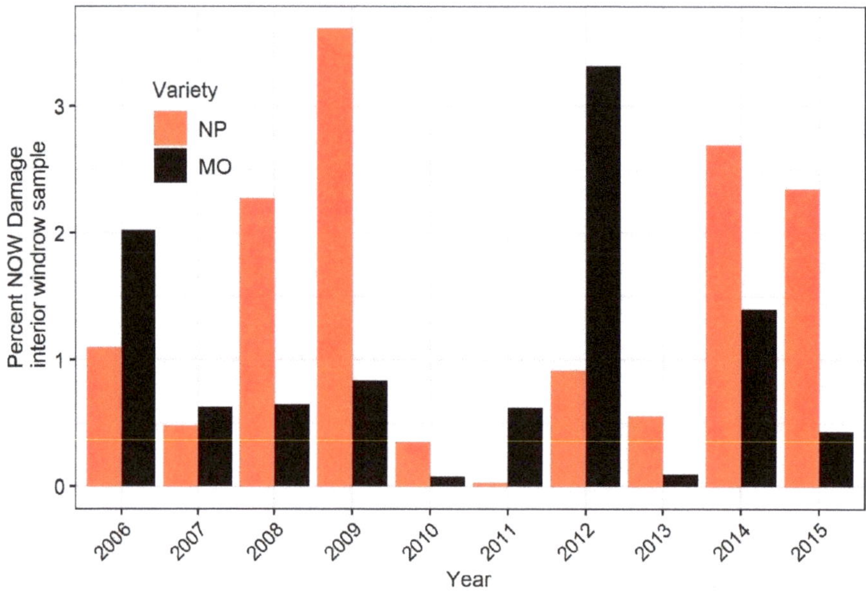

Figure 3. Percent navel orangeworm damage by year in Nonpareil and Monterey almonds in plots treated only with insecticide.

4. Discussion

In this study, windrow samples taken from the interior of the block and the edges were both strongly associated with percent damage from huller samples. Huller samples provide a more immediately relevant assessment of treatment effects on damage to almonds compared to windrow samples more typically used in research projects [15,35,41,43,45]. There are, however, trade-offs between the two approaches. Previously, industry and extension personnel have estimated that windrow samples have about twice the navel orangeworm infestation and damage compared to huller samples [35]. The difference between these two sampling methods is attributed to loss of damaged almonds between the two sampling points [35]. Air legs at various points in the harvest and processing pathway separate almonds from lighter field trash. After extensive feeding, damaged almonds can partition with this field trash; therefore, yield loss from navel orangeworm can be underestimated by the yield data typically received by a grower from a huller-processer. In the present study, we were able to take the extra step of further examination of damaged kernels from the huller to determine which were damaged by navel orangeworm. That option is rarely available to researchers. However, in addition to capturing more of the true loss to navel orangeworm damage, the researcher-gathered windrow samples have the advantages of being collected for the purpose of analysis. Unlike huller reports, which provide a percent damage for a large but imprecisely known sample, the windrow data provide integers representing nut damages and nuts examined, and therefore work better with analysis by generalized linear models. In contrast to the 2× estimates used in previous studies, we found that windrow samples taken from the interior of the orchard has 1.42× the damage of huller samples.

Navel orangeworm damage in edge samples also correlated well with huller samples, albeit not as well as internal windrow samples. Navel orangeworm damage in the edge samples was higher for all treatments. This higher damage is likely due both to mated females immigrating from sources of higher abundance outside the study site, and to faster maturation of almonds on the edges of the orchard. This faster maturation makes infestation of the almonds in the outside portion of the orchard a leading indicator of infestation in the orchard. Sampling edge site has therefore been tested as part of a multi-factor monitoring program to improve prediction of damage from navel orangeworm [39]. The association reported here of navel orangeworm damage with huller damage further supports the utility of this multi-factor monitoring approach.

Because mating disruption is often used with high value crops, studies of its efficacy are often done as an "overlay" in which plots treated with mating disruption and insecticide are compared to plots treated with insecticide alone [53]. Early mating disruption studies for navel orangeworm were unusual in using large untreated control plots without insecticide [41]. Subsequent studies have used the overlay approach [15,35]. The value of the crop at Lost Hills (10s of millions of dollars annually) precluded further untreated controls. The present approach allowed a valuable comparison of the contributions of an established and a new technology in integrated management of navel orangeworm.

Analysis of data for the first three years, when typical use of mating disruption and insecticide was compared using either alone or both together, found consistently similar damage between mating disruption and insecticide, and significant reduction when both were used together (Table 4, Figure 2). Broad analysis found this pattern through the 10-year study (Table 3), but with much variation (Figure 2). Mechanistically, the superior performance of mating disruption and insecticide together to either alone is plausible because they act at different points in the pest's lifecycle. Mating disruption acts against adults, reducing or preventing fertility [2,4]. Methoxyfenozide, the predominant insecticide in this study, primarily kills eggs and neonates [29]. Navel orangeworm eggs are laid on almonds close to the suture, and once the neonate larva enters the almond, it is no longer exposed to insecticide.

Experiments from 2008 to 2015 sought to improve cost-effectiveness of the aerosol dispensers. Aerosol dispensers initially proved to perform better than the hand-applied devices then tested [41]. Subsequent data indicates similar crop protection from several aerosol formulations now marketed, and from a meso-dispenser formulation [35]. Mechanisms of mating disruption invoked by the aerosol formulations [15,45] are likely broadly similar to those invoked by the meso-dispenser formulation (CB, unpublished data). Microencapsulated (sprayable) formulations are also commercially available, but do not offer the season-long suppression provided by the aerosol and meso-dispenser products (BH, unpublished data). The data from 2008 to 2011, comparing 2.5 or 5 dispensers per acre, found a marginally significant effect from the number of dispensers, but a stronger effect from using insecticide in addition to 2.5 dispensers per acre (Table 5).

The amount and year-to-year variation in the huller damage from navel orangeworm illustrates additional difficulties in obtaining full IPM benefits from mating disruption. Another recent study found between 0.9% and 1.1% huller damage as a breakpoint above which mating disruption for navel orangeworm in almonds increases grower return [35]. That analysis was based on the premium and penalty of another large almond processor, and may not be entirely applicable to a vertically integrated company such as Wonderful Farming. Processors and industry groups like the Almond Board of California (ABC) [54] tend to be concerned with providing the cleanest product possible in order to maintain the broadest market possible for almonds, and they may receive benefits from extra pest management input that would not accrue to a grower selling to a major processor. It is also illustrative that variable and occasionally high damage occurred despite overall favorable conditions for control of navel orangeworm. The orchards were managed by a well-capitalized company, and there was an ongoing commitment to orchard sanitation (winter removal and destruction of unharvested almonds) to a far higher degree than is common practice [19].

The aforementioned observations are consistent with the recent suggestion that mating disruption for navel orangeworm is a prudent insurance against high damage [35]. However, these observations also demonstrate the difficulty of fully realizing the potential for mating disruption to reduce insecticide input, such as the 25% reduction called for by ABC between 2020 and 2025 [54], and realizing such reduction will require improved monitoring methods, and greater adoption and confidence in such measures. Data from this and other commercial orchards over part of this period were used to determine which components of a multipart monitoring system best predicted damage [39]. That study found that pre-harvest sampling of almonds and trapping for females provided the best available prediction of damage, with an r^2 of approximately 0.5. Alternative attractants provide improved detection of navel orangeworm and are less impacted by mating disruption [42,55–57]. It is unclear, however, whether captures baited with these attractants which capture both sexes in traps [42] are as directly related to damage as the female traps in the previously mentioned study, and it appears there may be a trade-off between prediction power and detection sensitivity. Further, monitoring gains to provide greater confidence in the ability to base insecticide applications on in-season data may come from improved female attractants and trapping systems, possibly aided by improvements in trap automation and information [58].

The year-to-year variation in relative damage in Nonpareil further illustrates the complexity of navel orangeworm damage. Monterey is widely planted, and was the most prevalent variety in the current study site after Nonpareil. Nonpareil is the most commercially valuable almond variety, and has the poorest shell seal and therefore is most exposed to navel orangeworm [21]. Monterey has a much tighter shell seal, and is therefore thought of as less susceptible to navel orangeworm [21]. However, Monterey matures six weeks later than Nonpareil and, therefore, navel orangeworm populations are often in another generation and more abundant by the time Monterey is susceptible. This might be why, in a previous three-year study that found a negative correlation between shell seal and navel orangeworm infestation across varieties, Nonpareil and Monterey had similar damage [21]. The more long-term data from this study indicate greater variation than observed in this previous three-year study [21]. The hypothesis that a tighter shell seal in Monterey is offset by greater navel orangeworm abundance suggests that greater damage in Monterey than Nonpareil might come in cooler years, when Nonpareil would be less exposed to navel orangeworm. The comparison of degree-day accumulation at 15 June suggests that degree-day accumulation does not predict relative damage between Nonpareil and Monterey. It is possible that phenology of the nut is as important to damage patterns across varieties as phenology of the moth: for example, years in which poor conditions at bloom and pollination (in February) might impact Monterey more than Nonpareil. Such conditions cause more uneven maturation and delay harvest, therefore causing greater exposure. This hypothetical explanation is speculative, but illustrates that research to improve prediction of navel orangeworm damage needs to consider both the phenology of the navel orangeworm and that of the host.

5. Conclusions

Navel orangeworm damage trends in this 10-year case study showed a consistent trend of lower damage in almond plots treated with both insecticide for navel orangeworm and mating disruption compared to either alone. This study also provided a more quantitative estimate of the relationship between field and processor damage from navel orangeworm, confirming that the processor data understate loss from navel orangeworm. Variation from year to year in the relative navel orangeworm damage between two widely planted varieties with different maturities demonstrates the importance of protecting all varieties, and considering all varieties when comparing tactics for reduction of navel orangeworm damage. Year-to-year variation in navel orangeworm damage despite stringent management illustrates the challenge in taking mating disruption for navel orangeworm from a tool to lower risk of navel orangeworm damage to a tool to advance the industry goal of lowering insecticide input.

Supplementary Materials: The following are available online at https://www.mdpi.com/2075-445 0/12/2/188/s1, Table S1: Insecticides used in Lost Hills 2006–2011, Table S2: Insecticide applications in Lost Hills 2012–2015.

Author Contributions: Conceptualization, B.S.H.; methodology, B.S.H.; software, B.S.H. and C.S.B.; validation, B.S.H.; formal analysis, C.S.B.; investigation, B.S.H.; resources, B.S.H.; data curation, C.S.B.; writing—original draft preparation, C.S.B.; writing—review and editing, B.S.H.; visualization, C.S.B.; supervision, B.S.H.; project administration, B.S.H.; funding acquisition, B.S.H. All authors have read and agreed to the published version of the manuscript.

Funding: This research was funded by USDA-ARS project numbers 5325-42000-037 and 6036-22000-028, and with funding from the Almond Board of California (ABC) agreement number 58-5325-4-042.

Data Availability Statement: The almond damage and navel orangeworm degree (°F) data used in this paper are available in a public repository (doi 10.5281/zenodo.4553809; see https://doi.org/10.5281/zenodo.4553809 accessed on 21 February 2021).

Acknowledgments: The authors wish to thank J. Rosenheim (UC Davis) for valuable discussions, S. Gooder, A. Pedro, C. Harris, and E. Higuera (Wonderful Orchards) and W. Gee (USDA-ARS) for invaluable technical assistance, as well as Wonderful Orchards for the data set for our analyses, and placement of the kairomone blend-baited traps. Mention of trade names or commercial products in this publication is solely for the purpose of providing specific information and does not imply recommendation or endorsement by the U.S. Department of Agriculture. USDA is an equal-opportunity provider and employer.

Conflicts of Interest: B.H. worked for Wonderful Orchards at the time of the study, and subsequently has worked for Trece Inc., another company that sells mating disruption products for the navel orangeworm. He is not biased by these associations. Neither company had a role in the design of the study; in analyses or interpretation of data; in the writing of the manuscript, or in the decision to publish the results.

References

1. Cardé, R.T.; Minks, A.K. Control of moth pests by mating disruption: Successes and constraints. *Annu. Rev. Entomol.* **1995**, *40*, 559–585. [CrossRef]
2. Miller, J.R.; Gut, L.J. Mating disruption for the 21st century: Matching technology with mechanism. *Environ. Entomol.* **2015**, *44*, 427–453. [CrossRef]
3. Abd El-Ghany, N.M. Semiochemicals for controlling insect pests. *J. Plant Prot. Res.* **2019**, *59*, 1–11.
4. Evenden, M. Mating disruption of moth pests in integrated pest management. In *Pheromone Communication in Moths. Evolution, Behavior, and Application*; Allison, J.D., Carde, R.T., Eds.; University of California Press: Oakland, CA, USA, 2016; pp. 365–393.
5. Witzgall, P.; Kirsch, P.; Cork, A. Sex pheromones and their impact on pest management. *J. Chem. Ecol.* **2010**, *36*, 80–100. [CrossRef]
6. Peterson, R.K.D.; Higley, L.G.; Pedigo, L.P. Whatever Happened to IPM? *Am. Entomol.* **2018**, *64*, 146–150. [CrossRef]
7. UC ANR Statewide IPM Program. What Is Integrated Pest Management. 2020, Describes and Defines IPM. Available online: https://www2.ipm.ucanr.edu/What-is-IPM/ (accessed on 31 May 2020).
8. Wilson, H.; Daane, K.M. Review of ecologically-based pest management in California vineyards. *Insects* **2017**, *8*, 108. [CrossRef] [PubMed]
9. Lucchi, A.; Suma, P.; Ladurner, E.; Iodice, A.; Savino, F.; Ricciardi, R.; Cosci, F.; Marchesini, E.; Conte, G.; Benelli, G. Managing the vine mealybug, *Planococcus ficus*, through pheromone-mediated mating disruption. *Environ. Sci. Pollut. Res.* **2019**, *26*, 10708–10718. [CrossRef] [PubMed]
10. Rodriguez-Saona, C.R.; Polk, D.; Holdcraft, R.; Koppenhöfer, A.M. Long-term evaluation of field-wide oriental beetle (Col., Scarabaeidae) mating disruption in blueberries using female-mimic pheromone lures. *J. Appl. Entomol.* **2014**, *138*, 120–132. [CrossRef]
11. Arakaki, N.; Nagayama, A.; Kobayashi, A.; Hokama, Y.; Sadoyama, Y.; Mogi, N.; Kishita, M.; Adaniya, K.; Ueda, K.; Higa, M.; et al. Mating disruption for control of *Melanotus okinawensis* (Coleoptera: Elateridae) with synthetic sex pheromone. *J. Econ. Entomol.* **2008**, *101*, 1568–1574. [CrossRef]
12. Mahroof, R.M.; Phillips, T.W. Mating disruption of *Lasioderma serricorne* (Coleoptera: Anobiidae) in stored product habitats using the synthetic pheromone serricornin. *J. Appl. Entomol.* **2014**, *138*, 378–386. [CrossRef]
13. Martini, A.; Baldassari, N.; Baronio, P.; Anderbrant, O.; Hedenström, E.; Högberg, H.E.; Rocchetta, G. Mating disruption of the pine sawfly *Neodiprion sertifer* (Hymenoptera: Diprionidae) in isolated pine stands. *Agric. For. Entomol.* **2002**, *4*, 195–201. [CrossRef]
14. McGhee, P.S.; Miller, J.R.; Thomson, D.R.; Gut, L.J. Optimizing aerosol dispensers for mating disruption of codling moth, *Cydia pomonella* L. *J. Chem. Ecol.* **2016**, *42*, 612–616. [CrossRef]
15. Burks, C.S.; Thomson, D.R. Optimizing efficiency of aerosol mating disruption for navel orangeworm (Lepidoptera: Pyralidae). *J. Econ. Entomol.* **2019**, *112*, 763–771. [CrossRef]
16. Wilson, H.; Burks, C.S.; Reger, J.E.; Wenger, J.A. Biology and management of navel orangeworm (Lepidoptera: Pyralidae) in California. *J. Integr. Pest. Manag.* **2020**, *11*, 25. [CrossRef]

17. Wade, W.H. Biology of the navel orangeworm, *Paramyelois transitella* (Walker), on almonds and walnuts in northern California. *Hilgardia* **1961**, *31*, 129–171. [CrossRef]
18. Andrews, K.L.; Barnes, M.M. Invasion of pistachio orchards by navel orangeworm moths from almond orchards. *Environ. Entomol.* **1982**, *11*, 278–279. [CrossRef]
19. Higbee, B.S.; Siegel, J.P. New navel orangeworm sanitation standards could reduce almond damage. *Calif. Agric.* **2009**, *63*, 24–28. [CrossRef]
20. Sappington, T.W.; Burks, C.S. Patterns of flight behavior and capacity of unmated navel orangeworm adults (Lepidoptera: Pyralidae) related to age, gender, and wing size. *Environ. Entomol.* **2014**, *43*, 696–705. [CrossRef]
21. Hamby, K.; Gao, L.W.; Lampinen, B.; Gradziel, T.; Zalom, F. Hull split date and shell seal in relation to navel orangeworm (Lepidoptera: Pyralidae) infestation of almonds. *J. Econ. Entomol.* **2011**, *104*, 965–969. [CrossRef]
22. Rosenheim, J.A.; Higbee, B.S.; Akerman, J.; Meisner, M.H. Ecoinformatics can infer causal effects of crop variety on insect attack by capitalizing on 'pseudoexperiments' created when different crop varieties are interspersed: A case study in almonds. *J. Econ. Entomol.* **2017**, *110*, 2647–2654. [CrossRef] [PubMed]
23. Rice, R.E. A comparison of monitoring techniques for the navel orangeworm. *J. Econ. Entomol.* **1976**, *69*, 25–28. [CrossRef]
24. Sanderson, J.P.; Barnes, M.M.; Seaman, W.S. Synthesis and validation of a degree-day model for navel orangeworm (Lepidoptera: Pyralidae) development in California almond orchards. *Environ. Entomol.* **1989**, *18*, 612–617. [CrossRef]
25. Sanderson, J.P.; Barnes, M.M. Ability of egg traps to detect the onset of second-generation navel orangeworm (Lepidoptera: Pyralidae) moth activity in almond orchards. *J. Econ. Entomol.* **1990**, *83*, 570–573. [CrossRef]
26. Pathak, T.B.; Maskey, M.L.; Rijal, J.P. Impact of climate change on navel orangeworm, a major pest of tree nuts in California. *Sci. Total Environ.* **2020**, *755*, 142657. [CrossRef]
27. Siegel, J.P.; Bas Kuenen, L.P.S.; Ledbetter, C. Variable development rate and survival of navel orangeworm (Lepidoptera: Pyralidae) on wheat bran diet and almonds. *J. Econ. Entomol.* **2010**, *103*, 1250–1257. [CrossRef]
28. Kuenen, L.P.S.; Siegel, J.P. Protracted emergence of overwintering *Amyelois transitella* (Lepidoptera: Pyralidae) from pistachios and almonds in California. *Environ. Entomol.* **2010**, *39*, 1059–1067. [CrossRef]
29. Higbee, B.S.; Siegel, J.P. Field efficacy and application timing of methoxyfenozide, a reduced-risk treatment for control of navel orangeworm (Lepidoptera: Pyralidae) in almond. *J. Econ. Entomol.* **2012**, *105*, 1702–1711. [CrossRef]
30. Siegel, J.P.; Strmiska, M.M.; Niederholzer, F.J.A.; Giles, D.K.; Walse, S.S. Evaluating insecticide coverage in almond and pistachio for control of navel orangeworm (*Amyelois transitella*) (Lepidoptera: Pyralidae). *Pest. Manag. Sci.* **2019**, *75*, 1435–1442. [CrossRef]
31. Demkovich, M.R.; Siegel, J.P.; Walse, S.S.; Berenbaum, M.R. Impact of agricultural adjuvants on the toxicity of the diamide insecticides chlorantraniliprole and flubendiamide on different life stages of the navel orangeworm (*Amyelois transitella*). *J. Pest. Sci.* **2018**, *91*, 1127–1136. [CrossRef]
32. Bagchi, V.A.; Siegel, J.P.; Demkovich, M.R.; Zehr, L.N.; Berenbaum, M.R. Impact of pesticide resistance on toxicity and tolerance of hostplant phytochemicals in *Amyelois transitella* (Lepidoptera: Pyralidae). *J. Insect Sci.* **2016**, *16*, 62. [CrossRef]
33. Demkovich, M.; Siegel, J.P.; Higbee, B.S.; Berenbaum, M.R. Mechanism of resistance acquisition and potential associated fitness costs in *Amyelois transitella* (Lepidoptera: Pyralidae) exposed to pyrethroid insecticides. *Environ. Entomol.* **2015**, *44*, 855–863. [CrossRef] [PubMed]
34. Reger, J.; Wenger, J.; Brar, G.; Burks, C.; Willson, H. Evaluating response of mass-reared and irradiated navel orangeworm, *Amyelois transitella* (Lepidoptera: Pyralidae) to crude female pheromone extract. *Insects* **2020**, *11*, 703. [CrossRef]
35. Haviland, D.R.; Rijal, J.P.; Rill, S.M.; Higbee, B.S.; Gordon, C.A.; Burks, C.S. Management of navel orangeworm (Lepidoptera: Pyralidae) using four commercial mating disruption systems in California almonds. *J. Econ. Entomol.* **2020**. [CrossRef]
36. Niu, G.; Pollock, H.S.; Lawrance, A.; Siegel, J.P.; Berenbaum, M.R. Effects of a naturally occurring and a synthetic synergist on toxicity of three insecticides and a phytochemical to navel orangeworm (Lepidoptera: Pyralidae). *J. Econ. Entomol.* **2012**, *105*, 410–417. [CrossRef]
37. Niu, G.; Rupasinghe, S.G.; Zangerl, A.R.; Siegel, J.P.; Schuler, M.A.; Berenbaum, M.R. A substrate-specific cytochrome P450 monooxygenase, CYP6AB11, from the polyphagous navel orangeworm (*Amyelois transitella*). *Insect Biochem. Mol. Biol.* **2011**, *41*, 244–253. [CrossRef] [PubMed]
38. Hamby, K.A.; Nicola, N.L.; Niederholzer, F.J.A.; Zalom, F.G. Timing spring insecticide applications to target both *Amyelois transitella* (Lepidoptera: Pyralidae) and *Anarsia lineatella* (Lepidoptera: Gelechiidae) in almond orchards. *J. Econ. Entomol.* **2015**, *108*, 683–693. [CrossRef] [PubMed]
39. Rosenheim, J.A.; Higbee, B.S.; Ackerman, J.; Meisner, M. Predicting nut damage at harvest using different in-season density estimates of *Amyelois transitella*: Analysis of data from commercial almond production. *J. Econ. Entomol.* **2017**, *110*, 2692–2698. [CrossRef]
40. Shorey, H.H.; Gerber, R.G. Use of puffers for disruption of sex pheromone communication among navel orangeworm moth (Lepidoptera: Pyralidae) in almonds, pistachios, and walnuts. *Environ. Entomol.* **1996**, *25*, 1154–1157. [CrossRef]
41. Higbee, B.S.; Burks, C.S. Effects of mating disruption treatments on navel orangeworm (Lepidoptera: Pyralidae) sexual communication and damage in almonds and pistachios. *J. Econ. Entomol.* **2008**, *101*, 1633–1642. [CrossRef]
42. Burks, C.S.; Higbee, B.S.; Beck, J.J. Traps and attractants for monitoring navel orangeworm (Lepidoptera: Pyralidae) in the presence of mating disruption. *J. Econ. Entomol.* **2020**, *113*, 1270–1278. [CrossRef]

43. Higbee, B.S.; Burks, C.S.; Cardé, R.T. Mating disruption of the navel orangeworm (Lepidoptera: Pyralidae) using widely spaced aerosol dispensers: Is the pheromone blend the most efficacious disruptant? *J. Econ. Entomol.* **2017**, *110*, 2056–2061. [CrossRef]

44. Benelli, G.; Lucchi, A.; Thomson, D.; Ioriatti, C. Sex pheromone aerosol devices for mating disruption: Challenges for a brighter future. *Insects* **2019**, *10*, 308. [CrossRef]

45. Burks, C.S.; Thomson, D.R. Factors affecting disruption of navel orangeworm (Lepidoptera: Pyralidae) using aerosol dispensers. *J. Econ. Entomol.* **2020**, *113*, 1290–1298. [CrossRef]

46. USDA Agricultural Research Service. An areawide control program for navel orangeworm. 2012. Available online: https://portal.nifa.usda.gov/web/crisprojectpages/0412631-an-areawide-control-program-for-navel-orangeworm.html (accessed on 13 December 2020).

47. Wikipedia contributors. Section (United States Land Surveying). Wikipedia, The Free Encyclopedia. 2020. Available online: https://en.wikipedia.org/w/index.php?title=Section(United_States_land_surveying)&oldid=992229483 (accessed on 13 December 2020).

48. R Core Team. *R: A Language and Environment for Statistical Computing*; R Foundation for Statistical Computing: Vienna, Austria, 2020.

49. SAS Institute Inc. *SAS/STAT 14.2 User's Guide*; SAS Institute, Inc.: Cary, NC, USA, 2016.

50. Kenward, M.G.; Roger, J.H. Small sample inference for fixed effects from restricted maximum likelihood. *Biometrics* **1997**, *53*, 983–997. [CrossRef]

51. University of California State-Wide, I.P.M. Degree-Days: Navel Orangeworm. 2021. Available online: http://www.ipm.ucdavis.edu/calludt.cgi/DDMODEL?MODEL=NOW (accessed on 6 February 2021).

52. CIMIS. California Irrigation Management Information System. 2021. Available online: https://cimis.water.ca.gov/Default.aspx (accessed on 6 February 2021).

53. Gut, L.J.; Stelinski, L.L.; Thomson, D.R.; Miller, J.R. Behaviour-modifying chemicals: Prospect and constraints in IPM. In *Integrated Pest. Management: Potential, Constraints and Challenges*; Koul, O., Dhaliwal, G.S., Cuperus, G.W., Eds.; CAB International: Wallingford, UK, 2004; pp. 73–121.

54. Almond Board of California. Almond Orchard 2025 Goals. 2019, p. 8. Available online: http://www.almonds.com/sites/default/files/GoalsRoadmap_2019_web.pdf (accessed on 6 December 2020).

55. Beck, J.J.; Higbee, B.S. Volatile Blends and the Effects Thereof on the Navel Orangeworm Moth. U.S. Patent No 9,220,261, 29 December 2015.

56. Beck, J.J.; Light, D.M.; Gee, W.S. Electrophysiological responses of male and female *Amyelois transitella* antennae to pistachio and almond host plant volatiles. *Entomol. Exp. Appl.* **2014**, *153*, 217–230. [CrossRef]

57. Burks, C.S. Combination phenyl propionate/pheromone traps for monitoring navel orangeworm (Lepidoptera: Pyralidae) in almonds in the vicinity of mating disruption. *J. Econ. Entomol.* **2017**, *110*, 438–446. [CrossRef]

58. Lima, C.F.M.; Leandro, M.E.D.d.A.; Valero, C.; Coronel, L.C.P.; Bazzo, C.O.G. Automatic detection and monitoring of insect pests—A review. *Agriculture* **2020**, *10*, 161. [CrossRef]

 *insects*

Article

Can Mating Disruption Be a Possible Route to Control Plum Fruit Moth in Mediterranean Environments?

Gabriella Lo Verde [1], **Salvatore Guarino** [2], **Stefano Barone** [1] **and Roberto Rizzo** [3,*]

1 Department of Agricultural, Food and Forest Sciences, University of Palermo Viale delle Scienze, 90128 Palermo, Italy; gabriella.loverde@unipa.it (G.L.V.); stefano.barone@unipa.it (S.B.)
2 Institute of Biosciences and Bioresources (IBBR), National Research Council of Italy (CNR), Corso Calatafimi 414, 90129 Palermo, Italy; salvatore.guarino@ibbr.cnr.it
3 CREA Research Centre for Plant Protection and Certification, SS.113, Km 245,5, 90011 Bagheria, Palermo, Italy
* Correspondence: roberto.rizzo@crea.gov.it

Received: 28 July 2020; Accepted: 28 August 2020; Published: 1 September 2020

Simple Summary: *Grapholita funebrana* is a main pest of plum throughout the Palearctic region. The management of this pest is generally carried out with chemical insecticides. In this study we investigated the suitability of the mating disruption as alternative method of control of this pest. Experiments were carried out in organic plum orchards during 2012 and 2014. Trap catches and fruit sampling were carried out to estimate the efficacy of this technique in reducing males catch and fruit infestation. The results indicated that the males caught in traps placed in the treatment plots was always significantly lower than untreated plots. The chemical analysis of the pheromone emission from the dispenser, carried out by solid-phase micro-extraction followed by gas chromatography, indicated an optimal duration of these tool for at least 60 days of field exposure. Fruit sampling evidenced that pheromone treatment significantly reduced fruit infestation, but not economic damage, particularly on the cultivar for which a high susceptibility to the moth infestation is known.

Abstract: Control of the plum fruit moth, *Grapholita funebrana* Treitschke (Lepidoptera: Tortricidae), has been mainly based on the use of chemical insecticides, which can cause undesirable side effects, leading to a growing interest towards alternative sustainable strategies. The aim of this study was to evaluate the effect of the mating disruption technique on *G. funebrana* infestation in plum orchards, by comparing the number of male captures in pheromone-baited traps, and evaluating the damage to fruits in plots treated with the pheromone dispersers and in control plots. The study was carried out in 2012 and 2014 in three organic plum orchards, on the cultivars Angeleno, Friar, President and Stanley. To evaluate the pheromone emission curve of the dispensers from the openings to the end of the trials, a chemical analysis was carried out by solid phase micro-extraction followed by gas chromatography, followed by mass spectrometry. In all years and orchards the mean number of males caught in traps placed in the treatment plots was always significantly lower than untreated plots. Pheromone emission from the dispensers was highest at the opening, and was still considerable at 54 days of field exposure, while it significantly decreased after 72 days of field exposure. Cultivar was confirmed to be an essential factor in determining the fruit infestation level. Pheromone treatment significantly reduced fruit infestation, but not economic damage.

Keywords: *Grapholita funebrana*; Tortricidae; sex pheromones; integrated pest management

---

1. Introduction

The growing demand for plums is due to their nutritional peculiarities, as they seem to provide a variety of health benefits, thanks to their healthful compounds [1–4]. As a consequence, there has been a worldwide increase in plum cultivated areas (*Prunus domestica* L. and *Prunus salicina* Lindl.), reaching 2.7 million hectares in 2018 [5]. The global production of plums increased at an average annual rate of 2.3% in the period 2007–2018, and reached the peak of plum fruits production in 2018, with 12.6 million tons [5].

Grapholita funebrana Treitschke (Lepidoptera Tortricidae), commonly called the plum fruit moth (PFM), is an oligophagous species, feeding on the fruits of several hosts within the family Rosaceae [6–8]. The species is considered the main pest of plums throughout the Palearctic region [9].

The number of PFM generations per year varies depending on climate: in warmer areas, as in central and southern Italy, three generations per year can occur [8,10,11]. Females of PFM lay eggs on the exocarp surface of developing fruits of host plants [12]. Neonate larvae bore into fruits, where they feed and develop. Damage is due to the feeding activity of the larvae inside the fruits, leading to changes in fruit coloration, early ripening and fruit fall, with consequent yield decreases [8,10]. Furthermore, infested fruits show penetration holes (characterized by the presence of gum) made by neonate larvae, and exit holes made by mature larvae leaving the fruit [13]. The losses of production of plums determined by PFM infestations can be quite significant, especially in plum cultivars particularly susceptible to this pest [8,11].

The control of PFM in plum orchards has been mainly based on the use of chemical insecticides, which can cause side effects on beneficial insects, both pollinators and natural enemies, and can have serious implications for human health and the environment [14], or the development of resistance, as found in other tortricids, like *Grapholita molesta* Busck [15,16], *Cydia pomonella* L. [17–19] and *Lobesia botrana* (Denis & Schiffermuller) [20]. Moreover, in organic plum orchards, moth control is particularly difficult due to the small number of authorised products.

In this context, also in consideration that the European Union policy (Directive 2009/128/CE) is encouraging a reduction in the use of pesticides [21], growing research attention has been devoted to the development of alternative environmentally friendly and sustainable strategies to control insect pests of agricultural importance [22–27]. Among them, the manipulation of insect behavior with the use of pheromones is receiving increasing attention [28–31]. Pheromones have been effectively applied in the management of dangerous lepidopteran species, through the technique of the mating disruption, based on the release of high amounts of synthetic sex pheromones into a crop, thus interfering with the mate finding processes of a given pest species [32–35].

In the case of PFM, the sex pheromone is characterized by two main active compounds, i.e., (Z)-8-dodecenyl acetate and (E)-8-dodecenyl acetate [36], with (Z)-8-dodecen-1-ol as a minor component [37]. The sexual behavior exhibited by PFM is similar to other tortricid pest species that are controlled by mating disruption, thus suggesting the possibility of the application of this technique for PFM too. To date, despite the high economic importance of this insect pest, limited information is available about the application of mating disruption for its control [38,39]. The objective of this research was to evaluate the efficacy of the mating disruption technique in Sicilian organic plum orchards. The effect of the mating disruption method on PFM population and fruit damage was assessed by comparing the captures of PFM males in pheromone-baited traps and evaluating the infestation of fruits of four different plum cultivars in plots treated with the pheromone and in control plots.

2. Materials and Methods

2.1. Study Areas and Dispenser Placement

The research was performed in three different Sicilian organic plum orchards located in San Giuseppe Jato, Palermo province, Italy, in two different years. In 2012, field trials were carried out in a plum orchard of 3 hectares (37°99′87″ N, 13°21′12″ E), in which pheromone dispensers were placed (treated plot), while an untreated plum orchard of 1 ha, about 500 m away from the first one, was used as the control plot. In 2014, trials were carried out in two different orchards. In the first one (Orchard A, 37°99′31″ N, 13°22′47″ E), the pheromone dispensers were placed in 4 ha, while a surface of 1 ha, about 400 m far from it, was used as the control plot. In the second one, dispensers were placed in a 3 ha plum orchard (Orchard B, 37°99′57″ N, 13°20′74″ E), and a surface of 1 ha, about 200 m away, was used as the control.

All the orchards were on flat land and regularly shaped, planted with 8-years-old plum trees grafted on Mirabolano (*Prunus cerasifera* Ehrh.) rootstocks trained to a vase shape. Plum trees were spaced with 6 m between rows and 3 m between trees within a row. The four cultivars chosen for the study were Angeleno and Friar (*Prunus salicina* Lindl.) and Stanley and President (*Prunus domestica* L.). In all plum orchards, the four cultivars were distributed along eight rows (two rows for each cultivar). This distribution pattern of the cultivars was repeated over the entire surface of each orchard. The trees were managed using routine organic cultural practices; during the research, no insecticide treatments were carried out in the orchards.

Isomate® OFM Rosso Flex (Shin-Etsu Chemical Co. Ltd., Ohtemachi Chiyoda-ku, Tokyo, Japan) pheromone dispensers were placed in the field once during the season (on 30 March in all years), before the adult of PFM of the wintering generation emerged. Dispensers were hung in the upper third of tree canopy with a density of 550 pieces per hectare. Two dispensers per tree were placed in the outer row that delimited each experimental plot, while one dispenser per tree was placed in the rows following the latter and one dispenser every two trees within its perimeter. According to the manufacturer, each dispenser was loaded with 254 mg of pheromone mixture. Pheromone dispensers consisted of two parallel capillary tubes made of polyethylene sealed at the ends, filled with the PFM pheromone blend, consisting of (Z)-8-dodecenyl acetate (89.6%), (E)-8-dodecenyl acetate (5.4%) and (Z)-8-dodecen-1-ol (1%). The gap in the middle allows each dispenser to form a loop that can be easily deployed by placing the dispenser on a branch.

2.2. Trap Captures

The monitoring of PFM flight was carried out by placing three pheromone-baited (Isagro, Milano, Italy) sticky traps in each of the six experimental plots from the second half of March. Traps were checked every 10 days. Male genitalia extraction and observation of insects caught at each date were done to confirm the specific identification of the tortricid species.

2.3. Fruit Infestation

To compare the PFM field infestation occurring in the different cultivars present in the pheromone treated plots and in the control plots, fruit samplings were carried out, followed by dissection under a microscope to assess the presence of larvae. In detail, in 2012, for each cultivar, four groups of three plum trees, randomly chosen in each plot, were used for fruit sampling, while in 2014, three groups of three plum trees were used for each orchard and each cultivar. On these trees, starting from the first catches in the traps (on 9 April in 2012 and on 5 April in 2014), field observations were carried out on 100 fruits from each cultivar in each experimental treatment plot, in order to detect the first eggs laid by PFM. Afterwards, fruit sampling was done every two weeks, from 28 May to 24 July 2012, and from 23 May to 30 July 2014. In 2012, at each sampling date, 8 fruits per tree (96 per cultivar) were collected randomly around the canopy, whereas in 2014, 10 fruits per tree (90 per cultivar) were collected. In 2012, the cultivar Friar was not sampled due to inadequate fruit production. All fruits were then kept in the Department of Agricultural, Food, and Forest Science (University of Palermo, Palermo, Italy), and were dissected under a stereomicroscope to record the presence of PFM larvae. Fruit was considered infested when larvae or their penetration and exit holes were present.

2.4. Estimation of Pheromone Release from Dispensers

In 2014, the residual emission of sex pheromone by field dispensers periodically collected at each fruit sampling was evaluated, in comparison with a new dispenser. Three dispensers were periodically collected at each fruit sampling, extracted and analyzed by GC-MS. The pheromone emission rate from the dispenser was analyzed by headspace using a solid phase micro-extraction (SPME) in static air [40], an equilibrium process involving the headspace and the polymeric fiber stationary phase [41]. The stationary phase used as the coatings was poly(dimethylsiloxane) (PDMS, 100 µm). A manual SPME holder from the same manufacturer was used for injections. Fibers were conditioned in a gas chromatograph injector port as recommended by the manufacturer: PDMS at 250 °C for 30 min. SPME extractions were performed in climatic chambers (27 ± 2 °C and 50 ± 5% RH). For pheromone collection, the releasers' samples were placed into 40 mL vials, which were sealed with a poly(tetrafluoroethylene) silicon septum-lined cap (Supelco, Bellefonte, PA, USA). An SPME needle was then inserted through the septum and headspace volatiles were absorbed on the exposed fiber for 30 min in a conditioned room (29 ± 1 °C; 40 ± 5% RH). The release rates of the dispensers were measured at the opening and after 54, 72, 89, and 122 days of field exposure. Experiments were replicated three times for each day of sampling from the releaser opening and field exposure. In order to perform a chemical analysis on the collected pheromone, immediately after the end of the sampling time, the loaded fiber was desorbed in the gas chromatograph inlet port for 2 min. Coupled gas chromatography-mass spectrometry (GC-MS) analyses of the headspace extracts from the pheromone releasers were performed on an Agilent 6890 GC system interfaced with an MS5973 quadruple mass spectrometer, which was injected onto a DB5-MS column in 1/50 split mode. Injector and detector temperatures were 260 °C and 280 °C respectively. Helium was used as the carrier gas. The GC oven temperature was set at 40 °C for 5 min, and then increased by 10 °C/min to 250 °C. Electron impact ionization spectra were obtained at 70 eV, recording mass spectra from 40 to 550 amu. The measurements of the pheromone emission rate were accomplished by integrating the pheromones' peaks.

2.5. Statistical Analysis

Data on the number of males of PFM caught in the traps were analyzed, after a root square transformation, using a general linear model (GLM) procedure, in which the factors' treatment (pheromone treated plot/control plot), date and their interaction were included. Tukey's HSD test ($p < 0.05$) was then applied to assess the significant difference between treatments.

As infestation data were recorded as the presence/absence on fruits, a binary logistic regression analysis was performed, separately for each year and orchard, to assess the significance of the independent factors. Treatment (placement of pheromone dispenser vs. control), cultivar and date were included in the models as factors. Moreover, as variable susceptibility of the studied cultivars was known from a previous study [8], the interaction effect between the treatment and cultivar was included in the analysis.

The mean of the chromatographic areas obtained from headspace SPME pheromone collection of releaser with different field age were integrated and compared by using one-way ANOVA, followed by Tukey's HSD test ($p < 0.05$).

MINITAB software was used for all statistical analyses (Minitab Inc., State College, PA, USA).

3. Results

3.1. Trap Captures

Overall, 2246 males were caught in the traps, of which 2214 were *G. funebrana*, while 32 were belonging to *G. molesta*. Figure 1 shows the trend of PFM male captures per trap per week. First catches in the traps were recorded on 9 April in 2012 and on 5 April in 2014, in both Orchards A and B. An increase in catches was recorded from the second half of April, corresponding to the first PFM generation. After this date, in all years, the number of males in the traps was clearly higher in the control plots (Figure 1). In 2012, the highest number of captures was recorded on 27 June, with a mean number of 81.3 males per trap. In 2014, in Orchard A, the peak was recorded on 13 July (60.3 males per trap); in Orchard B it was recorded on 19 July (21 males per trap). In pheromone plots, the maximum number of males in the traps were recorded on 27 June 2012 (6 males per trap), and on 19 July 2014 (4.3 males per trap) in Orchard A, and on the same date in Orchard B (5.8 males per trap). Statistical analysis showed that the effect of treatment, date and their interaction were significant in all years (Table 1). The average number of males caught in traps placed in the treatment plots was always significantly lower than in untreated plots (Figure 1). Moreover, treatment was the most important of the factors included in the analyses (Table 1). Overall, the GLM models allowed us to effectively explain the variability in the caught number of PFM males, as shown by the values of adjusted R^2 values: 80.50 in 2012 and 92.40 in 2014 for Orchard A and 78.97 in 2014 for Orchard B.

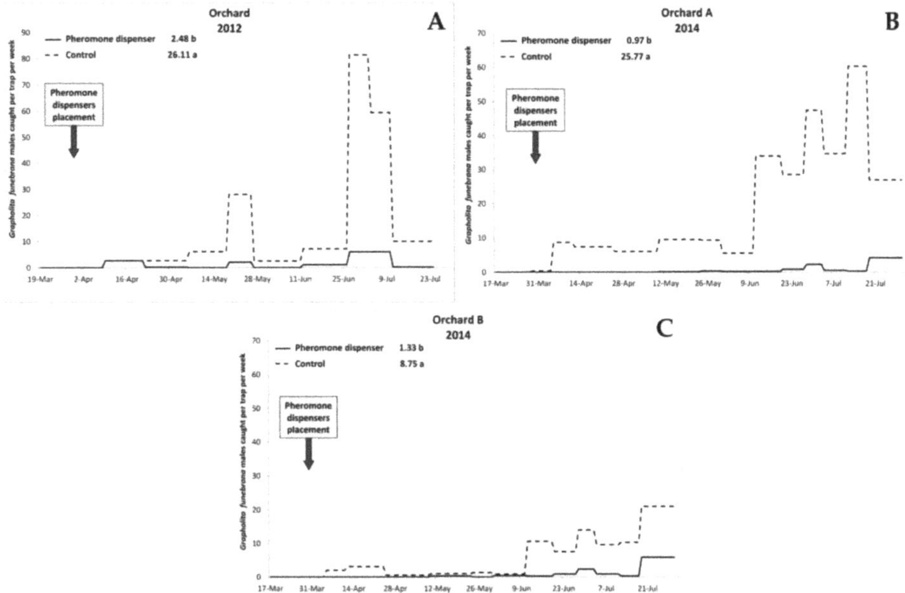

Figure 1. Trend of *Grapholita funebrana* males catches per trap per week in (**A**) Orchard 2012, (**B**) Orchard A 2014 and (**C**) Orchard B 2014. The mean number of captures per trap over the entire study period is reported in the legend; different letters indicate significant differences between the means (GLM followed by Tukey's HSD test, $p < 0.05$).

Table 1. Results of the general linear model (GLM) analyses, performed each year separately on the number of *Grepholita funebrana* males caught in the traps, and effect of the different factors considered (asterisks indicate significant factors within each year, * $p < 0.01$).

Factors	Orchard 2012		Orchard A 2014		Orchard B 2014	
	df	*F*-Values	*df*	*F*-Values	*df*	*F*-Values
Treatment	1	101.41 *	1	620.15 *	1	110.91 *
Date	8	12.16 *	12	18.73 *	11	13.55 *
Treatment * Date	8	4.64 *	12	9.65 *	11	2.70 *
Error	36		52		48	
Total	53		77		71	

3.2. Fruits Infestation

The first eggs on plums were found on 18 May 2012 and on 9 May 2014 in both orchards. In 2012, the first infested fruits were recorded on 28 May 2012 on Angeleno fruits, in both treated and control plots, and on President on fruits sampled in the control plot, whereas the first infested fruits on the Stanley cultivar were found on 11 June in the control plot (Figure 2). In 2014, in Orchard A, the first infested fruits were recorded on 23 May in all cultivars and treatments, with the only exception of the treated plot of the Stanley cultivar, in which infested fruits were recorded starting from 27 June (Figure 3). In Orchard B, the first infested fruits were recorded on 23 May, in both treated and control plots of Angeleno, Friar and President, while in the Stanley cultivar, infested fruits were recorded starting from 10 June in the control plot and from 27 June in the pheromone treated plot (Figure 4). During all years and orchards, on most of the sampling dates, the mean number of infested fruits per tree was lower in the treated plots. Angeleno was always the most infested cultivar and the only one in which a lower number of infested fruits per tree was recorded in the treated plot for all years, orchards and sampling dates (Figures 2–4).

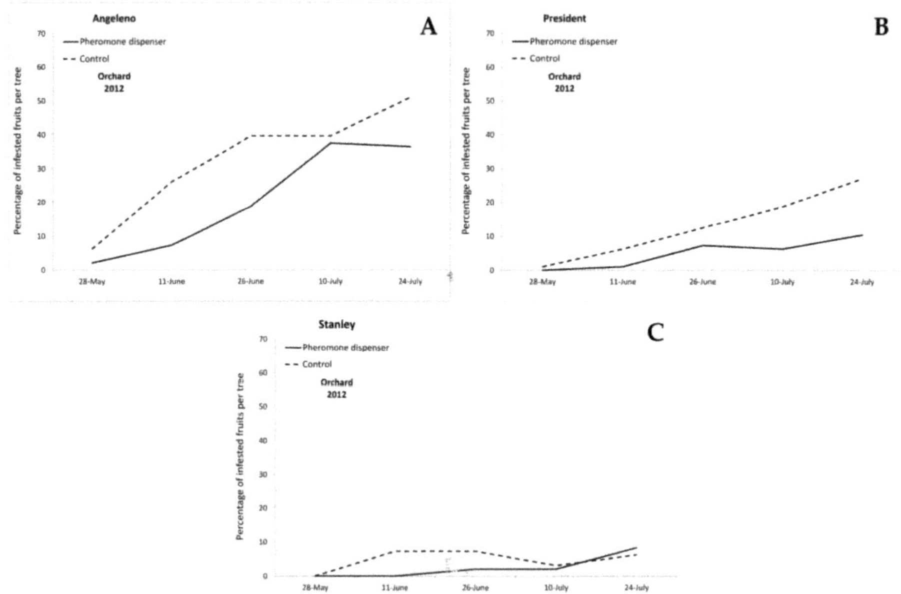

Figure 2. Infestation trend (mean percentage of infested fruits per tree) recorded in 2012 in the cultivars (**A**) Angeleno, (**B**) President and (**C**) Stanley.

Statistical analysis showed in all years and orchards a significant effect of the factors treatment, cultivar and date (Table 2). The interaction between cultivar and treatment was not significant, indicating that the effect of the treatment does not depend on the cultivar, and at the same time that the differences among the cultivars are not due to the treatment. Overall, the values of the adjusted R^2 values obtained in the GLM models allowed us to explain 56.15%, 71.11% and 71.02% of the variability in the infestation found in 2012 and in 2014 in Orchard A, and in 2014 in Orchard B, respectively. Moreover, the most important among the factors included in the analyses was the date, always followed by the cultivar (Table 2).

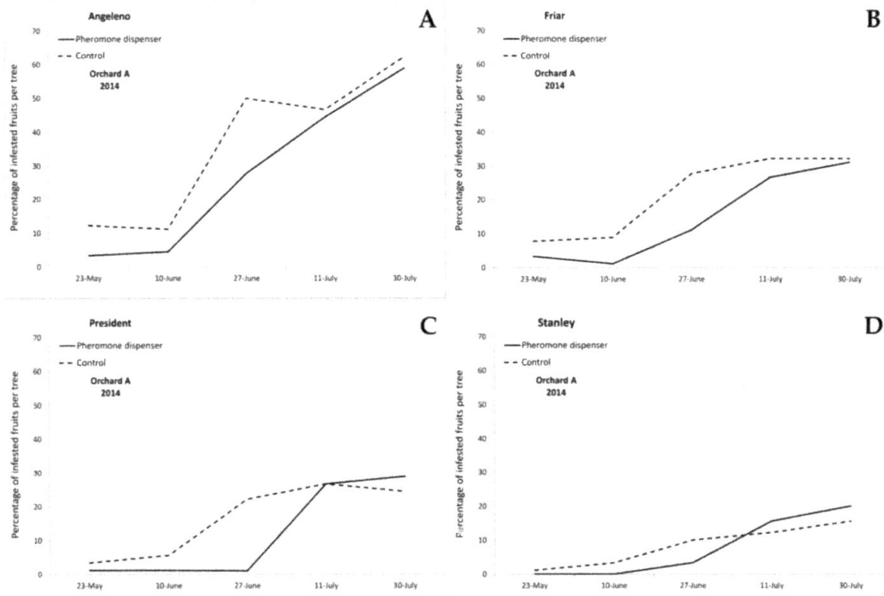

Figure 3. Infestation trend (mean percentage of infested fruits per tree) recorded in 2014 in Orchard A in the cultivars (**A**) Angeleno, (**B**) Friar, (**C**) President, and (**D**) Stanley.

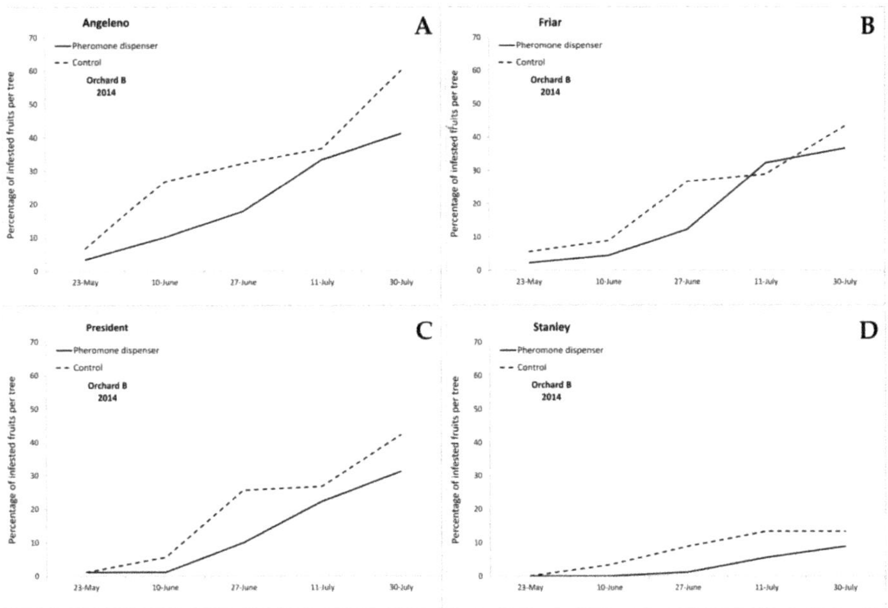

Figure 4. Infestation trend (mean percentage of infested fruits per tree) recorded in 2014 in Orchard B in the cultivars (**A**) Angeleno, (**B**) Friar, (**C**) President, and (**D**) Stanley.

Table 2. Results of the binary logistic regression, separately performed for each year and orchard, on the infestation data and effect of the different factors considered (asterisks indicate significant factors within each year, * $p < 0.01$). All of the goodness-of-fit tests of the reported models were found to be non-significant.

Factors	Orchard 2012		Orchard A 2014		Orchard B 2014	
	df	Chi-Squared	*df*	Chi-Squared	*df*	Chi-Squared
Regression	9	470.642 *	11	635.21 *	11	597.45 *
Treatment	1	20.42 *	1	9.54 *	1	17.39 *
Cultivar	2	107.49 *	3	77.21 *	3	89.35 *
Date	4	190.87 *	4	436.95 *	4	391.02 *
Treatment * Cultivar	2	1.83	3	2.96	3	3.60
Error	350		348		348	
Total	359		359		359	

However, the infestation percentages found on the last sampling date (corresponding to fruit harvesting) showed a reduction, due to mating disruption in treated plots of Angeleno and Friar in all years (Table 3). In the President cultivar, a reduced infestation due to mating disruption was found only in 2012 and 2014 in Orchard B, while for Stanley, a reduction was observed only in 2014 in Orchard B. For Angeleno, the most susceptible cultivar [8], the infestation in treated plots was quite high in all years (Table 3). In Orchard A, no differences were recorded between treatment and control with regard to fruit infestation.

Table 3. Percentage of infestation recorded at harvest, and differences between treated and control in the different cultivars and years.

Cultivar	Orchard 2012			Orchard A 2014			Orchard B 2014		
	Pheromone Dispenser	Control	Difference	Pheromone Dispenser	Control	Difference	Pheromone Dispenser	Control	Difference
Angeleno	36.46	51.04	−14.58	58.89	62.22	−3.33	41.11	60	−18.89
Friar	-	-	-	31.11	32.22	−1.11	36.67	43.33	−6.66
President	10.42	27.08	−16.66	28.89	24.44	+4.45	31.11	42.22	−11.11
Stanley	8.33	6.25	+2.08	20.00	15.56	+4.44	8.89	13.33	−4.44

3.3. Estimation of Pheromone Release from Dispensers

The mean release rates of pheromone from dispensers, measured by SPME in static air, are shown in Figure 5. The pheromone emission significantly decreased over the field-ageing period for both dispensers placed in the field ($F = 46.09$; $df = 5$; $p < 0.001$; ANOVA). Pheromone emission was higher at releaser opening than at all the other times of sampling ($p < 0.05$, ANOVA followed by Tukey's test). Releasers with 54 days of field exposure emitted more pheromone emission than the releasers with a longer duration of field exposure ($p < 0.01$; ANOVA followed Tukey's HSD test). However, no statistical differences were observed in pheromone emission from releasers recovered from 72 to 122 days of field exposure.

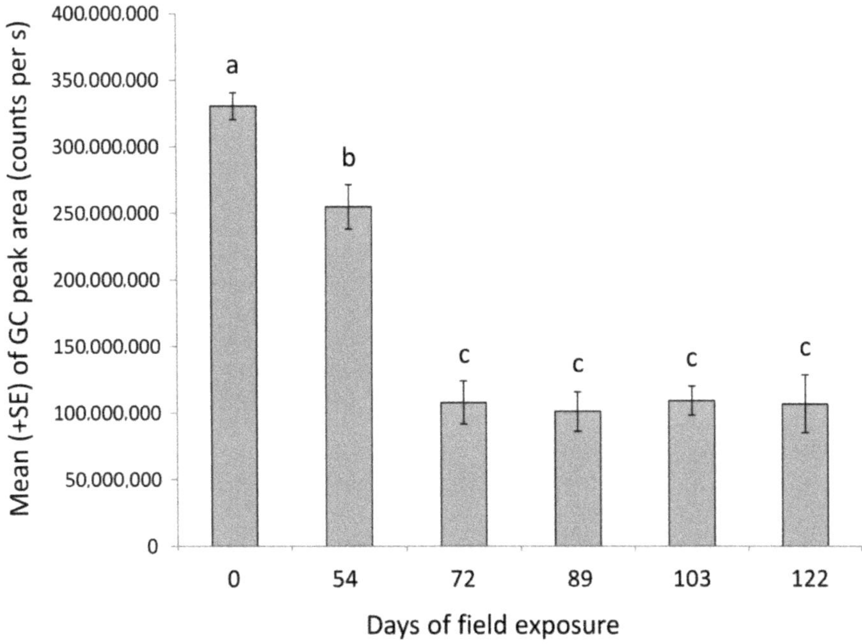

Figure 5. Mean (+SE) release rates of *Grapholyta funebrana* pheromone from dispensers measured by collecting the volatiles, by using the SPME headspace method at 22 °C. No letter in common indicates significant differences for $p < 0.05$ (ANOVA followed by Tukey's HSD test).

4. Discussion

Pheromone-mediated mating disruption aims to interfere with mate finding, reduce insect population growth, and prevent crop damage [42]. The main system for mate finding in moths is due to the female release of small amounts of sex pheromone, which is detected by males through their highly sensitive neurosensory structures [43,44]. In addition to the prevention of mating, pheromone treatment can also result in a delay in mating, thus impacting the fitness and subsequent population dynamics of the target insect pest [42]. Mating disruption using pheromone dispensers, distributed into a crop, has had great success for the control of harmful tortricids like *L. botrana* [45–47], *G. molesta* [48,49] and *C. pomonella* [50,51]. Nevertheless, some difficulties can occur when high populations of pests are present, and it may be necessary to reduce the population of the first moth generation, as found for PFM by [52]. For this reason, mating disruption should be considered as one control method within a structured strategy of integrated pest management, including other control methods [52,53]. With regard to PFM, few studies on its control by pheromone-based methods have been carried out in central Europe [39,52–54]. In Italy, the effect of pheromone-based methods on PFM was assessed in studies carried out in central and northern Italian regions on the Stanley cultivar, adopting the false-trail or the mating disruption methods [6,9]; no information is available for plum orchards in Mediterranean environments.

Our results show that catches of male PFM moths in pheromone traps were strongly reduced in the treated plots in comparison with the control plots, as found in other studies [9,49,53]. In the control plots, the trend of PFM males catches clearly shows the occurrence of peaks corresponding to at least two different PFM generations in the study period. Despite the reduction in male captures, the population trend in treated plots showed capture peaks on the same dates as the control plot. The pheromone emission rate from the dispensers significantly decreased with field aging until day 72; specifically, the emission of pheromone from the dispenser after opening decreased by about

25% on the 54th day of field exposure, and by 66% on the 72nd day of field exposure after opening. Afterwards, the pheromone emission was stable. The lower levels of male catches during the entire sampling period in treated plots led us to surmise that the pheromone, despite a lowering release rate, was still effective in reducing the captures of PFM males until the end of July (about 120 days after dispenser placement in the field), when the insect population increased to its highest levels. However, the pheromone emission results also suggest that a second application of pheromone dispensers or the use of automatic aerosol devices releasing pheromone puffs at programmed time intervals [31,46] might give a better effect on PFM mating disruption.

Fruit infestation recorded in control plots was higher compared to pheromone treated plots in the entire sampling period in all cultivars and years, with the only exceptions being samples of the cultivars Friar (Orchard B) and Stanley (Orchard A) collected on 11 July 2014. It should be noted that, in control plots, the infestation increase followed the trend of catches in the traps, which is related to an increase of the insect population due to the development of the second generation. In the pheromone treated plots, the reduction in trap catches was more evident than the reduction in the fruit infestation level. This led us to suppose that, despite the fact that the pheromone level in the field was effective in reducing trap attractiveness for PFM males, it did not inhibit the insect mating and oviposition to the same extent. This was particularly evident in Orchard A, in which no differences in the fruit infestation were found at harvest, despite the very low PFM catches recorded in the pheromone treated plot. Moreover, in Orchard B, the infestation level in both treated and control plots was higher than in 2012, although the number of PFM males in traps in the control plot was very low in 2014 compared to 2012. This suggests that the use of pheromone traps to evaluate the efficacy of the mating disruption technique can give information about changes in moth flights, but cannot be considered reliable as a stand-alone monitoring tool in areas treated with mating disruption. This has been observed also in other studies carried out on tortricid moth control in apple orchards and vineyards in Italy, as the information provided by traps needs frequent field scouting to evaluate the effective control of target species through mating disruption [46].

In our study, the reduction in fruit infestation recorded in treated plots compared to control plots cannot be considered suitable from an economic point of view, particularly in the most susceptible cultivar, Angeleno, in which the infestation in treated plots was 36% (2012), 59% (Orchard A) and 41% (Orchard B). In contrast, other studies carried out in central Europe showed that the application of mating disruption significantly reduced fruit damage and maintained a percentage of infested fruits below the economic injury level [49].

Our study confirmed that cultivar represents an important factor in determining the fruit infestation level, as already demonstrated by [8]. In particular, Stanley, a cultivar previously studied in Italy by [6,9], was less infested than Angeleno, Friar and President [8,11]. This complicates the comparison between our results and most of the literature, in which the cultivar is not reported.

The probability of fruit infestation is related both to the cultivar and to the density of the insect pest population. The PFM population density can be affected by environmental factors, such as the occurrence of wild host plants, like *Prunus spinosa* L. [9,52]. It could be useful to lower the density of the PFM generation, by chemical or biological methods [52,55], like the use of entomopathogenic nematodes against diapausing larvae [53].

5. Conclusions

This study showed that mating disruption application in the management of PFM can reduce the population level and the fruit infestation level. However, the use of mating disruption as a stand-alone method for controlling the PFM in Mediterranean plum orchards did not prove to be sufficient to contain the infestation below the economic damage threshold, especially on cultivars that are particularly susceptible to the PFM attack and when a high population density of the insect occurs. Therefore, it will require supplemental control methods to reduce fruit infestation. From this perspective, accurate economic assessments relating to possible combinations of control methods should be carried out, in order to achieve a sustainable use of pesticides, as promoted by the European Union (Directive 2009/128/CE). However, using the mating disruption method requires know-how and experience, while at present, just a few publications are available for PFM compared to other tortricids like *L. botrana* or *G. molesta*.

Interesting issues to be investigated in future studies could be aimed at improving knowledge about the susceptibility of the most widely cultivated cultivars, which is an essential factor in determining the fruit infestation level. Moreover, more efforts should be carried out to assess the effectiveness of the mating disruption method in the same orchards in consecutive years, to evaluate the additive effect on the reduction of the population. However, it should be useful to optimise the emission of an adequate amount of pheromone in the field that can provide control of PFM throughout the entire season, for example, through the use of automatic aerosol devices [31,46]. Furthermore, improved knowledge about distribution and forecast models [56,57] could provide further opportunities for the integrated management of this key pest in plum orchards.

Author Contributions: Conceptualization, experimental design, field sampling, and infestation recording: R.R. and G.L.V. Laboratory tests: S.G.; Statistics: S.B., G.L.V., R.R.; Writing: original draft preparation, review and editing; R.R., G.L.V. and S.G. All authors have read and agreed to the published version of the manuscript.

Funding: This study was funded by the University of Palermo through a research grant (Roberto Rizzo) and FFR 2012/2013 funds (G. Lo Verde).

Acknowledgments: The authors are grateful to Emanuele Savona, Giorgio Ferrara and Giovanni Ferrara, owners of the plum orchards, to Valentina Mirabella and Francesco Intravaia for their help in the field and laboratory work.

Conflicts of Interest: The authors declare no conflict of interest.

References

1. Wallace, T.C. Dried plums, prunes and bone health: A comprehensive review. *Nutrients* **2017**, *9*, 401. [CrossRef] [PubMed]

2. Stacewicz-Sapuntzakis, M. Dried plums and their products: Composition and health effects an updated review. *Crit. Rev. Food Sci. Nutr.* **2013**, *53*, 1277–1302. [CrossRef] [PubMed]

3. Kim, D.O.; Jeong, S.W.; Lee, C.Y. Antioxidant capacity of phenolic phytochemicals from various cultivars of plums. *Food Chem.* **2003**, *81*, 321–326. [CrossRef]

4. Prajapati, P.M.; Solanki, A.S.; Sen, D.J. Nutrition value of plum tree for health. *Int. Res. J. Pharm.* **2012**, *3*, 53–56.

5. FAO (Food and Agricultural Organization of United Nation). Available online: http://www.fao.org/faostat/ en/#data/QC (accessed on 30 May 2020).

6. Riolo, P.; Bruni, R.; Cappella, L.; Rama, F.; Isidoro, N. Control of the plum fruit moth, *Grapholita funebrana* (Treitsch.) (Lepidoptera, Tortricidae), by false-trail following. *IOBC/WPRS Bull.* **2010**, *54*, 401–404.

7. Negahban, M.; Sedaratian-Jahromi, A.; Ghanee-Jahromi, M.; Haghani, M. Monitoring of an Iranian population of *Grapholita funebrana* Treitschke, 1835 (Lepidoptera: Tortricidae) using sex pheromone traps: An applicable procedure for sustainable management. *Entomofauna Z. Für Entomol.* **2016**, *37*, 241–252.

8. Rizzo, R.; Farina, V.; Saiano, F.; Lombardo, A.; Ragusa, E.; Lo Verde, G. Do *Grapholita funebrana* Infestation Rely on Specific Plum Fruit Features? *Insects* **2019**, *10*, 444. [CrossRef]

9. Sciarretta, A. Auto-confusion trials for mating disruption of *Grapholita funebrana* in plum orchards. In Proceedings of the Giornate Fitopatologiche 2010, Cervia (RA), Italy, 9–12 March 2010; Atti, Volume primo, pp. 179–184.

10. Molinari, F. Notes on biology and monitoring of *Cydia funebrana* (Treitschke). *IOBC/WPRS Bull.* **1995**, *18*, 39–42.

11. Rizzo, R.; Lo Verde, G.; Lombardo, A. Effectiveness of spinosad and mineral oil for control of *Grapholita funebrana* Treitschke in organic plum orchards. *New Medit* **2012**, *11*, 70–72.

12. Jackson, D.M. Searching behavior and survival of 1st-instar codling moth. *Ann. Entomol. Soc. Am.* **1982**, *75*, 284–289. [CrossRef]

13. Alford, D.V. *Farbatlas der Obstschädlinge*; Ferdinand Enke: Stuttgart, Germany, 1987; pp. 168–169.

14. Aktar, W.; Sengupta, D.; Chowdhury, A. Impact of pesticides use in agriculture: Their benefits and hazards. *Interdiscip. Toxicol.* **2009**, *2*, 1–12. [CrossRef] [PubMed]

15. Kanga, L.H.; Pree, D.J.; Van Lier, J.L.; Walker, G.M. Management of insecticide resistance in oriental fruit moth (*Grapholita molesta*; Lepidoptera: Tortricidae) populations from Ontario. *Pest Manag. Sci.* **2003**, *59*, 921–927. [CrossRef] [PubMed]

16. Guo, Y.; Chai, Y.; Zhang, L.; Zhao, Z.; Gao, L.L.; Ma, R. Transcriptome analysis and identification of major detoxification gene families and insecticide targets in *Grapholita molesta* (Busck) (Lepidoptera: Tortricidae). *J. Insect Sci.* **2017**, *17*, 43. [CrossRef]

17. Ioriatti, C.; Tasin, M.; Charmillot, P.J.; Reyes, M.; Sauphanor, B. Early detection of resistance to tebufenozide in field populations of *Cydia pomonella* L.: Methods and mechanisms. *J. Appl. Entomol.* **2007**, *131*, 453–459. [CrossRef]

18. Reyes, M.; Barros-Parada, W.; Ramírez, C.C.; Fuentes-Contreras, E. Organophosphate resistance and its main mechanism in populations of codling moth (Lepidoptera: Tortricidae) from Central Chile. *J. Econ. Entomol.* **2015**, *108*, 277–285. [CrossRef] [PubMed]

19. Joshi, N.K.; Hull, L.A.; Krawczyk, G. Insecticide Baseline Sensitivity in Codling Moth (Lepidoptera: Tortricidae) Populations From Orchards Under Different Management Practices. *J. Entomol. Sci.* **2020**, *55*, 105–116. [CrossRef]

20. Tacoli, F.; Cargnus, E.; Moosavi, F.K.; Zandigiacomo, P.; Pavan, F. Efficacy and mode of action of kaolin and its interaction with bunch-zone leaf removal against *Lobesia botrana* on grapevines. *J. Pest Sci.* **2019**, *92*, 465–475. [CrossRef]

21. Hillocks, R.J. Farming with fewer pesticides: EU pesticide review and resulting challenges for UK agriculture. *Crop Protect.* **2012**, *31*, 85–93. [CrossRef]

22. Todd, J.H.; Barratt, B.I.; Tooman, L.; Beggs, J.R.; Malone, L.A. Selecting non-target species for risk assessment of entomophagous biological control agents: Evaluation of the PRONTI decision-support tool. *Biol. Control* **2015**, *80*, 77–88. [CrossRef]

23. González-Chang, M.; Wratten, S.D.; Lefort, M.C.; Boyer, S. Food webs and biological control: A review of molecular tools used to reveal trophic interactions in agricultural systems. *Food Webs* **2016**, *9*, 4–11. [CrossRef]

24. Holland, J.M.; Bianchi, F.J.; Entling, M.H.; Moonen, A.C.; Smith, B.M.; Jeanneret, P. Structure, function and management of semi-natural habitats for conservation biological control: A review of European studies. *Pest Manag. Sci.* **2016**, *72*, 1638–1651. [CrossRef] [PubMed]

25. Benelli, G.; Pavela, R.; Maggi, F.; Petrelli, R.; Nicoletti, M. Commentary: Making green pesticides greener? The potential of plant products for nanosynthesis and pest control. *J. Clust. Sci.* **2017**, *28*, 3–10. [CrossRef]

26. Benelli, G. Plant-borne compounds and nanoparticles: Challenges for medicine, parasitology and entomology—Green-Nanopest DRUGS. *Environ. Sci. Poll. Res.* **2018**, *25*, 10149–10150. [CrossRef] [PubMed]

27. Athanassiou, C.G.; Kavallieratos, N.G.; Benelli, G.; Losic, D.; Usha Rani, P.; Desneux, N. Nanoparticles for pest control: Current status and future perspectives. *J. Pest Sci.* **2018**, *91*, 1–15. [CrossRef]

28. Witzgall, P.; Kirsch, P.; Cork, A. Sex pheromones and their impact on pest management. *J. Chem. Ecol.* **2010**, *36*, 80–100. [CrossRef] [PubMed]

29. Kaplan, I. Attracting carnivorous arthropods with plant volatiles: The future of biocontrol or playing with fire? *Biol. Control* **2012**, *60*, 77–89. [CrossRef]

30. Pérez-Staples, D.; Shelly, T.E.; Yuval, B. Female mating failure and the failure of 'mating' in sterile insect programs. *Entomol. Exp. Appl.* **2013**, *146*, 66–78. [CrossRef]

31. Lucchi, A.; Ladurner, E.; Iodice, A.; Savino, F.; Ricciardi, R.; Cosci, F.; Conte, G.; Benelli, G. Eco-friendly pheromone dispensers—A green route to manage the European grapevine moth? *Environ. Sci. Pollut. R.* **2018**, *25*, 9426–9442. [CrossRef]

32. Cardé, R.T. *Principles of mating disruption. Behavior-Modifying Chemicals for Pest Management: Applications of Pheromones and Other Attractants*; Marcel Dekker: New York, NY, USA, 1990; pp. 47–71.

33. Cardé, R.T.; Minks, A.K. Control of moth pests by mating disruption: Successes and constraints. *Annu. Rev. Entomol.* **1995**, *40*, 559–585. [CrossRef]

34. Suckling, D.M. Issues affecting the use of pheromones and other semiochemicals in orchards. *Crop Prot.* **2000**, *19*, 677–683. [CrossRef]

35. Miller, J.R.; Gut, L.J.; De Lame, F.M.; Stelinski, L.L. Differentiation of competitive vs. non-competitive mechanisms mediating disruption of moth sexual communication by point sources of sex pheromone (part 2): Case studies. *J. Chem. Ecol.* **2006**, *32*, 2115–2143. [CrossRef] [PubMed]

36. Arn, H.; Delley, B.; Baggiolini, M.; Charmillot, P.J. Communication disruption with sex attractant for control of the plum fruit moth, *Grapholitha funebrana*: A two-year field study. *Entomol. Exp. Appl.* **1976**, *19*, 139–147. [CrossRef]

37. Guerin, P.M.; Arn, H.; Buser, H.R.; Charmillot, P.; Tóth, M.; Sziráki, G. Sex pheromone of *Grapholita funebrana*. Occurrence of Z-8- and Z-10-tetradecenyl acetate as secondary components. *J. Chem. Ecol.* **1986**, *12*, 1361–1368. [CrossRef] [PubMed]

38. Charmillot, P.J.; Blaser, C.; Maggiolini, M.; Arn, H.; Delley, B. Confusion sexuelle contre le Carpocapse de prunes (Grapholita funebrana Tr.): I. Essais de lutte in verges. *Bull. Soc. Ent. Suisse* **1982**, *55*, 55–63.

39. Stefanova, D.; Vasilev, P.; Kutinkova, H.; Andreev, R.; Palagacheva, N.; Tityanov, M. Possibility for control of plum fruit moth *Grapholita funebrana* Tr. by pheromone dispensers. *J. Biopestic.* **2019**, *12*, 153–156.

40. Angeli, G.; Anfora, G.; Baldessari, M.; Germinara, G.S.; Rama, F.; De Cristofaro, A.; Ioriatti, C. Mating disruption of codling moth *Cydia pomonella* with high densities of Ecodian sex pheromone dispensers. *J. Appl. Entomol.* **2007**, *131*, 311–318. [CrossRef]

41. Pawliszyn, J. *Solid Phase Microextraction: Theory and Practice*; John Wiley & Sons: New York, NY, USA, 1997; p. 264.

42. Mori, B.A.; Evenden, M.L. When mating disruption does not disrupt mating: Fitness consequences of delayed mating in moths. *Entomol. Exp. Appl.* **2013**, *146*, 50–65. [CrossRef]

43. Cardé, R.T.; Haynes, K.F. Structure of the pheromone communication channel in moths. In *Advances in Insect Chemical Ecology*; Cardé, R.T., Millar, J.G., Eds.; Cambridge University Press: Cambridge, UK, 2004; pp. 283–332.

44. Cardé, R.T.; Willis, M.A. Navigational strategies used by insects to find distant, wind-borne sources of odor. *J. Chem. Ecol.* **2008**, *34*, 854–866. [CrossRef]

45. Cooper, M.; Varela, L.G.; Smith, R.J.; Whitmer, D.R.; Simmons, G.A.; Lucchi, A.; Broadway, R.; Steinhauer, R. Growers, scientists and regulators collaborate on European grapevine moth program. *Calif. Agric.* **2014**, *4*, 125–133. [CrossRef]

46. Ioriatti, C.; Lucchi, A. Semiochemical strategies for tortricid moth control in apple orchards and vineyards in Italy. *J. Chem. Ecol.* **2016**, *42*, 571–583. [CrossRef]

47. Hummel, H.E. A brief review on *Lobesia botrana* mating disruption by mechanically distributing and releasing sex pheromones from biodegradable mesofiber dispensers. *Biochem. Mol. Biol. J.* **2017**, *3*, 1. [CrossRef]

48. Kovanci, O.B.; Walgenbach, J.F.; Kennedy, G.G.; Schal, C. Effects of application rate and interval on the efficacy of sprayable pheromone for mating disruption of the oriental fruit moth *Grapholita molesta*. *Phytoparasitica* **2005**, *33*, 334–342. [CrossRef]

49. Kutinkova, H.Y.; Arnaudov, V.A.; Dzhuvinov, V.T. Sustainable Control of Oriental Fruit Moth, *Cydia molesta* Busck, Using Isomate OFM Rosso Dispensers in Peach Orchards in Bulgaria. *Chem. Eng. Trans.* **2015**, *44*.

50. Knight, A.L.; Stelinski, L.L.; Hebert, V.; Gut, L.; Light, D.; Brunner, J. Evaluation of novel semiochemical dispensers simultaneously releasing pear ester and sex pheromone for mating disruption of codling moth (Lepidoptera: Tortricidae). *J. Appl. Entomol.* **2012**, *136*, 79–86. [CrossRef]

51. Kovanci, O.B. Co-application of microencapsulated pear ester and codlemone for mating disruption of *Cydia pomonella*. *J. Pest Sci.* **2015**, *88*, 311–319. [CrossRef]

52. Pyatnova, Y.B.; Kislitsyna, T.I.; Vojnova, V.N. Testing the pheromone of *Grapholitha molesta* and *Grapholitha funebrana* for the control of the population numbers by mating disruption. *Zashchita I Karantin Rastenii* **2013**, *8*, 33–35.

53. Toups, I.; Zimmer, J.; Pfeiffer, B.; Schmückle-Tränkle, G. Regulation of plum moth (*Cydia funebrana*) with mating disruption and entomopathogenic nematodes in organic orchards. In *Fördergemeinschaft Ökologischer Obstbau eV (FÖKO), Proceedings of the Ecofruit. 14th International Conference on Organic Fruit-Growing*; Hohenheim, Germany, 22–24 February 2010, Fördergemeinschaft Ökologischer Obstbau e.V.: Weinsberg, Germany, 2010; pp. 169–175.

54. Kutinkova, H.; Dzhuvinov, V.; Samietz, J.; Veronelli, V.; Iodice, A.; Bassanetti, C. Control of plum fruit moth, *Grapholita funebrana*, by Isomate OFM rosso dispensers, in plum orchards of Bulgaria. *IOBC/WPRS Bull.* **2011**, *72*, 53–57.

55. Ardizzoni, M.; Iodice, A.; Caruso, S. Four years of mating disruption for the control of plum fruit moth *Cydia funebrana* (Treitschke), in plum orchard in Emilia-Romagna region (North Italy). *IOBC/WPRS Bull.* **2013**, *91*, 205–208.

56. Butturini, A.; Tiso, R.; Molinari, F. Phenological forecasting model for *Cydia funebrana*. *EPPO Bull.* **2000**, *30*, 131–136. [CrossRef]

57. Sciarretta, A.; Trematerra, P.; Baumgärtner, J. Geostatistical analysis of *Cydia funebrana* (Lepidoptera: Tortricidae) pheromone trap catches at two spatial scales. *Am. Entomol.* **2001**, *47*, 174–185. [CrossRef]

 *insects*

Article

Mating Disruption of *Helicoverpa armigera* (Lepidoptera: Noctuidae) on Processing Tomato: First Applications in Northern Italy

Giovanni Burgio *, Fabio Ravaglia, Stefano Maini, Giovanni Giorgio Bazzocchi, Antonio Masetti and Alberto Lanzoni

Department of Agricultural and Food Sciences, Alma Mater Studiorum-Università di Bologna, Viale G. Fanin, 42, 40127 Bologna, Italy; fabio.ravaglia@studio.unibo.it (F.R.); stefano.maini@unibo.it (S.M.); giovanni.bazzocchi@unibo.it (G.G.B.); antonio.masetti@unibo.it (A.M.); alberto.lanzoni2@unibo.it (A.L.)
* Correspondence: giovanni.burgio@unibo.it; Tel.: +39-051-209-6289

Received: 20 December 2019; Accepted: 19 March 2020; Published: 26 March 2020

Abstract: *Helicoverpa armigera* is a polyphagous and globally distributed pest. In Italy, this species causes severe damage on processing tomato. We compared the efficacy of mating disruption with a standard integrated pest management strategy (IPM) in a two-year experiment carried out in Northern Italy. Mating disruption registered a very high suppression of male captures (>95%) in both growing seasons. Geostatistical analysis of trap catches was shown to be a useful tool to estimate the efficacy of the technique through representation of the spatial pattern of captures. Lower fruit damage was recorded in mating disruption than in the untreated control plots, with a variable efficacy depending on season and sampling date. Mating disruption showed a higher efficacy than standard IPM in controlling *H. armigera* infestation in the second season experiment. Mating disruption showed the potential to optimize the *H. armigera* control. Geostatistical maps were suitable to draw the pheromone drift out of the pheromone-treated area in order to evaluate the efficacy of the technique and to detect the weak points in a pheromone treated field. Mating disruption and standard IPM against *H. armigera* were demonstrated to be only partially effective in comparison with the untreated plots because both strategies were not able to fully avoid fruit damage.

Keywords: mating disruption; cotton bollworm; processing tomato; geostatistics

1. Introduction

Helicoverpa armigera (Hubner) (Lepidoptera: Noctuidae), the African bollworm or cotton bollworm, is distributed worldwide with the exception of North America. This polyphagous pest causes severe damage to many crops including tomato, cotton, pea, chickpea, sorghum, and cowpea [1]. The severity of cotton bollworm damage varies by crop and region and is influenced by the temporal scale [2]. Due to its dispersive and migratory behavior, the incidence of this pest is often unpredictable. In Italy, the severity of damage caused by *H. armigera* has increased in recent years [3], especially on processing tomato. *Helicoverpa armigera* is listed as a quarantine pest by the European and Mediterranean Plant Protection Organization [4,5].

Italy is the most important tomato (*Licopersicon exculentum* Mill) producer in Europe, and the Emilia-Romagna region (Northern Italy) accounts for approximately 30% of all Italian production [6]. Emilia-Romagna is also the leader in organic tomato cultivation (69% of Italian production) because of favorable pedoclimatic conditions and the presence of important processing companies. The area of organic tomato production is steadily increasing due to the farmers' profit and the high demand for organic tomatoes by consumers.

A high level of resistance to chemical insecticides by *H. armigera* (i.e., carbamates) has led to control failures in many parts of the world including Europe [7]. For this reason, there is an increasing interest in alternative approaches to controlling this pest. Pheromones have been utilized in a variety of ways including mass trapping, disruption of mating communication, monitoring, and surveying [8,9]. Mating disruption has been tested on the Noctuidae species including *Spodoptera* spp. infesting vegetable crops, onions, cotton, herbs, and ornamentals, both in open fields and greenhouses [10–16], and has been shown to be an effective method to control pest populations. However, even if pheromone application has been found to disrupt the males' ability to locate a pheromone source, in some cases, the larval populations were not reduced to the point at which insecticide sprays could be eliminated [14]. Mating disruption for *H. armigera* management has been demonstrated to be effective in reducing field infestations [17–19]. However, with this species, the efficacy of mating disruption also showed a certain degree of variability and in some cases, a lower level of control was achieved in comparison to the chemical sprays. A number of factors including field size, crop species, receiving environment, local climatic characteristics (e.g., dominant winds), behaviors of insect target species (e.g., pre-oviposition flight behavior of mated female [19]), can influence the field efficacy of this technique.

Mating disruption seems to be particularly suitable for processing tomato in Northern Italy where this crop is cultivated on wide areas characterized by an intensive cropping system. Moreover, this technique is characterized by the lack of negative impacts and can be integrated with reduced risk insecticides like microbial products.

The objective of the study was to evaluate the efficacy and feasibility of mating disruption to control *H. armigera* infestations in an area where processing tomato are intensively grown by analyzing both the reduced trap capture and subsequent fruit damage reduction. The efficacy of this technique was also compared with a standard integrated pest management strategy (IPM) by means of fruit damage evaluation [20]. The field trials were carried out on a farm representative of tomato cultivation in the Emilia-Romagna region.

2. Materials and Methods

2.1. Site Description

The study was conducted in an area cropped with processing tomato located in Ravenna Province, the Emilia-Romagna region, Italy. A pilot farm, managed using integrated pest management (IPM) methods and representative of the tomato cultivation conditions of the region, was selected. The IPM method consisted of a strategy according to which insecticide need was determined on the basis of insect density monitoring performed twice a week.

2.2. Mating Disruption Trials Planning

In each year, two treatments were compared: (i) mating disruption (i.e., pheromone-treated) and (ii) the control (i.e., pheromone-untreated), where pheromones were not applied. The field experiments were carried out in two consecutive growing seasons, 2011 and 2012, in two nearby fields. In both years, insecticide sprays were applied across the whole tomato field, according to the IPM strategy (Table 1). In 2011, a 1-ha pheromone-treated area and a 1-ha pheromone-untreated control area (both approximately 100 × 100 m) were delimited into the western half of an 18-ha tomato field (44°30′31″N, 12°11′43″E). In 2012, two areas (5 ha each, both approximately 180 × 270 m) were delimited within a 15-ha tomato field (44°30′12″N, 12°11′13″E), and designated pheromone-treated and pheromone-untreated (control). In each year, the pheromone untreated control area was located upwind, approximately 300 m away from the pheromone-treated area to avoid pheromone drift into the pheromone-untreated control area. Pheromone-treated and pheromone-untreated control areas were divided into four quadrants for replication purposes.

Table 1. Insecticide sprays applied to tomato fields for *H. armigera* control.

Year	Day	Commercial Name	Active Ingredient	Dose kg c.p./ha
2011	04 July	Steward®	Indoxacarb (30 g/L)	0.125
	21 July	Affirm®	Emamectin Benzoate (0.95%)	1.5
2012	19 July	Affirm®	Emamectin Benzoate (0.95%)	1.5
	10 August	Steward®	Indoxacarb (30 g/L)	0.125

c.p., commercial product.

2.3. Pheromone Treatments

BioSelibate HA dispensers (Suterra Europe, Valencia, Spain), consisting of a sawdust type material each containing 0.29 g a.i. (Z-11 hexadecenal and Z-9 hexadecenal in a ratio of 91:9), were used. The target application rate was 100 dispensers/ha (=29 g a.i./ha) that were manually hung on 0.8 m high rods in a 10 × 10 m grid.

In 2011, the experiment started on May 27 (dispensers were hung on May 27; tomato was transplanted on May 25 and 26) and continued until August 16. The tomato was harvested from August 19 to 22. In 2012, the experiment started on June 20 (the dispensers were hung on June 20; tomato was transplanted from June 6 to 8) and tomato was harvested on September 7. Currently, the pheromones for the mating disruption of *H. armigera* are not yet commercial and were provided by Suterra to cover a total of 6 ha.

2.4. Male Capture Evaluation

Pheromone-baited traps (AgriSense funnel trap green/yellow/transparent) were used to verify if male cotton bollworm moths were able to locate a pheromone source in the pheromone-treated area. Trap catches in the pheromone-treated area were compared with those recorded in the pheromone-untreated control area. Four traps in the 2011 trial and eight in the 2012 trial were placed in each of the pheromone-treated and untreated control areas. Traps were baited with *H. armigera* pheromone lures (Septa pheromone lure, Suterra). Baited traps were hung at about 80–100 cm above the ground and at least 10 m inside the area and 35 m far apart from each other. Each trap had an insect killing strip (a.i., 15% diazinon) at the bottom of the trap. Male moths were collected from the traps weekly. In addition to these traps, a grid of pheromone baited traps (26 traps in total in 2011 and 42 traps in 2012) were also placed to cover all the area of the tomato field where the areas were delimited with the aim to map the area where mating disruption may have been effective. All the traps were georeferenced using a handheld Magellan SporTrak Map® GPS unit.

2.5. Fruit Damage Estimation

Within the pheromone-treated and pheromone-untreated control areas, four plots left without insecticide sprays (one per quadrant) were delimited with the aim of sampling for the evaluation of fruit damage. Likewise, four plots were also selected within an area of the field receiving only the chemical spray to control *H. armigera*.

Cotton bollworm damage on tomato (proportion of damaged fruits) in the pheromone-treated, pheromone-untreated control, and insecticide treated areas was estimated by visual inspection using a sample of fruits from within each of four plots nested within each treatment, over six consecutive samplings. In both years, samplings 1–6 corresponded to weeks 7–12 after transplanting (WAT). In each of the four plots, the samplings were performed by checking the fruits for 30 s on 10 randomly selected plants. Fruit damage estimation in each treatment was taken on a weekly basis. On each sampling date, a different set of plants were sampled. Fruit damage estimation followed a stratified design, with treatments (n = 3) nested into years (n = 2), plots (n = 4) nested into each treatment, and sampling dates (n = 6) in each year.

2.6. Data Analysis

2.6.1. Male Capture Analysis

The comparison of the male catches in the pheromone-treated and pheromone-untreated areas was analyzed by the Mann–Whitney U test ($p < 0.05$). The ratio of the catch reduction in the pheromone-treated area with respect to the pheromone-untreated area was calculated as follows:

$$\text{Catch reduction ratio} = \frac{\text{Pheromone untreated area catch} - \text{pheromone treated area catch}}{\text{Pheromone untreated area catch}} \times 100$$

Data from the pheromone traps were also analyzed using geostatistics, in order to compare the spatial pattern of the male captures between the pheromone-treated (mating disruption) area and the pheromone-untreated (control) area. One of the main objectives of geostatistical studies is to provide a spatial representation of data by estimating variable values at unsampled locations. Geostatistics offers a great variety of interpolation methods including stochastic techniques like kriging, and deterministic methods like inverse distance weighting (IDW) [21,22]. IDW was selected as the interpolation tool to provide a visual representation of the population pattern in pheromone-treated and pheromone-untreated areas. Maps estimated by IDW were validated by cross-validation analysis, comparing the predicted and observed trap catch values using linear correlation analysis [23]; in addition, the mean prediction errors of the estimates were calculated. Geostatistical analysis was employed using ArcGIS, with the geostatistics ARCMAP extension (ESRI, Redlands, CA, USA).

2.6.2. Fruit Damage Assessment

In each sampling date, fruit damage was calculated as the ratio of damaged fruits on the total of fruit sampled; standard errors of the damage ratio were calculated according to a binomial distribution [24]. The ratio of damaged fruits was analyzed using log linear analysis, a method that mimics a factorial analysis of variance [25] and allows for simultaneous evaluation of multiple interactions among categorical variables. Log linear analysis uses a likelihood ratio statistic χ^2 that has an approximate χ^2 distribution. In our log linear analysis, the response variable was the ratio of damaged fruits, while the design (or independent) variables were: years (2011–12), treatments (mating disruption–chemical–untreated control), and sampling dates (n = 6). Additionally, a model involving "plot" (n = 4) as the design variable was tested, but this variable was removed because it did not show significant interaction with the response variable and the other design variable ($p > 0.05$). Although all interactions between variables were calculated by log linear analysis, only associations of the response variable (proportion of fruit damage) with design variables were taken into account for data interpretation.

The fruit damage recorded in the twelfth week after transplant (WAT 12) was the most relevant for the final evaluation of the treatment efficacy because it was the last sampling date before the harvest. For this reason, the frequency of damaged fruit at WAT 12 was analyzed by the χ^2 test followed by a z-test to compare the column proportions and rank the efficacy of the treatments [26]. Bonferroni correction was implemented to adjust the p-level of the z-test. This procedure was performed, separately for each year, using the IBM SPSS 23 statistics package (IBM corporation, Armonk, NY, USA).

3. Results

3.1. Mating Disruption Evaluation

The suppression of male captures was very high in both growing seasons (Figure 1, Table 2). In particular, the suppression ratio calculated for total capture was 99.2% and 98.4% in 2011 and 2012, respectively (Table 2). It is remarkable that capture suppression was also higher than 97% on the last sampling date at the end of August, corresponding to the harvest (Table 2).

The validation analysis of IDW maps is reported in Table 3 including the correlation analysis of the predicted vs. observed values and the calculation of the mean prediction errors. Eleven out of 12 of the contour maps were statistically supported by cross-validation analysis. In Figures 2 and 3, the IDW maps of *H. armigera* male distribution, calculated from the total catch per trap during the sampling period, are shown. Only the maps of the total catch per year are reported, because they properly describe the spatial pattern of the catches during both full field seasons. In each year, the male catch within the pheromone-treated area can be visualized and compared with those of the remaining part of the pheromone-untreated field. The gaps of catches within the treated areas in the maps can be considered as a demonstration of the effectiveness of the male capture reduction due to mating disruption (Figures 2 and 3). The catch patches, visualized as the darker filled contours, indicate the areas of the fields where male disruption was less or not effective. These areas of reduced efficacy correspond to the hedges and to the east zone of the mating disruption area, which is adjacent to the pheromone-untreated tomato. Moreover, the catch patches reached the highest values in the northeast area of the maps (up to 150 and 350 male captures in the 2011 and 2012 maps, respectively), corresponding to the pheromone untreated (control) area.

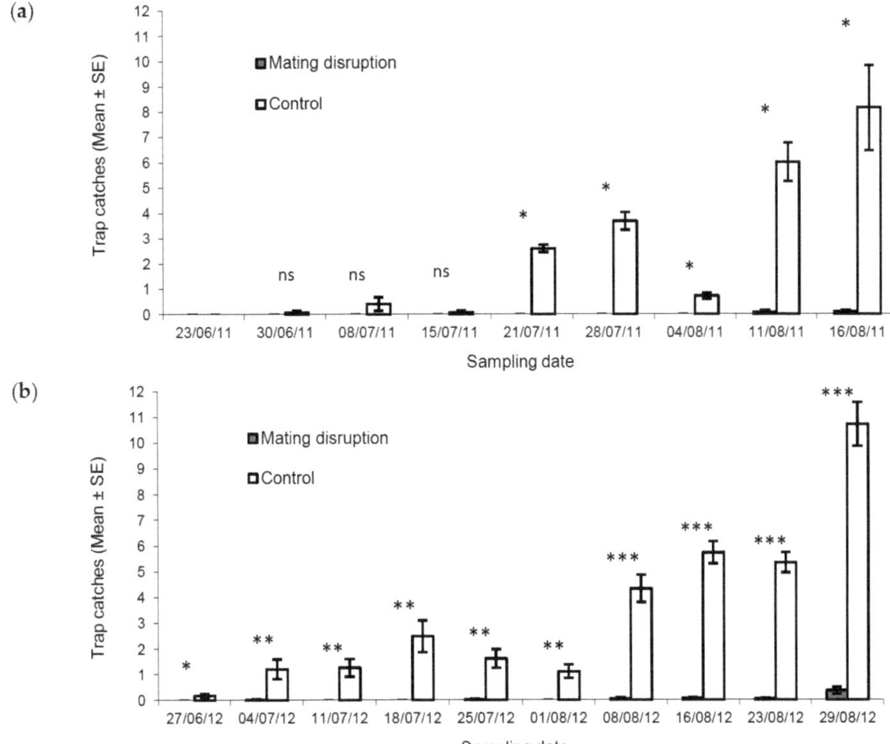

Figure 1. Mean number of *Helicoverpa armigera* males trapped per night in pheromone-treated (mating disruption) and pheromone-untreated (control) areas in 2011 (**a**) and 2012 (**b**). Bars represent the standard errors of the means. Male catches were compared using the Mann–Whitney U test: ns = not significant; * $p < 0.05$; ** $p < 0.01$; *** $p < 0.001$.

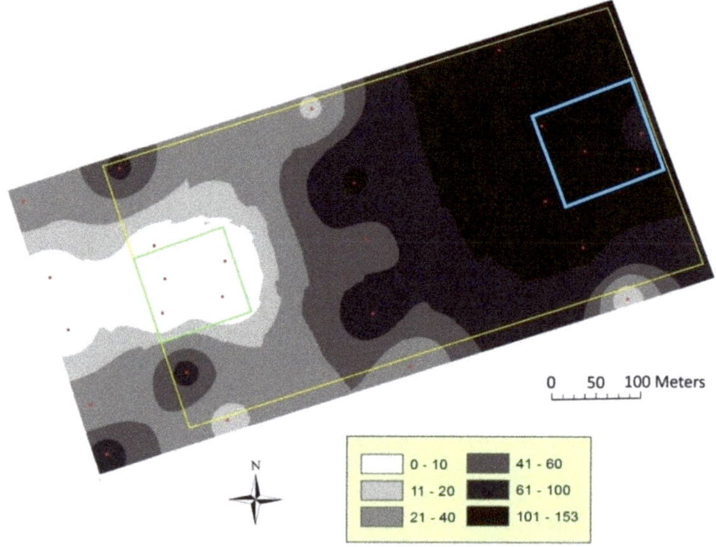

Figure 2. Distribution map of *Helicoverpa armigera* males in 2011 calculated from the total catch per trap during the sampling period (23 June to 16 August). The mating disruption area (1 ha) is represented by the small square on the left; the control area (1 ha) is represented by the small square on the right. Sampling points (pheromone baited traps) are represented by dots.

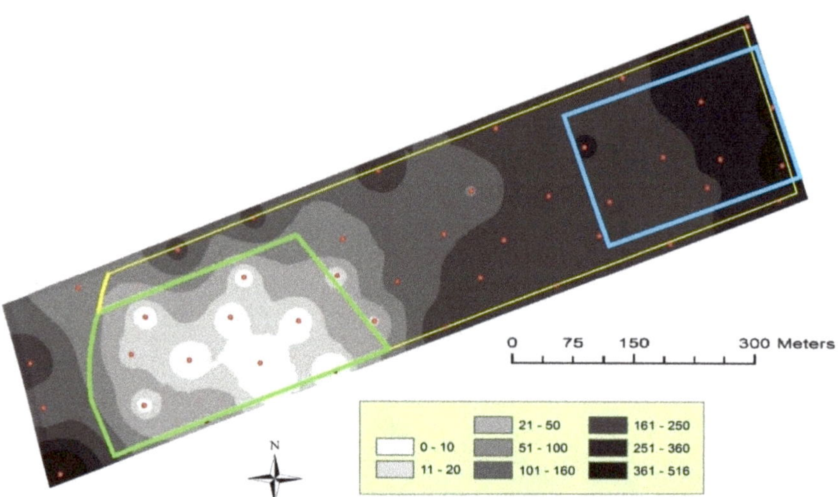

Figure 3. Distribution map of *Helicoverpa armigera* males in 2012 calculated from the total catch per trap during the sampling period (27 June to 29 August). The mating disruption area (5 ha) is represented by the area on the left; the control area is represented by the rectangle on the right. Sampling points (pheromone baited traps) are represented by dots.

Table 2. Ratio of catches reduction (%) in the pheromone-treated (mating disruption) area with respect to the pheromone-untreated (control) area in 2011 and 2012.

Year	Weeks after Dispenser Position												
	1	2	3	4	5	6	7	8	9	10	11	12	Total
2011	NA	NA	NA	-	100	100	100	100	100	100	98.8	98.8	99.1
2012	100	98.5	100	100	98.9	100	98.8	98.6	99.3	96.7	NA [1]	NA [1]	98.4

NA = not assessed; [1] Tomato harvested; - = no male catches.

3.2. Fruit Damage Estimation

Lower fruit damage was found in the mating disruption and chemical spray areas than in the control plots for most sampling dates (Figure 4). No damage by other pests was detected during the field trials (e.g., *Tuta absoluta* (Meyrick) (Lepidoptera Gelechiidae). Data analysis showed single and multiple significant interactions between fruit damage and the design variables (Table 4) fruit damage between treatments ($p < 0.001$), years ($p < 0.001$), and sampling dates ($p < 0.001$). Overall, the damage on fruits was higher during 2012 (7.72%) than in 2011 (4.8%). Fruit damage had significant multiple interactions between variables and, for this reason, data were split in each field season and sampling date to provide an interpretation of the seasonal trend of fruit damage in each treatment.

Table 3. Results of the cross-validation analysis of the inverse distance weighting (IDW) maps. The linear correlation of the predicted against measured values are reported.

Year	Map	R	p	Mean Prediction Error
2011	July, 21	0.80	<0.001	0.01
	July, 28	0.75	<0.001	−0.74
	August, 4	0.24	>0.05	−0.01
	August, 11	0.45	<0.05	−0.44
	August, 16	0.45	<0.05	0.34
	Total catches	0.71	<0.001	−0.91
2012	August, 1	0.42	<0.05	−0.62
	August, 8	0.81	<0.001	−0.99
	August, 16	0.69	<0.001	−2.88
	August, 23	0.57	<0.001	−1.79
	August, 29	0.70	<0.001	−3.67
	Total catches	0.73	<0.001	−13.6

Table 4. Log linear analysis of fruit damage. Only the interaction of response variable (fruit damage) with design variables (treatments–years–plots–sampling dates) are considered in the analysis.

Effect	χ^2	d.f.	p
Treatments * Damage	103.2	2	<0.01
Years * Damage	80.1	1	<0.01
Date * Damage	77.3	5	<0.01
Treatments * Years * Damage	66.7	2	<0.01
Treatments * Date * Damage	48.5	10	<0.01
Years * Date * Damage	39.7	5	<0.01

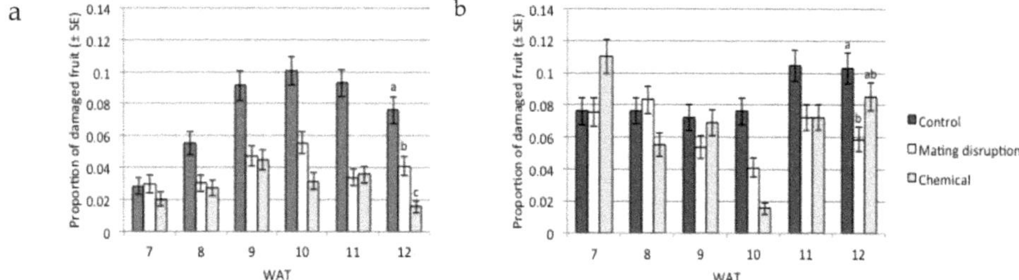

Figure 4. Proportion of damaged fruit (±binomial SE) for the treatments in the 2011 (**a**) and 2012 (**b**) seasons at each sampling date. WAT = weeks after transplanting. WAT 12 corresponded to the last sampling, one day before harvest. Bars bearing different letters are significantly different for $p < 0.05$ (χ^2 test followed by the z-test).

It is remarkable that the relative efficacy of both control methods varied during the year, and this evidence was corroborated by the significant multiple interaction among "treatments * years * fruit damage" (Table 4); in particular, chemicals were more effective in 2011 than in 2012. Fruit damage was shown to also be dependent by treatments * sampling date and by treatment * year. Within each year and corresponding to the harvest (WAT 12), a χ^2 test followed by a z-test was used to rank the efficacy of the treatments. Using this method, chemicals were scored as more effective than mating disruption in 2011 (Figure 4A), resulting in the following rank of efficacy: chemical > mating disruption > untreated control. On the other hand, chemical control was less effective in WAT 12 in the 2012 season, resulting in non-significant differences in comparison with the untreated control (Figure 4B).

4. Discussion

Mating disruption demonstrated a variable efficacy in controlling *H. armigera*, measured by the analysis of fruit damage. A significant difference between control and mating disruption was obtained in both seasons, thus showing a robustness of the data obtained in the two-year replication of this study. In the 2011 season, the efficacy of mating disruption was lower than the chemical control, but in 2012, this trend was reversed. Overall, both control techniques against *H. armigera* were demonstrated to be only partially effective in comparison with the untreated control because the strategies were not able to fully avoid fruit damage. It is remarkable that mating disruption was more effective in the 2012 season when applied on a 5 ha field; in contrast, mating disruption applied on 1 ha (2011 season) resulted in a lower reduction of fruit damage. A wide area approach is a cornerstone of the mating disruption approach [9] and it could be hypothesized that the increased area of application in 2012 led to a higher efficacy of mating disruption.

The geostatistical analysis of trap catch reported in this study was a useful tool to evaluate the efficacy of mating disruption through the representation of the spatial pattern of catches. Catch gaps and patches can be interpreted as areas where catch reduction is optimal or ineffective, respectively. In particular, maps were used to verify how catch reduction was affected by the position of field dispensers and by dominant winds, in order to highlight and interpret potential border effects. The maps seem suitable to visualize the effects on male catches as a result of potential pheromone drift out of the pheromone-treated area, showing a partial efficacy of the capture reduction, in the downwind borders of the pheromone-treated area. Geostatistical techniques have been used to characterize the spatial and temporal variability of male *H. armigera* catches in Spain, affecting pest management actions, and as a powerful tool in precision agriculture systems [27]. That study demonstrated that moths were aggregated at the borders of tomato field, gradually colonizing the inner area on cloudless days when northeastern winds were predominant. A geostatistical analysis of the spatial heterogeneity of bollworm eggs was studied in China using semi-variance and kriging interpolation, providing a population risk analysis [21,28]. Authors showed that there was a high risk at early pest population stages (mid-June). Geostatistical maps of the spatial distribution of male catches proved to be suitable to analyze the efficacy of *Spodoptera littoralis* (Boisduval) (Lepidoptera Noctuidae) mating disruption [16] and to study the spatial distribution of vegetable pests including *Phthorimaea operculella* (Zeller) (Lepidoptera Gelechiidae) and the western corn rootworm *Diabrotica virgifera virgifera* LeConte, in Northern Italy [29,30]. Moreover, georeferencing tools can be used to decide on the best installation site according to topography and wind direction, when pheromone aerosol devices for mating disruption are intended to be employed [20]. A number of practical applications of spatial analysis in managing pests are reviewed and discussed by Sciarretta and Trematerra [22].

The reason for the low efficacy of the chemical treatment in 2012 is unknown, but this fact is in agreement with the field data of the extension services of the Emilia-Romagna region [31]. In particular, our data seem to corroborate the anecdotal findings that chemical control of *H. armigera* in Northern Italy tomato often achieves a partial and variable efficacy. In this context, the application of mating disruption in wide areas including adjacent fields treated with pheromones, seems to have the potential to enhance the efficacy of the technique. The crucial role of the pattern of dispenser distribution over a much greater area outside the cultivated field has been stressed by Kerns [14] and de Souza et al. [11], in order to make mating disruption effective. For these reasons, future application of the mating disruption technique should consider an increase in the number of dispensers in the areas adjacent to the hedges of the field in order to balance the reduced efficacy due to border effects [20,27].

5. Conclusions

Mating disruption could be applied as an IPM strategy to optimize *H. armigera* control in order to increase the efficacy of chemical or microbial control or to reduce the number of chemical sprays in highly infested areas. Pheromone dispensers for mating disruption of *H. armigera* are not commercially available yet, but we hope that this technique will be available soon for stakeholders involved in IPM and organic insect management. Further studies should take into account the application of mating disruption in combination with microbial agents like *Bacillus thuringiensis* subsp. *kurstaki/aizawai* and nuclear polyhedrosis virus (HaNPV), in order to promote a more ecological and sustainable control strategy to minimize the negative side effects of chemical control including the selection of insecticide-resistant strains and the harmful impact on beneficial insects occurring in agroecosystems.

Author Contributions: G.B., A.L. and S.M. conceived and designed the research; A.L., F.R., G.G.B. and G.B. conducted the experiments; G.B., A.M. and A.L. analyzed the data; G.B. wrote the first draft of the manuscript. All authors discussed the results and wrote the final manuscript. All authors have read and agreed to the published version of the manuscript.

Funding: Field trials were funded by the Emilia-Romagna Region and CRPV (Centro Ricerche Produzioni Vegetali). Traps and pheromone dispensers were provided by SUTERRA Europe, Valencia, Spain.

Acknowledgments: Special thanks go to Cristina Alfaro for their technical suggestions and to Davide Montanari for technical support in the field.

Conflicts of Interest: The authors declare no conflict of interest. The funders had no role in the design of the study; in the collection, analyses, or interpretation of data; in the writing of the manuscript, or in the decision to publish the results.

References

1. Lammers, J.W.; Macleod, A. Report of a Pest Risk Analysis: Helicoverpa Armigera (Hübner, 1808). Available online: https://secure.fera.defra.gov.uk/phiw/riskRegister/downloadExternalPra.cfm?id=3879 (accessed on 2 July 2018).
2. Cherry, A.; Cock, M.; van den Berg, H.; Kfir, R. Biological control of *Helicoverpa armigera* in Africa. In *Biological Control in IPM Systems in Africa*; Neuenschwander, P., Borgemeister, C., Langewald, J., Eds.; CAB International: Wallingford, UK, 2003; pp. 329–346.
3. Sannino, L.; Espinosa, B. *I Parassiti Animali Delle Solanacee*; Edagricole: Milano, Italy, 2009; p. 290.

4. EPPO European and Mediterranean Plant Protection Organization. Distribution Maps of Quarantine Pests. *Helicoverpa armigera*. Available online: https://gd.eppo.int/taxon/HELIAR/distribution (accessed on 2 July 2018).

5. EPPO European and Mediterranean Plant Protection Organization. Datasheets on Quarantine Pests–Helicoverpa Armigera. Available online: https://gd.eppo.int/download/doc/132_datasheet_HELIAR. pdf (accessed on 2 July 2018).

6. (Istat) Istituto Nazionale di Statistica. Census Database. 2018. Available online: http://dati-censimentoagricoltura.istat.it (accessed on 2 July 2018).

7. Buès, R.; Bouvier, J.C.; Boudinhon, L. Insecticide resistance and mechanisms of resistance to selected strains of *Helicoverpa armigera* (Lepidoptera: Noctuidae) in the south of France. *Crop Prot.* **2005**, *24*, 814–820. [CrossRef]

8. Mitchell, R.E. Pheromones: As the glamour and glitter fade: The real work begins. *Fla. Entomol.* **1986**, *69*, 132–139. [CrossRef]

9. Birch, M.C.; Haynes, K.F. *Insect Pheromones*, 1st ed.; Studies in Biology No. 147; Hodder Arnold, H & S.: London, UK, 1982; p. 58.

10. Kehat, M.; Dunkelblum, E.; Gothilf, S. Mating disruption of the cotton leafworm, *Spodoptera littoralis* (Lepidoptera: Noctuidae), by release of sex pheromone from widely separated Hercon-laminated dispensers. *Environ. Entomol.* **1983**, *12*, 1265–1269. [CrossRef]

11. De Souza, K.; McVeigh, L.J.; Downham, M.C.A.; Smith, J.; Moawad, G.M. Mating-disruption of *Spodoptera littoralis* in Egypt. *WPRS Bull.* **1993**, *16*, 212.

12. Mitchell, E.R.; Kehat, M.; McLaughin, J.R. Suppression of mating by beet armyworm (Noctuidae: Lepidoptera) in cotton with pheromone. *J. Agri. Entomol.* **1997**, *14*, 17–28.

13. Wakamura, S.; Takai, M. Communication disruption for control of the beet armyworm, *Spodoptera exigua* (Hübner), with synthetic sex pheromone. *Jpn. Agri. Res. Q.* **1995**, *29*, 125.

14. Kerns, D.L. Mating disruption of beet armyworm (Lepidoptera: Noctuidae) in vegetables by a synthetic pheromone. *Crop Prot.* **2000**, *19*, 327–334. [CrossRef]

15. Rama, F.; Reggiori, F.; Albertini, A. Control of *Spodoptera littoralis* (Bsdv.) by biodegradable, low-dosage, slow-release pheromone dispensers. *IOBC/Wprs Bull.* **2011**, *72*, 59–66.

16. Lanzoni, A.; Bazzocchi, G.G.; Reggiori, F.; Rama, F.; Sannino, L.; Maini, S.; Burgio, G. *Spodoptera littoralis* male capture suppression in processing spinach using two kinds of synthetic sex-pheromone dispensers. *Bull. Insectol.* **2012**, *65*, 311–318.

17. Kehat, M.; Dunkelblum, E. Sex pheromones: Achievements in monitoring and mating disruption of cotton pests in Israel. *Arch. Insect Biochem. Physiol.* **1993**, *22*, 425–431. [CrossRef]

18. Kehat, M.; Anshelevich, L.; Gordon, D.; Harel, M. Evaluation of Shin-Etsu twist-tie rope dispensers by the mating table technique for disrupting mating of the cotton bollworm, *Helicoverpa armigera* (Lepidoptera: Noctuidae), and the pink bollworm, *Pectinophora gossypiella* (Lepidoptera: Gelechiidae). *Bull. Entomol. Res.* **1989**, *88*, 141–148. [CrossRef]

19. Chamberlain, D.J.; Brown, N.J.; Jones, O.T.; Casagrande, E. Field evaluation of a slow release pheromone formulation to control the American bollworm, *Helicoverpa armigera* (Lepidoptera: Noctuidae) in Pakistan. *Bull. Entomol. Res.* **2000**, *90*, 183–190. [CrossRef] [PubMed]

20. Benelli, G.; Lucchi, A.; Thomson, D.; Ioriatti, C. Sex Pheromone Aerosol Devices for Mating Disruption: Challenges for a Brighter Future. *Insects* **2019**, *10*, 308. [CrossRef] [PubMed]

21. Rossi, R.E.; Borth, P.W.; Tollefson, J.J. Stochastic simulation for characterizing ecological spatial patterns and appraising risk. *Ecol. Appl.* **1993**, *3*, 719–735. [CrossRef] [PubMed]

22. Sciarretta, A.; Trematerra, P. Geostatistical tools for the study of insect spatial distribution: Practical implications in the integrated management of orchard and vineyard pests. *Plant Protect. Sci.* **2014**, *50*, 97–110. [CrossRef]

23. Isaaks, E.H.; Srivastava, R.M. *An Introduction to Applied Geostatistics*; Oxford University Press: New York, NY, USA, 1989; p. 592.

24. Zar, J.H. *Biostatistical Analysis*, 5nd ed.; Pearson Prentice Hall: Upper Saddle River, NJ, USA, 2010; p. 944.

25. Steel, R.G.D.; Torrie, J.H.; Dickey, D.A. *Principles and Procedures of Statistics. A Biometrical Approach*, 3rd ed.; WCB McGraw-Hill: New York, NY, USA, 1997; p. 666.

26. Sharpe, D. Your Chi-Square test is statistically significant: Now what? *Pract. Assess. Res. Eval.* **2015**, *20*, 1–10.

27. Garcia Moral, F.J. Analysis of the spatio—Temporal distribution of *Helicoverpa armigera* Hb. in a tomato Field using a stochastic approach. *Biosyst. Eng.* **2011**, *93*, 253–259. [CrossRef]

28. Ge, S.K.; Carruthers, R.I.; Ma, Z.F.; Zhang, G.X.; Li, D.M. Spatial heterogeneity and population risk analysis of cotton bollworm, *Helicoverpa armigera*, in China. *Insect Sci.* **2005**, *12*, 255–261. [CrossRef]

29. Masetti, A.; Butturini, A.; Lanzoni, A.; De Luigi, V.; Burgio, G. Area-wide monitoring of potato tuberworm (*Phthorimaea operculella*) by pheromone trapping in Northern Italy: Phenology, spatial distribution and relationships between catches and tuber damage. *Agric. For. Entomol.* **2015**, *17*, 138–145. [CrossRef]

30. De Luigi, V.; Furlan, L.; Palmieri, S.; Vettorazzo, M.; Zanini, G.; Edwards, C.R.; Burgio, G. Results of WCR monitoring plans and evaluation of an eradication program using GIS and Indicator Kriging. *J. Appl. Entomol.* **2011**, *135*, 38–46. [CrossRef]

31. Dradi, D.; ASTRA Innovazione e Sviluppo, Cesena, Italy. Personal communication, 2012.

Review

Tephritid Fruit Fly Semiochemicals: Current Knowledge and Future Perspectives

Francesca Scolari [1,*], Federica Valerio [2], Giovanni Benelli [3], Nikos T. Papadopoulos [4] and Lucie Vaníčková [5,6,*]

1   Institute of Molecular Genetics IGM-CNR "Luigi Luca Cavalli-Sforza", I-27100 Pavia, Italy
2   Department of Biology and Biotechnology, University of Pavia, I-27100 Pavia, Italy; federica.valerio02@universitadipavia.it
3   Department of Agriculture, Food and Environment, University of Pisa, Via del Borghetto 80, 56124 Pisa, Italy; giovanni.benelli@unipi.it
4   Department of Agriculture Crop Production and Rural Environment, University of Thessaly, Fytokou st., N. Ionia, 38446 Volos, Greece; nikopap@uth.gr
5   Department of Chemistry and Biochemistry, Faculty of AgriSciences Mendel University in Brno, Zemedelska 1, CZ-613 00 Brno, Czech Republic
6   Department of Forest Botany, Dendrology and Geobiocoenology, Faculty of Forestry and Wood Technology, Mendel University in Brno, Zemedelska 1, CZ-613 00 Brno, Czech Republic
*   Correspondence: francesca.scolari@igm.cnr.it (F.S.); lucie.vanickova@mendelu.cz (L.V.); Tel.: +39-0382-986421 (F.S.); +420-732-852-528 (L.V.)

Simple Summary: Tephritid fruit flies comprise pests of high agricultural relevance and species that have emerged as global invaders. Chemical signals play key roles in multiple steps of a fruit fly's life. The production and detection of chemical cues are critical in many behavioural interactions of tephritids, such as finding mating partners and hosts for oviposition. The characterisation of the molecules involved in these behaviours sheds light on understanding the biology and ecology of fruit flies and in addition provides a solid base for developing novel species-specific pest control tools by exploiting and/or interfering with chemical perception. Here we provide a comprehensive overview of the extensive literature on different types of chemical cues emitted by tephritids, with a focus on the most relevant fruit fly pest species. We describe the chemical identity, production modality and behavioural relevance of volatile pheromones, host-marking pheromones and cuticular hydrocarbons, as well as the technological advances available for their characterisation. The variegate set of approaches integrating the use of the identified chemical signals for the control of wild populations of key pests is also explored. Last but not least, key challenges for future basic to applied research regarding tephritids are outlined.

Citation: Scolari, F.; Valerio, F.; Benelli, G.; Papadopoulos, N.T.; Vaníčková, L. Tephritid Fruit Fly Semiochemicals: Current Knowledge and Future Perspectives. *Insects* **2021**, *12*, 408. https://doi.org/10.3390/insects12050408

Academic Editor: Eric W. Riddick

Received: 27 March 2021
Accepted: 27 April 2021
Published: 30 April 2021

Publisher's Note: MDPI stays neutral with regard to jurisdictional claims in published maps and institutional affiliations.

Abstract: The Dipteran family Tephritidae (true fruit flies) comprises more than 5000 species classified in 500 genera distributed worldwide. Tephritidae include devastating agricultural pests and highly invasive species whose spread is currently facilitated by globalization, international trade and human mobility. The ability to identify and exploit a wide range of host plants for oviposition, as well as effective and diversified reproductive strategies, are among the key features supporting tephritid biological success. Intraspecific communication involves the exchange of a complex set of sensory cues that are species- and sex-specific. Chemical signals, which are standing out in tephritid communication, comprise long-distance pheromones emitted by one or both sexes, cuticular hydrocarbons with limited volatility deposited on the surrounding substrate or on the insect body regulating medium- to short-distance communication, and host-marking compounds deposited on the fruit after oviposition. In this review, the current knowledge on tephritid chemical communication was analysed with a special emphasis on fruit fly pest species belonging to the *Anastrepha*, *Bactrocera*, *Ceratitis*, and *Rhagoletis* genera. The multidisciplinary approaches adopted for characterising tephritid semiochemicals, and the real-world applications and challenges for Integrated Pest Management (IPM) and biological control strategies are critically discussed. Future perspectives for targeted research on fruit fly chemical communication are highlighted.

Keywords: pheromone; olfactory cues; mating disruption; cuticular hydrocarbons; host-marking pheromone; true fruit flies; olfaction; odours

1. Introduction

Insect semiochemicals are compounds belonging to different chemical classes that regulate intra- and inter-specific communication, affecting major behavioural and physiological responses [1–3]. Based on the identity of the emitter and the receiver, semiochemicals can be classified as pheromones (i.e., molecules mediating communication between co-specifics) or allelochemicals (i.e., compounds involved in communication between individuals of different species). Allelochemicals include kairomones (beneficial to the receiver but producing disadvantages for the emitter), synomones (molecules that benefit both the emitter and the receiver), allomones (beneficial to the producer and with neutral effects to the receiver), and apneumones (chemicals of non-biological origin beneficial to the receiver) [4–6]. Semiochemicals mediate a number of behavioural processes, such as the identification of food sources, the location of mates and hosts for oviposition, and the avoidance of predators [7].

To achieve these different functions, insects use volatile, semi-volatile, and non-volatile chemicals that are involved in long-, medium-, and short-distance communication, respectively. These stimuli can be detected by sensory neurons of the olfactory system on the antennae and maxillary palps [8], such as in the case of volatile molecules, or by neurons of the gustatory system mainly on proboscis, ovipositor and legs, which are able to perceive non-volatile chemicals through contact chemoreception [9].

Among semiochemicals, pheromones (from the Greek words "φερειν"-transfer, and "ορμαν"-excite) can be classified in two main categories: (*i*) releaser pheromones that produce an immediate response in a recipient individual (e.g., a male fly orienting toward a female guided by sex pheromone), (*ii*) primer pheromones that trigger the initiation of a complex physiological response not immediately observable [10]. The complex functions pheromones exert are mediated by sex, aggregation, alarm, trails and host-marking compounds [11,12].

Since 1959, when the term 'pheromone' was proposed [13] and the first pheromone, bombykol, was chemically identified in the silkworm moth, *Bombyx mori* (L.) (Lepidoptera: Bombycidae) [14], an increasing number of studies have focused on unravelling the chemistry and biological roles of these substances in numerous species. In most cases, pheromones are blends of individual chemicals that can be shared among species, but that are mixed in species-specific combinations (i.e., quantitatively and qualitatively) [15,16]. So far, volatile pheromones have been described as being composed of two or three compounds in moths [17,18], one in *Drosophila melanogaster* Meigen (Diptera: Drosophilidae) ((Z)-undec-4-enal [19]), or by complex blends in honey bees (Hymenoptera: Apidae) [20]. The chemical diversity of pheromone blends is very high, including acetate esters, alcohols, aldehydes, carboxylic acids, hydrocarbons, epoxides, ketones, benzenoid compounds, isoprenoids, terpenoids, and triacylglycerides [21]. In the case of Lepidoptera, as well as other insects that mainly rely on long-distance sexual signalling, volatile pheromones are the primary semiochemicals adopted [22]. Other insects, such as *Drosophila* species, are characterised by complex courtship rituals and use cuticular hydrocarbons (CHs) of both high and low-volatility [23]. The body surface of many insect species is indeed covered by a thin film of wax, composed mainly of hydrocarbons. Complex mixtures of esters, alcohols and free fatty acids are components of the cuticular wax in some insects [24]. Beside their cuticle waterproofing function, the long-chain hydrocarbons of insects are involved in chemical communication, serving as sex pheromones, kairomones, species- and gender-recognition cues, nestmate recognition compounds, fertility and dominance cues, chemical mimicry, and primer pheromones. Such key roles boosted research efforts in the past several decades on many dipteran species, including fruit flies, house flies and mosquitoes [25,26]. Insect CHs are usually a mixture of compounds that may include *n*-alkanes, alkenes, terminally branched monomethylalkanes, internally branched monomethylalkanes, dimethylalkanes, trimethylalkanes and others. They are synthesised by an elongation-decarboxylation pathway in oenocytes, which are associated with epidermal cells or fat bodies. After synthesis, CHs are transported through haemolymph by lipophorin carrier [27,28].

An additional type of semiochemicals is used by several parasitic and phytophagous insects immediately after egg-laying, the host-marking pheromone (HMP). Its function is to affect the oviposition behaviour of conspecifics in a way that subsequent eggs are not deposited in their already utilised resource, thus reducing the time spent on the already exploited resource and the competition for limited host resources, with advantages for both the marker and the seeker [29]. The HMP can be synthetized by female fruit flies in the form of a complex molecule [30–32] or a simple compound [33–35]. The receptors located in the tarsi and mouthparts of females searching for an oviposition site allow the detection of HMPs [29,36,37].

In the last decades, progress has been made to determine the identity and composition of semiochemicals in insects, as well as the chemical specificity and functional properties of molecules mediating semiochemical perception such as odour and taste receptors, odorant and gustatory binding proteins. Insects developed extremely refined abilities to produce and discriminate among different arrays of chemicals. In this framework, an increasingly deeper knowledge of the mechanisms underlying semiochemicals' production and stimuli coding is being acquired, also due to advancements in analytical approaches, which, in turn, is providing multiple novel/improved tools for insect pest control.

In the scenario depicted above, true fruit flies (Diptera: Tephritidae) are excellent models to investigate the differentiation of semiochemicals' production and perception. Including more than 5000 species and 500 genera, this family is one of the largest among dipteran, with a worldwide distribution. Many tephritids are important pests of agricultural commodities infesting a wide range of fruits and vegetables. The most pestiferous species belong to genera *Anastrepha* (Schiner), *Bactrocera* (Macquart), *Ceratitis* (Macleay), and *Rhagoletis* (Loew) [38] (Figure 1).

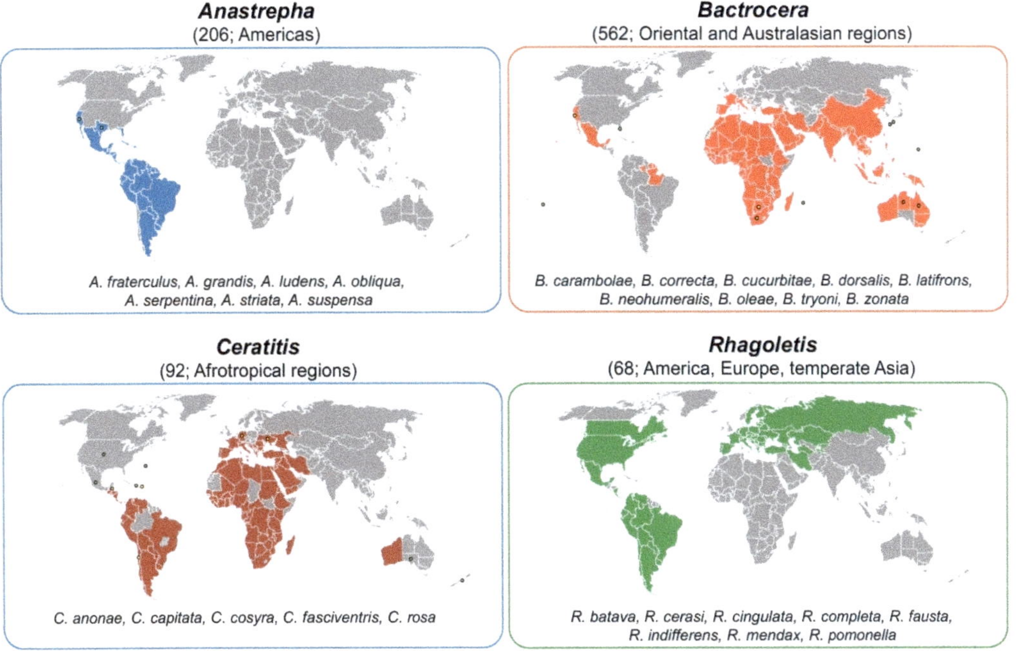

Figure 1. Worldwide distribution of species belonging to the *Anastrepha, Bactrocera, Ceratitis* and *Rhagoletis* genera. The map is based on current distribution data of species belonging to each genus retrieved from the EPPO Global Database (https://gd. eppo.int) (accessed on 13 November 2020) and integrated with information from CABI (https://www.cabi.org/isc/) (accessed on 13 November 2020) and the literature. In brackets, the total number of living species (obtained from Catalogue of Life: 2019 Annual Checklist [39]) for each genus and the original geographic range are reported. Lists of the major pests within each genus are also indicated. Background map was retrieved from https://freevectormaps.com/world-maps/WRLD-EPS-01-0011?ref=atr (accessed on 29 July 2020). Information about the status of invasion is reported in more details for the species *A. ludens, B. dorsalis* and *C. capitata*, for which yellow and green dots indicate transient presence or achieved eradication, respectively.

As an example, the damage caused annually in Africa by the Oriental fruit fly (*Bactrocera dorsalis* [Hendel]) has been estimated to USD 2 billion mainly due to export trade bans [40]. Moreover, due to a number of biological features including multivoltinism, long adult longevity, high fecundity, remarkable response to various stresses, increased capacity to overwinter [41], several species became aggressive global invaders, imposing strict quarantine regulations in several fruit-producing countries [42,43]. Because of their high importance, there is a list of ongoing programs worldwide that aim to eradicate, contain or suppress the populations of tephritid species [44].

Area-Wide Integrated Pest Management (AW-IPM) has been proven successful for the control of tephritid pest species and incorporates different components, such as thorough population monitoring employing sophisticated trapping systems, the Sterile Insect Technique (SIT), the Male Annihilation Technique (MAT) and often bait insecticidal sprayings [45]. To ensure effective application of the above components of the AW-IPM programs, an in-depth understanding of insect communication and mating strategies is required. This field has been widely investigated in tephritids and an expanding body of literature is available.

In this review, we critically discuss the current knowledge on pheromone-based communication in tephritid fruit flies, as well as its applied relevance for pest control. Starting from the role of these semiochemicals in tephritid reproductive behaviour, we analysed the most relevant, available literature focusing on (*i*) volatile pheromones released by males and/or females, (*ii*) HMPs, and (*iii*) CHs. For each of these three groups, the tissues involved in the production, their chemical identity, and the analytical methods applied for the identification, as well as the electrophysiological and behavioural tools employed to shed light on their ecological significance, are considered. Due to their high economic importance, insights are provided on pest species of the genera *Anastrepha, Bactrocera, Ceratitis* and *Rhagoletis*. Over the last decades, tephritid semiochemical research has not been equally dedicated to all three groups of semiochemicals. Thus, information related to volatile pheromones, mostly because of their more direct applied implications, is far more extended. This is also reflected in the present review, which additionally focuses on the role of semiochemical-based communication in species evolution. Tephritids display different levels of host specialisation, covering the whole spectrum from monophagy to polyphagy. Whether and how the adaptation to novel host plants drove the evolution of such an extreme diversity is still an open question and requires investigation of multiple factors (and their interactions), including those related to semiochemical-based communication. Finally, results focusing on applying the above-mentioned knowledge to fruit fly population monitoring and management, as well as to chemical taxonomy, are analysed, towards formulating major challenges for future research.

2. Semiochemicals and Reproductive Behaviour—An Overview

Semiochemicals are involved in the different phases of the reproductive behaviour of many tephritids, which include male lek formation and sexual signalling (sexual calling), courtship, acceptance of a mating partner and successful copulation (Figure 2).

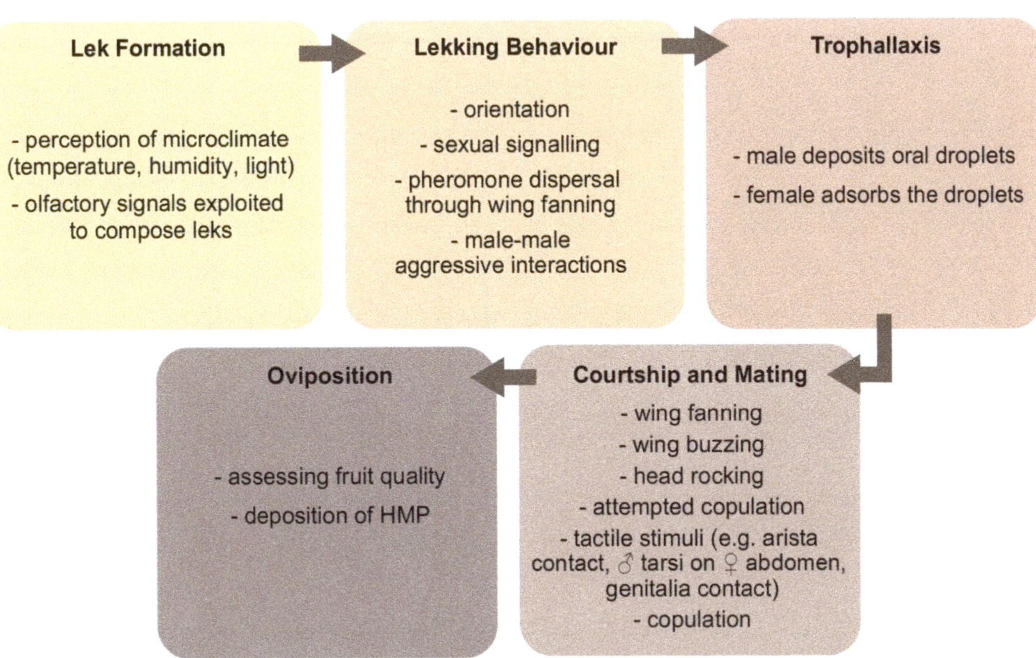

Figure 2. Diagram showing the main steps of mating behaviour in tephritids and the signals transmitted and perceived. Lekking behaviour is common in Tephritidae, but notably absent in *Rhagoletis* species. Similarly, pre-mating trophallaxis is common, but this phenomenon is known to occur also during and post mating [46]. HMP, host-marking pheromone.

The release and perception of semiochemicals are often accompanied by a range of intense behavioural interactions, including wing vibration and buzzing and head rocking [47–50].

There is a variety of mating systems in tephritid fruit flies, with lek-based ones being common in many species of economic importance [51]. Leks are mating arenas where males aggregate and perform sexual signalling without controlling the access to resources that may be critical for females or offering parental care, and provide only the sperm to females that freely choose their mates [52]. Males perform sexual signalling mainly on the under surface of leaves of preferred host trees that either bear fruits as in *Anastrepha fraterculus* (Wiedemann) and *B. dorsalis*, or do not bear fruits as in certain populations of the Mediterranean fruit fly (medfly) *Ceratitis capitata* (Wiedemann) (see [53] for a review). Instead, in *Zeugodacus cucurbitae* (Coquillett), lekking takes place more frequently on non-host plants [54,55]. Likewise, in *Bactrocera cacuminata* (Hering), the wild tobacco fruit fly, non-host plants containing the male attractant methyl eugenol (ME) serve as the main mating sites [56]. Lekking males, in groups of 2–10 individuals [53], perform sexual signalling that includes a set of visual, acoustic and olfactory signals, with volatile pheromone acting as the longer-distance cue.

On the other hand, at variance with many tephritids, most *Rhagoletis* spp. do not exhibit male lekking [51]. For example, males of *Rhagoletis pomonella* (Walsh), *Rhagoletis cerasi* (L.), *Rhagoletis rubicola* (Doane) and *Rhagoletis mendax* (Curran) individually search for potential mates mainly on host fruits and the top surface of leaves [57], mostly during the late afternoon and at dusk [51,58,59]. However, in *Rhagoletis batava* (Hering) individuals have been reported to form small groups in response to male-released pheromone that might be considered leks [60]. In certain species, such as *Anastrepha suspensa* (Loew) and *Bactrocera papayae* (Drew & Hancock), males use olfactory cues to locate the leks. However, other studies demonstrated that this is not the case for *Anastrepha obliqua* (Macquart) and *Anastrepha ludens* (Loew), and it is rather controversial for *C. capitata*, *B. dorsalis*, and *Z. cucurbitae* [53]. Whether the cues exploited by males to aggregate in leks derives from their own pheromone emissions, or from the plants where leks occurs is still under debate. Indeed, in *C. capitata* it has been proposed that plant volatiles, rather than male-emitted pheromones, contribute to male aggregation in leks [61,62], with α-copaene, a natural sesquiterpene widely present in plants, being regarded as the primary male cue for lek formation [63–66]. Male pheromone has been shown to attract conspecific males thus acting as aggregation pheromone in some *Bactrocera* species [67–72]. Nonetheless, Kobayashi and colleagues reported no male-male attraction in *B. dorsalis* [73].

Interestingly, sex pheromones are not uniquely produced by males in all tephritid species. In the *Ceratitis*, *Anastrepha* and *Rhagoletis* genera, the volatile chemicals identified in the adult headspace are of male origin. Instead, the role of females in emitting volatiles is evident in the genus *Bactrocera*. In all *Bactrocera* species studied so far, both sexes produce and release pheromone chemicals, albeit interspecific differences exist. In *Bactrocera musae* (Tryon), a polyphagous pest distributed in Australia and Papua New Guinea mainland [74], both sexes produce volatile pheromones, but its complexity is higher in females than males [75].

In the olive fruit fly, *Bactrocera oleae* (Rossi), females release the pheromone that plays a central role in the mating system of this species [76–79]. However, earlier and most recent studies suggested that male-emitted chemicals elicit female attraction [80–82]. The finding that both sexes emit and perceive chemicals during courtship [49,83], together with the identification of male wing vibration in this species [49], encourage further research to better understand the role of chemical communication in the mating behaviour of *B. oleae*. So far, males have been regarded to swarm before settling on olive canopy [41]. The modality of male aggregation to form leks in this species needs to be further clarified too, in order to understand the relevance of chemical communication in a species where female-borne pheromones have been well documented both as a short- and long-distance cue [41]. Recently it was proven that ethyl decanoate emitted by olive fruit fly females attracts other females, and hence may be involved in female-female aggressive interactions on oviposition sites [84].

Courtship implies a series of ritualized actions, and it is much more complex in the genus *Ceratitis*, particularly in *C. capitata* [85], than in other tephritids, such as many species of *Anastrepha* [86,87] and *Bactrocera* [88]. However, recent studies are revealing that, even in species where it was believed to be simplified, such as in *Bactrocera tryoni* (Froggatt), courtship is more complex than previously thought [48]. During courtship, volatile chemicals are perceived by either one of both sexes using the antennae, while less volatile compounds, such as CHs, have been suggested to be important in later stages, when male and female touch each other during mating attempts and mating *per se* [35,41,89]. Indeed, in several tephritids, fore- and hind-leg interactions, as well as male foreleg interaction with the female abdomen, 'kissing' (i.e., touching of the labella), and male tapping with the labellum on female thorax have been reported [41,48,90]. For example, in *Ceratitis rosa* (Karsch) and *A. fraterculus*, sex-specific differences in the quantitative composition of CH profile, together with reported mating incompatibilities, further suggest the role of CHs as short-range semiochemicals [89,91,92].

Trophallaxis (i.e., female provision with gifts -oral, genital or transdermal- by males [90,93,94]), a common behaviour in tephritids, is considered as a courtship signal as well (see [95] for a review). While in *Anastrepha* species a pre-mating transfer of male oral products to females by labrum-to-labrum contact is well known [90,96,97], this phenomenon has been described in medfly only recently [95]. In this species, the consumption of male-produced oral droplets by the female appeared to increase her receptivity to mating, suggesting that the chemical composition of these droplets may be a tool to assess male quality [35,95]. Further research in the medfly and in other major tephritid fruit fly species is thus essential to unravel how these substances are perceived and how this behaviour fits in the complex semiochemical-based communication frame. Interestingly, a recent work demonstrates that *A. ludens* males regurgitate more than females, and propose that regurgitation and deposition of series of droplets organized in lines or spirals by some fruit fly species may play multiple functions (e.g., collecting bacteria from the environment, or eliminating ingested toxicants), including the production of oral pheromones [98]. Moreover, signalling males in some *Anastrepha* spp. deposit pheromone on the leaf surface by abdominal tip dipping [99]. Some of these deposited components persist on leaves up to one hour after removal of signalling males and are able to attract females [99]. Abdominal dipping with deposition of a viscous substance from the male cercus (i.e., the external appendage close to the digestive tract) has also been observed in *Rhagoletis boycei* (Cresson) [100].

After mating, tephritid females undergo an almost immediate switch from response to male pheromone to host plant-oriented olfactory behaviour to seek for appropriate oviposition background. This phenomenon has been well documented for the medfly at the behavioural level [101] and molecular data suggest that genes related to olfaction and/or foraging are also changing in their transcriptional profiles [102]. Mating-related differential expression in genes involved in chemosensory perception has also been detected in *Bactrocera* [103,104], and *Anastrepha* spp. [105], with females of *B. tryoni* showing mating-induced switches in olfactory preference [106].

The oviposition behaviour of tephritids is highly heterogeneous and display species-specific differences, such as daily patterns of oviposition, clutch size, patterns of positioning the ovipositor, duration of the oviposition bout, and preferred plant site for the oviposition [107]. In the medfly, oviposition behaviour has been widely studied and described as being structured in four steps: arrival to the fruit, exploring (i.e., survey of fruit surface with head, labellum and ovipositor), ovipositor puncturing and drawing following oviposition conclusion [108,109]). In the last phase, the fruit surface is again explored by the females, with the aculeus of the ovipositor protracted, and a HMP is deposited [110].

The presence of conspecific eggs and developing larvae in breeding substrates may dramatically alter the oviposition behaviour in phytophagous insects. In tephritids, females often mark already used hosts with a pheromone to avoid overexploitation of the specific resource and hence reduce/eliminate competition [111,112]. To maximise the chances of survival and success of their progeny, phytophagous insects tend to avoid egg laying in already explored host resources [29,107]. Host marking is particularly important in endophytic species, such as tephritids, in which females oviposit inside fruits or other plant tissues, with no visible damage and no emission of specific plant volatiles in response to infestation and presence of fruit fly eggs [113]. HMPs are generally applied by female drawing the ovipositor following an egg-laying event [111].

Mechano-, hygro- and gustatory receptors are located on female ovipositor in tephritids [114]. Gustatory sensilla present on the tarsi of *R. pomonella* exhibit sensitivity to HMP [115]. Sensilla types on female ovipositor have been described in several *Bactrocera* and *Zeugodacus* species (i.e., *Z. cucurbitae*, *Bactrocera diaphora* [Hendel], *B. dorsalis*, *Bactrocera minax* [Enderlein], *Bactrocera scutellata* [Hendel] and *Bactrocera tau* [Walker] [116]; *B. tryoni* [117]), as well as in *R. pomonella* [118–120], but their characterisation is still patchy in species of the *Ceratitis* and *Anastrepha* genera [121].

Interestingly, HMPs have been shown to either deter or enhance oviposition, depending on concentration and other factors [122]. Deterrent effects induce different behaviours in responding females including suppression/disruption of oviposition, reduction of the number of egg clutches per fruit and of egg number per clutch, and dispersion to less infested (occupied) areas [107,113,123–127]. Although HMPs are predominantly recognised by individuals of the same species [29,128], interspecific perception of HMPs has also been described in tephritids. Cross-recognition has been demonstrated among species of the *Rhagoletis* [129,130], *Anastrepha* [131] and *Ceratitis* genera [33,34]. The host-marking behaviour may display different features even among species of the same genus. For example, small-sized fruit specialists (e.g., *Rhagoletis alternata* [Fallen], *Rhagoletis indifferens*

[Curran], *R. pomonella* and *R. cerasi*) often deposit HMPs [132]. Conversely, members of the *Rhagoletis suavis* group rarely mark the host targets [133,134] and commonly tend to lay eggs on infested fruits [133,135,136]. Two hypotheses have been proposed for the sporadic host-marking behaviour of the *suavis* group. According to the "no HMP deposition" hypothesis, all species of this group use as host walnut species (*Juglans* spp.), which are not infested by other flies of the genus in North America, and hence provide a competition-free resource for larvae [132]. The second hypothesis proposes the occurrence of a "male host-marking behaviour". *Rhagoletis boycei* males indeed usually touch the host fruit depositing a substance on its surface and females preferentially oviposit on the unmarked fruit [100]. According to this hypothesis, male host-mark replaces the female's one, causing a reduction in female marking behaviour. Male host-marking behaviour has been described in two species of the *suavis* group, *R. boycei* and *R. suavis* (Loew) [112].

The drawing of the aculeus after oviposition without an evident release of HMP has been described in *B. dorsalis* [137], *B. tryoni* and *B. jarvisi* (Tryon) [138], and *Z. cucurbitae* [139]. Instead of using their ovipositor right after oviposition, olive fly females spread the olive juice that leaks from the oviposition wound over the fruit surface using their labella. This behaviour appears to prevent other females from ovipositing on the same fruit [140–142].

3. Volatile Pheromones

3.1. Tissues Involved in Volatile Pheromone Production

Male pheromones seem to be released primarily from glands positioned in the rectum in *Ceratitis* [143,144], *Bactrocera* [73,78,145–147], and *Anastrepha* spp. [148–150]. However, interesting variations have been reported. This is the case of the goldenrod gall tephritid fly *Eurosta solidaginis* (Fitch), which possess no anal glands but enlarged rectum and pleural epidermis that have been suggested to be involved in pheromone production and/or storage [151].

In general, the rectal valve of *A. fraterculus* and *C. capitata* is located at the distal portion of the colon, while the intestinal canal enlarges in a chamber comprising four rectal papillae projecting in the lumen of the rectal ampulla (i.e., the anterior rectum) [152]. In its proximal region, the ampulla is lined with epithelial cells that increase in number and are organized in folds in the distal portion. In several *Anastrepha* species, the structure of the female rectum is similar, with differences in the features of the epithelium [153]. In *Bactrocera* spp. (e.g., *B. oleae*, *B. dorsalis*, *B. tryoni*, and *B. papayae*), the rectal glands of males and females display a remarkable sexual dimorphism [154–156]: muscles surrounding male glands are more abundant, suggesting a more complex contraction capacity to support the pheromone-storage and -release functions. The rectal sac (i.e., an evagination in the rectal gland) is present only in males. Because they are not surrounded by muscles and directly exposed to the haemolymph, the rectal pads, which extend into the gland as rectal papillae, have been suggested to be involved in transporting chemicals into the rectal gland [156].

During sexual calling, the anal tube is protruded to allow the epithelium of the distal rectum to be extruded. In medfly, when everted, the folded rectal epithelium is expanded in a balloon-like structure [143]. This evagination of the anal membranes at the abdominal tip, which appears as a droplet, has been described in *Anastrepha* and *Ceratitis* species, and it has been explained as a mean of expanding the evaporative surface, thus increasing pheromone emission and, consequently, attractiveness [99,149,157]. In addition, this behaviour is accompanied by protrusion of male expansion of the pleural abdominal region generating two lateral blisters [148,158].

Males of *C. capitata* and *C. rosa* display three types of sex-specific glands, (*i*) the anal glands, (*ii*) the pleural glands and (*iii*) the dimorphic salivary glands [159]. The pair of anal glands open onto the external cuticle close to the anal opening [159,160]. Males of *Anastrepha* species display only the pleural glands and dimorphic salivary glands [150,159]. In *Bactrocera* species (e.g., *B. oleae*, *B. tryoni* and *B. dorsalis*), both the anal glands and the dimorphic salivary glands are absent [145,159,161]. Currently, detailed information on *Rhagoletis* species are lacking, although the early work of Nation [159] found neither dimorphic salivary glands nor pleural glands in both *R. pomonella* and *R. juglandis* (Cresson).

In addition to the key function played by rectal tissues, pheromones have been later found to be also released orally. Salivary glands release pheromone components in *Anastrepha* spp. [148–150,162,163]. These glands are sexually dimorphic, with male salivary glands being ball-like structures associated with the crop [98,159,164]. Salivary glands have been suggested to be involved in the storage and, potentially, synthesis of pheromone components in medfly males [165]. This is supported by the fact that some chemicals of the pheromone blend identified in medfly male headspace were also detected in the extracts of salivary glands. Figure 3 summarizes the tissues involved in pheromone production in tephritids.

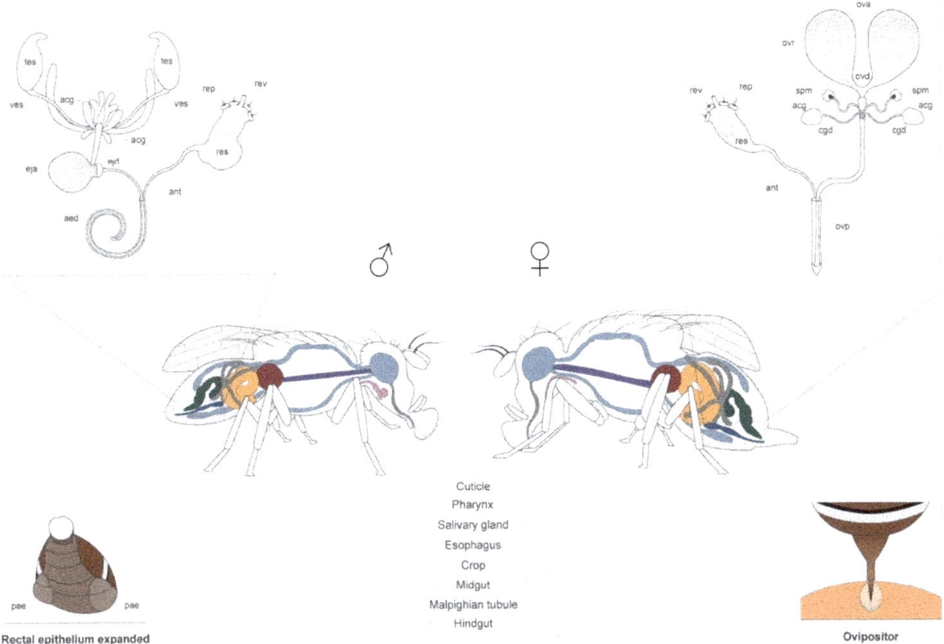

Figure 3. Tissues involved in semiochemicals' production in tephritids. Diagrammatic representation of a male (left) and female (right) generalized tephritid showing the tissues/body compartments involved in pheromone production. Reproductive tracts and rectum are shown in upper left (male) and right (female) boxes. Bottom left and right boxes show the abdomen of a calling male, showing the expanded rectal epithelium (pae, pleural abdominal expansion), and the ovipositor of a female. Abbreviations: Male reproductive system: tes (testes), ves (vas deferens), acg (accessory glands), ejd (ejaculatory duct), eja (ejaculatory apodeme), aed (aedeagus). Female reproductive system: ova (ovaries), ovr (ovariole), ovd (oviduct), spm (spermathecae), acg (accessory glands), cgd (colleterial gland duct), ovp (ovipositor). Rectum: rev (rectal valve), rep (rectal papilla), res (reservoir containing pheromone), ant (anal tube).

3.2. Composition of Volatile Pheromones

In tephritids, the volatile pheromone is a mixture of diverse chemical compounds with different isomers [166]. These chemicals are either newly synthesised or produced from precursors acquired from the diet. Plant-borne chemicals introduced with the diet become integrated into the pheromone mixture and contribute to male mating success. Diet-derived chemicals can unmodified be incorporated into the pheromone blend (e.g., raspberry ketone (RK) in *Z. cucurbitae* [167,168]) or can go through conversion in other compounds that are then used in the pheromone (e.g., ME in *B. dorsalis* that is converted in 2-allyl-4,5-dimethoxyphenol and *trans*-coniferyl alcohol [67,68,169,170]). *Bactrocera dorsalis* males are attracted to ME, and feeding on this compound was shown to directly benefit mating success [170,171].

The pheromone mixture comprises both major, minor and trace compounds, with the complete blend displaying stronger effects than individual compounds or a mix of a subset of compounds. Indeed, medfly females respond differentially to mixtures of major male pheromone compounds than to the complete blend [172]. In *B. oleae*, olean (1,7-dioxaspiro[5.5]undecane) is the major component of female sex pheromone, and it is more attractive to males than the other identified components (e.g., α-pinene, nonanal, and ethyl dodecanoate) [173]. However, the combination of all chemicals is more attractive than olean alone (see [170] for a review).

The composition of the pheromone mixture has been investigated using two main approaches: (i) sampling the headspace of calling males (or females in the case of Bactrocera species), (ii) extracting the chemicals from the rectal glands. Volatiles captured in the headspace have been so far collected and characterised in 18 tephritid species (Table 1; Table S1), while gland extracts have been obtained from 26 species (Table 2).

Table 1. List of tephritid species for which volatile pheromone has been chemically analysed from the headspace.

Genus	Species	Male/Female-Borne	References
Anastrepha	*A. fraterculus*	Male	[174–178]
	A. ludens	Male	[86,179–185]
	A. obliqua	Male	[186–191]
	A. serpentina	Male	[192]
	A. suspensa	Male	[148,179,184,193–202]
Bactrocera	*B. carambolae*	Male	[71]
	B. dorsalis s.s.	Male/Female	[67,203–205]
	B. musae	Male/Female	[75]
	B. oleae	Male/Female	[79,173,206,207]
	B. tryoni	Male/Female	[208–211]
	B. zonata	Male/Female	[212]
Zeugodacus	*Z. cucurbitae*	Male/Female	[203,204]
Ceratitis	*C. anonae*	Male	[213]
	C. capitata	Male	[165,172,214–224]
	C. fasciventris	Male	[213]
	C. rosa	Male	[213]
Rhagoletis	*R. batava*	Male	[60]
	R. cerasi	Male	[225]

Table 2. List of tephritid species for which pheromones have been derived from rectal glands extracts.

Genus	Species	Sex	References
Anastrepha	*A. fraterculus*	Male *	[163]
	A. ludens	Male *	[181]
	A. ludens	Male *	[226]
	A. ludens	Male *	[182]
	A. ludens	Male *	[184]
	A. ludens	Male *	[183]
	A. suspensa	Male *	[202]
Bactrocera	*B. albistrigatus*	Male	[227]
	B. cacuminatus	Male	[228]
	B. carambolae	Male	[71]
	B. cucumis	Male	[229]
	B. correcta	Male/Female	[230]

Table 2. *Cont.*

Genus	Species	Sex	References
	B. correcta	Male	[231]
	B. distincta	Male	[228]
	B. dorsalis	Male	[67]
	B. dorsalis	Male	[227]
	B. facialis	Male	[232]
	B. halfordiae	Male	[229]
	B. kirki	Male	[232]
	B. kraussi	Male	[232]
	B. latifrons	Male	[229]
	B. musae	Male/Female	[75]
	B. neohumeralis	Male	[208]
	B. nigrotibialis	Male	[227]
	B. occipitalis	Male	[229]
	B. oleae	Male/Female	[78]
	B. oleae	Male/Female	[84]
	B. oleae	Female	[228]
	B. oleae	Male/Female	[84]
	B. oleae	Male	[233]
	B. oleae	Female	[173]
	B. passiflorae	Male	[232]
	B. tryoni	Male	[208]
	B. tryoni	Female *	[209]
	B. tryoni	Female	[210]
	B. tryoni	Male	[211]
	B. tryoni	Male/Female	[234]
	B. umbrosa	Male	[235]
	B. xanthodes	Male	[232]
Zeugodacus	*Z. cucurbitae*	Male	[236]
	Z. cucurbitae	Male	[237]
	Z. cucurbitae	Male	[235]
	Z. tau	Male	[235]
Ceratitis	-	-	-
Rhagoletis	-	-	-

* abdominal extracts.

The two most represented chemical classes of the volatile compounds captured in the headspace in *Anastrepha*, *Bactrocera*, *Ceratitis* and *Rhagoletis* species are fatty acyls and organooxygen compounds. Prenol lipids, which represent most compounds in *Ceratitis* and *Anastrepha* species, and are abundant also in *Rhagoletis*, are instead poorly represented in *Bactrocera*. Lactones are particularly abundant in *Anastrepha* species (Table S1).

Within each genus, it is evident that only few compounds are shared among species (Figure 4). For example, in all four *Ceratitis* species investigated so far, namely *C. capitata*, *Ceratitis fasciventris* (Bezzi), *Ceratitis anonae* (Graham), and *C. rosa*, only three volatile chemicals, all belonging to the prenol lipid class, are emitted by males. These three compounds are linalool, (*E*)-β-ocimene, and (*Z*)-β-ocimene that also occur naturally in the host plants. In *C. capitata*, the perception of plant volatiles has been investigated by electroantennogram (EAG) and behavioural studies demonstrating that linalool, a compound representative of immature citrus fruit associated with high toxicity against immature stages of fruit flies and considered as an important compound conferring resistance against fruit fly larval development, has a significant deterrent effect [238]. Linalool was reported by numerous studies as an active constituent of medfly male sex pheromone that elicits a strong EAG response [217,220,239]. It also triggers antennal depolarisation in females of *C. fasciventris*, *C. anonae* and *C. rosa* (so-called *Ceratitis* FAR complex) [213].

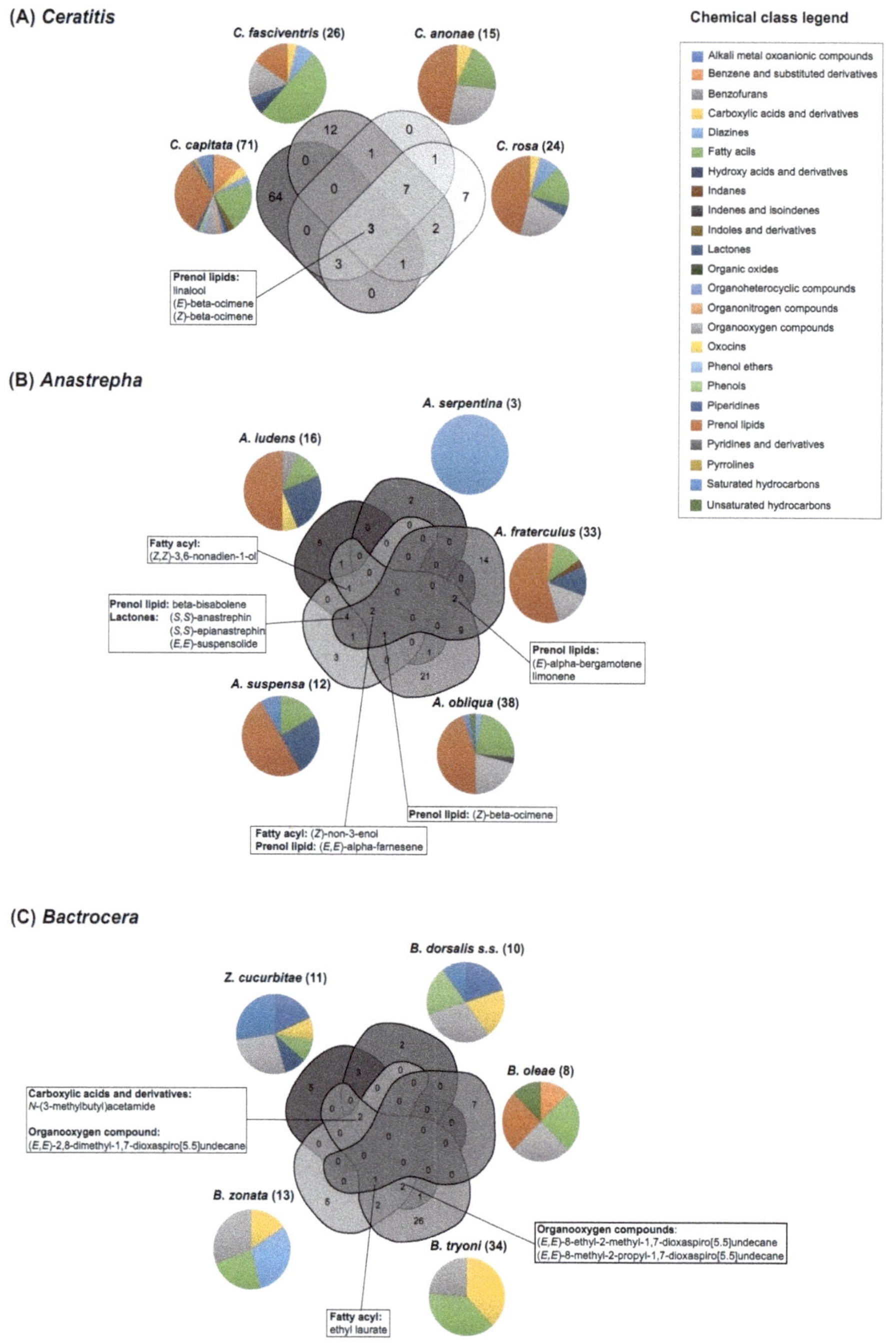

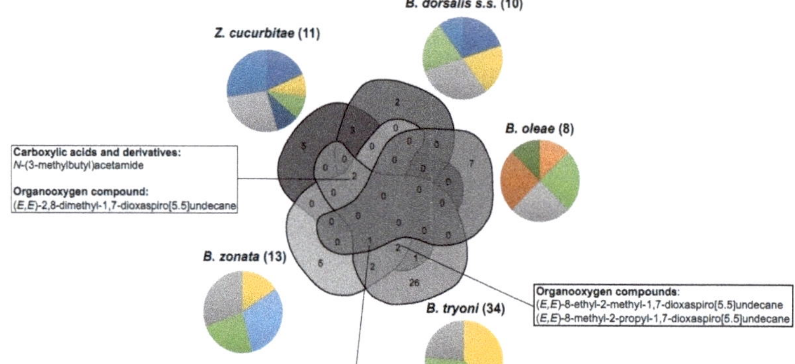

Figure 4. Overlap among chemicals identified in the headspace emissions of tephritids. Venn diagrams (http://bioinformatics. psb.ugent.be/webtools/Venn/) (accessed on 10 March 2021) illustrating the number of common and unique chemicals found among the compared species in the (**A**) *Ceratitis*, (**B**) *Anastrepha*, and (**C**) *Bactrocera* genera. Pie charts represent the proportion in chemical classes of the total number of compounds (in parenthesis) isolated for each species.

In *Anastrepha*, (Z)-non-3-en-1-ol (a member of the fatty acyl compound class) is the unique compound shared among *A. fraterculus*, *A. ludens*, *A. obliqua*, and *A. suspensa*. This compound is a typical host plant component and has been shown to elicit an active behavioural response in all four species [174,187,192,240,241]. For example, *A. fraterculus* females are attracted to (Z)-non-3-en-1-ol [174]. These behaviours may be interpreted as a first step in the complex mating process of this species, i.e., attracting females to the mating site. Since mating is strongly associated with host plants, the use of plant typical compounds (e.g., limonene and pinene, among others) would help females to find simultaneously mating and oviposition sites [242].

N-(3-methylbutyl)acetamide (carboxylic acids and derivatives class) and (E,E)-2,8-dimethyl-1,7-dioxaspiro[5.5]undecane (organooxygen compounds class) are shared among *Z. cucurbitae*, *B. dorsalis s.s.*, *B. tryoni*, and *B. zonata* (Saunders). N-(3-methylbutyl)acetamide elicits female attraction in *Z. cucurbitae*, *B. dorsalis* and *B. carambolae* [71,203,227]. The spiroacetal (E,E)-2,8-dimethyl-1,7-dioxaspiro[5.5]undecane elicits an antennal response in *B. musae* males, suggesting a biological role for this compound [75].

It is noteworthy that research efforts in investigating pheromone composition have been particularly intense in certain species such as the medfly, and this may reflect in the number of identified compounds. Future studies integrating multiple analytical approaches in all species will enable a more extensive description of the shared and unique chemical signatures of tephritid volatile pheromones.

The differences in the male pheromone composition that have been reported may be partially linked to technical aspects (e.g., different sampling techniques, experimental variables such as airflow, sampling duration, and column features), as well as to several factors affecting pheromone production and release, such as the time of the day [148,199], social context [148], food availability [198], and diet [219,243]. Larval diet (i.e., the different host fruits utilised) and adult age affect both the total amount of pheromone emitted by males, the relative quantity of major components, and the presence of specific minor compounds in the medfly [219]. Similarly, in *A. ludens*, the profile of the male pheromone blend and the relative abundance of the chemical components change in response to different host fruits used for larval development [243]. Moreover, qualitative and quantitative differences in the composition of male-emitted volatile pheromone have been detected among laboratory strains and different populations in *A. fraterculus* [175,176], and among laboratory strains and wild populations in *C. capitata* [220]. In addition, emission of male pheromone increases in *Anastrepha* species in response to treatments with methoprene (a synthetic analogue of the juvenile hormone), with important implications in gaining deeper insights into tephritid reproductive physiology and in enhancing the application of management methods such as the release of sterilised males [244,245].

The components of male pheromone also display different volatilities [246]. Six alkanes and related compounds have been identified in the headspace of medfly males reared on a standard wheat bran-based larval diet [219]. These compounds display poor volatility and do not belong to the published CHs identified in the cuticle of adult medflies. Thus, similar to *Drosophila* species [247,248], medfly seems to be able to deposit on the substrate and emit in the surrounding air CH-like compounds, which may serve as short/medium distance cues for mate localisation. The ability of males to deposit pheromone on the substrate, in addition to aerial pheromone release, has also been reported in *A. suspensa* [62,99]. Moreover, saturated C25, C27 and C29 hydrocarbons have been detected in the volatile pheromone of male melon fly *Z. cucurbitae* [204].

3.3. Does Host-Preference Affect the Volatile Pheromone Bouquet?

Pheromone precursors are acquired from four main sources: (i) *de novo* synthesis, (ii) conversion of precursors that insects acquire from host plants or substrate, (iii) direct incorporation from the host plants, and/or (iv) from endosymbionts [21,249–252]. Thus, it is likely that different variables influence the pheromone blend.

The chemical classes of their pheromone components and their specific identity are more similar among species of *Anastrepha*, *Ceratitis* and *Rhagoletis* genera, and rather different from species of the genus *Bactrocera*. The monophagous *B. oleae* is the most diverse species among the genus, with the majority of its pheromone volatiles being species-specific. To elucidate whether the composition of the pheromone components is related to feeding strategies of different species, we analysed the lists of volatile compounds identified in all species of the genera *Anastrepha*, *Ceratitis* and *Bactrocera*. This analysis could not be extended to *Rhagoletis* genus, as only the volatiles emitted by *R. cerasi* have been identified [225] so far, and a single compound has been isolated in *R. batava* [60].

The species with a monophagous or oligophagous feeding strategy mostly emit chemicals that are not shared with other species with the only exception of nonanal, α-pinene and *p*-cymene, which are in common between *B. oleae* and *R. cerasi*. Figure 5 shows the number of chemicals identified in each species as well as the intersections of overlapping compounds, represented by connected dots (see Table S2 for the complete lists of unique and overlapping chemicals).

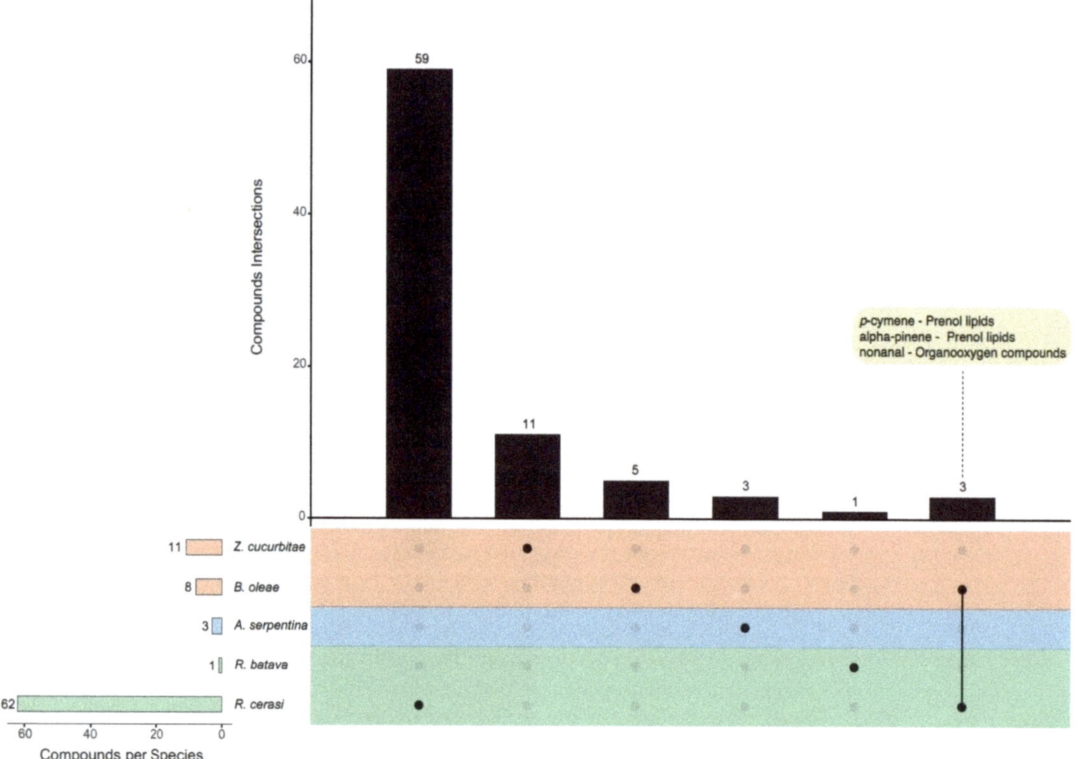

Figure 5. Comparison among the chemicals identified in the headspace emissions of monophagous and oligophagous tephritid species. UpSet plot showing unique and overlapping chemicals across five species belonging to the *Zeugodacus*, *Bactrocera*, *Anastrepha* and *Rhagoletis* genera. The intersection matrix is sorted in descending order. Connected dots represent intersections of overlapping chemicals and horizontal bars show the total number of compounds identified in each species headspace. The plot was generated using the UpSetR package in R [253].

The polyphagous species shared more compounds, as shown by the higher number of connected dots in Figure 6 (see Table S2 for the complete lists of unique and overlapping chemicals). This finding is particularly evident between species belonging to the same genus.

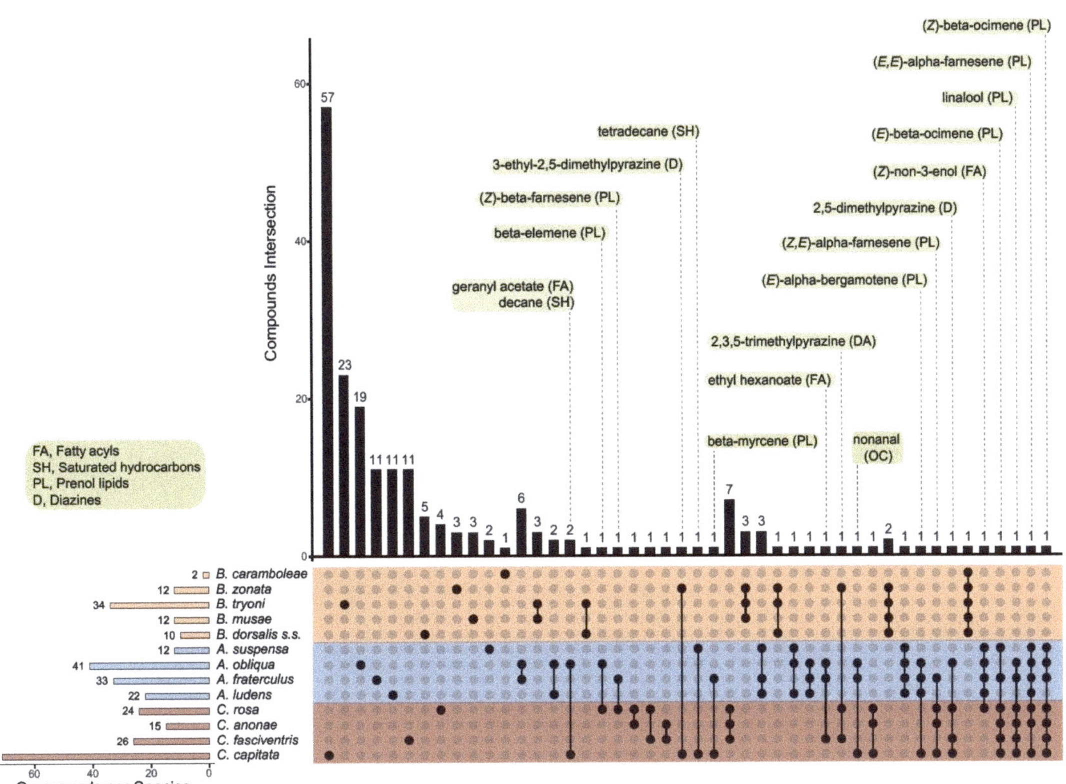

Figure 6. Comparison among the chemicals identified in the headspace emissions of polyphagous tephritid species. UpSet plot showing unique and overlapping chemicals across five species belonging to the *Bactrocera*, *Anastrepha* and *Ceratitis* genera. The intersection matrix is sorted in descending order. Different genera are indicated in different colors. Connected dots represent intersections of overlapping chemicals and horizontal bars show the total number of compounds identified in each species headspace. The plot was generated using the UpSetR package in R [253].

Figure 6 also reports the identity of the chemicals that are shared between different genera. It generally appears that the members of the *Anastrepha* and *Ceratitis* genera share a higher number of compounds, with chemicals belonging to the prenol lipids class being most frequently in common. Conversely, *Bactrocera* species display quite unique pheromone blend features. The only two compounds to be shared belong to the diazine class and are in common with *Ceratitis* species. This suggests that at least two major evolutionary forces, i.e., genomic background and host feeding strategy, interacting in a complex manner, have shaped the pheromone blend in these species. A comprehensive molecular phylogeny of tephritid species will be of great help in disentangling the effects of genome evolution and environmental selective pressures in shaping semiochemical-based behaviour in these species. Further research is required to deeply investigate how feeding strategies affect the production of semiochemicals, also with respect to the capacity of colonizing new hosts, a feature that characterises several invasive tephritid pests. A strong effect of larval host on the pheromone bouquet produced by the polyphagous species *A. ludens* and *A. obliqua* has been recently demonstrated [243].

3.4. Analytical Approaches to Unravel Pheromone Composition

The complexity of the signals involved in tephritid chemical communication, together with their presence in small amounts in natural systems, stimulated the adoption of very sensitive high-resolution analytical methods for their detection, identification and testing. These strategies include both conventional as well as reverse chemical ecology approaches. The workflow of conventional chemical ecology (CCE) approaches generally involves the following steps: (i) preliminary observation of the existence of pheromone-mediated communication in the target species, (ii) volatile pheromone sampling, (iii) characterisation of pheromone blends through analytical techniques, (iv) chemical synthesis of the identified compounds, (v) evaluation of their electrophysiological activity, (vi) behavioural assays (bioassays) to test the role of the isolated semiochemicals in laboratory and/or field set-ups in order to confirm their pheromone identity [254]. Recently, a different approach began to be utilised for the discovery of semiochemicals, reverse chemical ecology (RCE) [255]. Similarly to what occurs for receptor-based drug discovery, RCE exploits odorant binding proteins (OBPs) as molecular targets for the screening of behaviourally-active chemicals based on their binding affinity [256]. The workflow of RCE comprises the following steps: (i) identification of OBP targets through genomics and bioinformatics approaches, (ii) characterisation of OBP expression profile, (iii) purification and 3D-structural analysis, (iv) in vitro OBP: ligand binding assays using semiochemicals collected from insects or pure compounds, (v) in silico selection of test ligands, (vi) bioassays to verify the effects of the identified semiochemicals on insects *in vivo*. The key aspects of these approaches are described in the following sections.

3.4.1. Conventional Chemical Ecology Approaches

The components of the complex blends of tephritid pheromones can be isolated, identified and tested using interdisciplinary approaches involving bioassays, sensory physiology, analytical and organic chemistry, and biochemistry. Recent breakthroughs in analytical techniques allow the rapid screening of semiochemicals with more sensitive bioassays and their isolation and identification from relatively smaller amounts of material. Several non-destructive and artefact-free methods are available for collecting natural odorants from living organisms. Examples include the adsorption of odours on different polymer matrices contained in cartridges or filters. Trapped odours can be desorbed thermally or eluted with organic solvents followed by analysis using gas chromatography. The direct coupling of a chemical (flame ionization detector and/or mass spectrometer) and biological detector (e.g., the use of an insect antenna) permits simultaneous isolation and identification of bioactive components from trapped odours. Furthermore, the discovery of more efficient chemical synthetic methods now allows the state-of-art synthesis of semiochemicals of high purity whose field activity may provide answers to ecological and evolutionary questions associated with the importance of the chemical in the behaviour of the target insect [257].

Volatile Collection and Analytical Techniques for Their Identification

Chemical identification of tephritid volatiles requires a chromatographic separation followed by detection using spectrometric analytical methods. Although a wide range of methods is available in principle, the number which is suitable in practice depends upon the amount of insect material that can be obtained.

Solvent extraction of tephritid rectal glands using standard solvents such as heptane, hexane, methanol, ethanol, dichloromethane and ethyl ether is commonly used [165,212,230,231]. More recently, mixtures of solvents (acetonitrile/water and methanol/acetonitrile/water) were applied for the extraction of semiochemicals from the whole body of adult medflies [258].

Solid phase microextraction (SPME) is a solvent free, pre-concentration technique developed by Arthur and Pawliszyn [259] for application in solid, liquid, or gaseous samples (reviewed in [260]). The results obtained using SPME fibres are similar to those obtained with solvent extraction [261–263]. SPME is a known and effective alternative to liquid-liquid extraction. It provides some advantages over liquid-liquid extraction process because of reduced solvent consumption. In tephritids, different SPME coating materials have been used for volatile collections. A polydimethylsiloxane/divinylbenzene fibre (PDMS/DVB) was applied for pre-concentration of volatile compounds emitted by male and female medflies in different mating status (virgin or mated), and age (3 or 9 days old) [221]. In total, 70 compounds of diverse chemical classes such as alcohols, acids, aldehydes, terpenes, branched hydrocarbons and esters were reported and identified by gas chromatography coupled with mass spectrometric detection (GC-MS) [221]. Similarly, PDMS fibres have been used to compare the composition of the pheromone of medfly males from a standard laboratory strain reared as larvae on laboratory media and fresh fruits [219]. Five and 30 day-old males have been used, with 36 and 27 chemicals (mostly belonging to terpenes, amides, esters and alkanes) identified to be emitted by these two age-classes, respectively [219]. In a recent work on *C. capitata* volatiles, divinyl-benzene/carboxen/polydimethylsiloxane (DVB/CAR/PDMS) combined with GC-MS and gas chromatography-flame ionization detection (GC-FID) techniques resulted in the identification of 27, 23 and 29 compounds from larvae, pupae and adults, respectively [264]. PDMS/DVB SPME fibres have been applied to collect volatiles produced by *B. zonata* males and females and *R. batava* males, respectively [60,212].

Dynamic headspace is a common method for the collection of volatiles produced by fruit flies. The volatiles emitted by virgin-calling flies are collected using a modified technique of an air-collection apparatus as described by Nation [148]. In this technique, a purified airstream is blown over living flies enclosed in a glass chamber. The volatiles are collected onto traps containing adsorbents such as SuperQ, Tenax, Activated Carbon, etc. [212,220,222]. The air flow directed through the apparatus is controlled by flowmeter. Volatile collections are usually performed for 24 h. Afterwards, the traps are washed with heptane, hexane, diethyl ether or ethanol and the obtained extract is analysed by GC-MS, electroantennography, gas chromatography coupled to electroantennographic detection (GC-EAD) methods and used for bioassays.

GC-MS is one of the most useful tools for chemical analysis of volatile semiochemicals. The gas chromatography provides high-resolution separation of components within a complex mixture, and the mass spectrometric detection supplies structural information in addition to its role as a sensitive detector (the current limit of detection is at femtomole levels). By selection of the appropriate capillary column, practically all volatile organic compounds can be separated, including carboxylic acids, ketones, aldehydes, alcohols, aromatic compounds, and hydrocarbons [179,214,225,265–268].

For identification of the absolute configuration of tephritid volatile semiochemicals, the chiral column or chiral GC-MS can be used, as recently applied for pheromone identification of *R. batava* [60]. Most commercial chiral GC phases currently available are composed of modified cyclodextrins, which give a wide range of enantiomeric separations, but have the disadvantage of being thermally unstable above 200 °C, and are therefore useful for relatively volatile compounds only. Like chiral nuclear magnetic resonance (NMR) studies, chiral GC requires homochiral or enantioenriched synthetic standards, but unlike the NMR technique, only nanograms of impure material are necessary [269]. Two-dimensional gas chromatography with time-of-flight mass spectrometric detection (GC × GC-TOFMS) is a recently developed analytical technique that offers a solution to the chromatographic co-elution and provides high sensitivity and selectivity. In principle, the method consists of two GC systems (GC × GC) equipped with columns of different polarity connected by an interface with an integrated cryogenic trap. The cryogenic trap repeatedly condenses compounds eluting from the primary column and releases them periodically as short pulses to the secondary column. Parameters like duration and frequency of both condensation and injection pulses are variable and allow precise tuning of the instrument according to the requirements of the analysis. Since the GC × GC produces very narrow peaks (down to 50 ms, depending on the frequency of the cryogenic modulation) a time-of-flight mass spectrometric detector (TOFMS) with a high acquisition rate (up to 500 spectra per second) is required. The pulsed nature of the TOFMS source of ionisation further enhances the system accuracy by avoiding the spectral skewing common in a continuous ionisation mode. GC × GC with TOFMS detection thus operates with a high precision independent of the concentration range [270,271]. This method has been applied for analyses of fruit fly semiochemicals produced by species of the genera *Anastrepha*, *Bactrocera* and *Ceratitis* [176,213,220,272].

Gas chromatography coupled with Fourier transformed infrared spectroscopy (GC-FTIR) is relatively sensitive (detection threshold 10–100 mg) and particularly useful for identifying geometrical and positional isomers or functional groups. Thus GC-FTIR is an ideal instrument for the study of volatile organic compounds. Medfly synthetic attractants composed of *trans*-trimedlure isomers and *cis*-trimedlure isomers were analysed by GC-FTIR spectroscopy [273].

Matrix-assisted laser desorption/ionization with time-of-flight mass spectrometry (MALDI-TOFMS) and desorption electrospray ionization (DESI) mass spectrometry are soft ionization techniques developed for the analysis of biomolecules (biopolymers such as proteins, peptides and sugars) and large organic molecules (such as polymers, dendrimers and other macromolecules). However, they have been applied very sporadically for studies of the fruit fly pheromone components [274].

High performance liquid chromatography (HPLC) coupled to a MS detector is at present the favoured technique for the study of polar non-volatile tephritid semiochemicals. Other types of detectors that are commonly used with HPLC separation are UV and ELSD (evaporative light scattering) detectors [275–278]. Normal phase and reversed phase HPLC separation are still popular in lipid analysis. The mobile phases usually consist of methanol, propan-2-ol, or *n*-hexane. Chromatographic columns with C18-phases still prevail. A few separations were reported on C8-columns. In a recent metabolomic study of *B. dorsalis* larvae, liquid chromatography-mass spectrometry (LC-MS) and GC-MS were applied for the characterisation of endogenous metabolite changes and biochemical effects of azadirachtin [279].

NMR spectroscopy is one of the most informative, but least sensitive, modern spectrometric methods. Fourier transform ^{1}H and ^{13}C NMR spectroscopy require approximately 100 µg and 10 mg pure material, respectively (50 ng – 1 µg with very expensive NMR-nanoprobe). Since many insect pheromones are only present in nanogram amounts, a very high number of individuals are required. NMR is widely used to determine the absolute configuration of optically active compounds (semiochemicals), either using chiral shift reagents, or by converting enantiomers to diastereoisomers prior to spectroscopic study, or using chiral solvent. The use of NMR for absolute identification of semiochemicals has two main disadvantages: isolation of the pure substance is required, and an enantioenriched or a homochiral synthetic standard is needed. In modern semiochemical research, NMR spectroscopy is normally used only when insufficient structural information is provided by more sensitive methods [280]. Baker and Heath [281] applied NMR spectroscopy for the identification of lactone pheromone components emitted by *A. suspensa* and *A. ludens*.

Chemical Synthesis

The chemical synthesis of tephritid pheromones is a key step towards producing pheromone compounds for use in bioassays (usually very low amounts), and for insect pest monitoring and management purposes. Moreover, their absolute configuration is an important determinant of their biological activities [282]. In 1973, Jacobson and colleagues isolated and synthetized the first medfly pheromone components, i.e., methyl (*E*)-non-6-enoate and (*E*)-non-6-en-1-ol that were attractive eliciting also sexually excitatory response to females in laboratory trials [214]. Three additional major components of the medfly pheromone, namely ethyl (*E*)-oct-3-enoate, geranyl acetate, and (*E*,*E*)-α-farnesene were isolated and synthetized later [224]. In 2010, Olszewski and Grison [283] reported a novel and versatile synthetic approach to the preparation of (*E*)-non-6-en-1-ol.

Since the first successful studies on the isolation and synthesis of the medfly pheromone components, several efforts have been made to characterise the volatiles emitted by *Bactrocera* and *Rhagoletis* species, mainly using the enantioselective approach. Moreover, in the case of chiral pheromones, which occur in tephritids, the synthesis and following bioassays using stereoisomers are essential for recognition of the stereochemistry-bioactivity relationships. This knowledge also has practical implications for pest control as in most chiral pheromones only one enantiomer is biologically active [282]. In the case of *B. oleae*, olean (1,7-dioxaspiro[5.5]undecane) was isolated and identified as the major component of the female-produced sex pheromone in 1980 [78]. (4*R**,6*R**)-4-hydroxyoleane and (3*R**,6*R**)-3-hydroxyoleane (the asterisks indicate the chiral centers) were then isolated as minor components [78]. Bioassays involving the synthetic enantiomers of oleane showed that (*R*)-oleane is active on males, while (*S*)-oleane on females, and GC analyses showed that females can produce (±)-oleane [206]. The synthesis of enantiomers of 4-hydroxyoleanes was conducted by Mori and co-workers [284] and a recent study identified and synthesised 11 new *B. oleae* female-specific components [285].

Knowledge of pheromone composition in *Rhagoletis* is still limited to *R. cerasi* and *R. batava*. In this last species, Buda and colleagues recently applied GC-MS analyses of fly headspace and found that males emit (-)-δ-heptalactone. The authors then synthesised the two enantiomers of (-)-δ-heptalactone using enantioselective synthesis and found that only (-)-δ-heptalactone elicited an electrophysiological response in both sexes, proposing that this chemical may act as aggregation pheromone [60].

Despite these research efforts, similar studies focusing on the optimisation of chemical synthesis of pheromone components are completely missing for the genus *Anastrepha*, as well as for many other species belonging to the *Bactrocera*, *Ceratitis* and *Rhagoletis* genera.

Identification of Electrophysiologically-Active Volatiles

GC-EAD uses the electrophysiological response from a dissected insect antenna to assign activity to a gas chromatographic peak. The GC eluent is normally split between a flame ionization detector (FID) and the insect antenna, which is connected to an amplifier and recorded by silver or platinum electrodes. The response from each detector is recorded simultaneously so that an EAD response can be correlated with a specific peak in the FID chromatogram. This technique is particularly useful for the assignment of activity to chemicals which may be present as minor components in complex mixtures, although it provides no information about the function of the active compounds. GC-EAD analyses of male pheromone or rectal glands extracts have been performed on several tephritid species, including *C. fasciventris*, *C. anonae* and *C. rosa* [213], *C. capitata* [220,239], *B. oleae* [286], *A. serpentina* [192], *A. fraterculus* [176,287–289], *A. obliqua* [241] and *A. striata* [290]. Additional studies performed GC-EAD analyses on the headspace volatiles of both sexes, in the species *R. batava* [60], *Bactrocera frauenfeldi* (Schiner) [291], *B. musae* [75], and *C. capitata* [222]. In addition to GC-EAD, gas chromatography-electropalpogram detection (GC-EPD) has begun to be used to test the responses of tephritid maxillary palps to the pheromone emissions [291].

Furthermore, electroantennography-based experiments can be performed independently from gas chromatography. In this case, pure compounds are employed to assess the response to a certain stimulus and the required dose to elicit a detectable response. In this case, chemicals are delivered to insect antennae/palps in controlled conditions by pumping the odour from a reservoir [254]. Since early works in the '80s [242,292,293], numerous EAG-based studies have been performed in tephritids to test the electrophysiological response to volatiles of fruits and flowers [294–297], artificial attractants [298–302], as well as pheromone components [303,304]. The use of GC-EAD brought the EAG-based approaches to a higher level of sophistication by using the antenna/palp of the target insect as a detector for the gas chromatograph. However, while electroantennography is an excellent technique to quickly measures the change in the electrical potential between distal and proximal sections of the antenna/palp provoked by olfactory stimulation, EAG amplitude depends on the position of the electrode, the strength of the connection, and insect vitality [303,304]. Thus, EAG is considered a qualitative indicator of olfactory reception [305]. Also in tephritids, EAG signals are indeed known to vary in relation to the relative density of sensilla, which display a specific distribution on the funiculus, at the electrode location [306,307]. To increase the recording specificity, a novel method was developed integrating EAG recordings at multiple antennal positions with current source density (CSD) modelling [308,309] useful to map the functional activation of individual antennae. This method was applied to six tephritid species, i.e., *C. capitata*, *C. catoirii* (Bezzi), *Neoceratitis cyanescens* (Bezzi), *B. zonata*, *Z. cucurbitae*, and *D. demmerezi* to measure the response to seven volatile chemicals emitted by fruits and plants at four position in the funiculus.

A more quantitative measurement of the olfactory response, although requiring significant experimental effort on multiple individuals, can be achieved using single sensillum recording (SSR). SSR allows to measure the action potentials generated by olfactory sensory neurons (OSNs) within individual sensilla on antennae or palps using an electrode in contact with the lymph of the extracellular receptor [305]. To the best of our knowledge, this approach has been used only in two tephritids so far, but involving, in both cases, fruit odours as stimuli. The sensilla located on the tarsi of *C. capitata* [310], and the antennae of *R. pomonella* [311] were investigated.

Behavioural Assays to Identify Active Compounds

Bioassays play key roles in the study of tephritid semiochemicals [269]. Bioassays allow to assess the behavioural effect(s) (e.g., lek formation, male signalling, courtship, copulation, host finding for oviposition, HMP deposition) of synthesised active compounds (pure or in mixture) with respect to host or mate interactions, differently from EAG assays that provide the electrophysiological responses of isolated organs, structures or olfactory receptors (ORs). The biological response measured by a bioassay is essential to attribute to a molecule or a mixture of chemicals the pheromone identity. Behavioural effects in tephritid species can be evaluated in laboratory, semi-field and field conditions where both the stimuli and the background in which the stimuli are presented are tightly controlled [312]. Techniques for insect bioassays have been widely reviewed (see for example [254,313–315]) and widely applied to tephritid species, for which most experiments have been so far performed in laboratory conditions. Initial studies were done in close-range cage bioassays estimating medfly female attraction to male pheromone components based on landings on filter paper soaked with olfactory stimuli [217]. Currently used setups adopted for the study of tephritid olfaction include (i) two-choice systems, (ii) flight tunnels, and (iii) multi-arm olfactometers. Two-choice systems, using arenas (i.e., observation chambers) [173,316,317], as well as Y-tubes/T-maze [84,230,291,318], have been used to evaluate attractive or repulsive responses following exposure to pheromone (isolated from headspace or rectal glands); in this framework, dynamic systems like the Y-tubes, allowing an air flow carrying the chemicals to be evaluated for the chemo-ecological role should be preferred, since they avoid the risk of saturating receptors of the tested flies, which is common in still-air arenas. Flight tunnels, also known as wind tunnels, are extremely useful to monitor medium-distance flight responses to mate-derived chemicals [200,268,319]. Their use, for example, allowed to assess the attraction of females to male emissions in *C. capitata* [268] and *A. serpentina* [192], or of both sexes to ME-fed *B. papayae* males [69]. Wind tunnels also permitted to prove that pheromone components can exert behavioural effects in *A. ludens* females [180]. These systems can integrate different components for the simultaneous identification of released volatile chemicals and the assessment of their attractiveness, as well as recording environmental parameters and fly activity [200], including the observation of flight patterns in females responding to male-derived volatiles [319]. Lastly, multi-arm olfactometers have been mainly used so far to test fruit odours and with flies released into an area composed of multiple chambers from which airflow-containing odour flows [320].

Moving laboratory results to the field for real-world applications is a timely challenge in tephritid research. Therefore, field assays to evaluate tephritid attraction should be considered after successful laboratory evaluation of a given compound. Field studies can be performed either using live insects/glands extracts as a source of pheromone or traps with different types of dispensers releasing pheromone pure chemicals or mixtures, as described especially for the medfly and the olive fruit fly. In the case of *C. capitata*, a synthetic blend releasing three male pheromone compounds (i.e., ethyl (*E*)-oct-3-enoate, geranyl acetate and (*E,E*)-α-farnesene), in a ratio similar to that observed in natural conditions, was effective in attracting females, as shown by trap catches [224]. Another study found that trimedlure was more effective than pheromone individual components or mixtures in trapping flies [321]. For *B. oleae*, field experiments aimed at determining the attractive effect of the four major pheromone components stressed the importance of finding the ideal combination between attractant formulation and trap type/colour [322]. The use of polyethylene vials as dispensers of either the complete pheromone blend, racemic mixtures of the major components, or individual synthetic chemicals resulted effective in trapping *B. oleae* flies [317,323]. Open field tests were performed either with wild medfly females [224] or released *C. capitata* and *A. suspensa* females [324–327]. Tests in field cages with potted trees have also been performed, providing valuable information about behavioural responses to live conspecifics or male extracts in seminatural conditions for *A. obliqua*, *A. ludens* and *A. suspensa* [187,242,328].

3.4.2. Reverse Chemical Ecology Approaches

In addition to the above-described techniques to identify tephritid semiochemicals, recent studies are increasingly showing that RCE has great importance in understanding the molecular basis of insect chemical perception and identifying the active volatile semio-chemicals [255]. Studies began to be performed to identify genes involved in chemosensory perception in tephritids, including olfactory, ionotrophic and gustatory receptors (ORs, IRs and GRs, respectively), OBPs, odorant degrading enzymes (ODEs), and chemosensory proteins (CSPs) [329]. Insect OBPs are small soluble proteins mostly found in the chemosen-sillar lymph of sensory organs where they bind molecules of odorants and pheromones (see [330] for a review). Thus, OBPs are considered ideal molecular targets for binding assays to identify chemicals with a potential behaviourally-active role in tephritid biology. The expression of OBPs, followed by their purification and structural analysis, is indeed adopted to perform OBP ligand binding studies; ligands are screened from sets of volatiles emitted from host plants, pheromones, or synthetic attractants used in field applications, and once identified, behavioural responses are evaluated *in vivo* [254].

Identification and Functional Analysis of OBP Genes

Putative OBP genes have been initially identified through expressed sequence tag (EST) approaches in *C. capitata* [331], *B. dorsalis* [332,333], *R. suavis* [334], and *R. pomonella* [335] and their transcriptional profile started to be explored. Subsequently, with the advent of next generation sequencing, more OBPs have been discovered through mining of RNA-seq data and whole genome sequencing in several tephritid species, including *B. dorsalis*, *B. minax* [103,336–339], *A. fraterculus*, *A. obliqua* [105,340,341], and *C. capitata* [342]. Antennal proteomics profiling has been applied to *B. dorsalis* to identify differentially expressed genes, including OBP genes, in ME-responsive males [343]. Functional studies have been performed to assess their role in odour perception. These include tracing OBP genes expres-sion profiles in different tissues, developmental stages and in response to maturation and mating, RNA interference (RNAi) combined to electrophysiology to assess the involvement of target OBPs in odour detection, followed by behavioural assays [333,339,343–345].

Purification and 3D-Structural Analysis of Identified/Expressed OBPs

The binding specificity of OBPs expressed in the main chemosensory organs, i.e., antennae and maxillary palps, may help in the identification of pheromone/pheromone components that are still uncharacterised. Thus, OBPs that are (i) abundant in olfactory tissues, or, ideally, specific to these tissues, and (ii) showing sequence similarities to already characterised proteins known to be involved in chemical communication in other insects can be functionally characterised by using ligand-binding assays. OBPs are expressed in bacterial or yeast systems and the recombinant proteins purified with chromatographic steps using different techniques, including anion-exchange or gel filtration chromatography, or affinity chromatography on nickel columns when histidine-tags are added to the OBP sequence (see [346] for a review). Purified proteins can then be used for ligand-binding experiments and to solve their structure through X-ray crystallography or NMR spectroscopy [346].

The first, and, so far, the only available, structural characterisation of a tephritid OBP was recently obtained in the medfly using X-ray crystallography [347]. The structure of CcapOBP22 is characterised by six α-helical elements, a typical feature of insects' OBPs, interconnected by three disulphide bridges. Differently from other insect OBP structures, CcapOBP22 also carries a 7th α-helix at the C-terminus, which contributes to delimit the ligand-binding pocket. CcapOBP22 was co-crystallised with (E,E)-α-farnesene as ligand, further supporting the potential role of this protein in semiochemical perception in this species.

In vitro and *in silico* OBP: Ligand Binding Assays

Several approaches have been adopted to measure the affinity of OBPs to odorants (see [348] for a review). The most common method, which is fast and requires a limited amount of protein, is based on the use of fluorescent reporters, such as 8-anilinonaphthalene-1-sulfonic acid (ANS) and N-phenylnaphthalen-1-amine (1-NPN) in competitive binding experiments [349–351]. 1-NPN is a lipophilic crystalline solid that strongly binds insect OBPs [352]; when increasing amounts of a tested ligand are added to the OBP/1-NPN system, decreasing 1-NPN fluorescence emission is inferred as 1-NPN displacement since the ligand is assumed to compete for the binding pocket initially occupied by the fluorescent reporter.

This approach has been used in the medfly to evaluate the binding affinity of CcapOBP22 and CcapOBP24 to electrophysiologically-active components of the male pheromone, as well as to the two synthetic attractants trimedlure and ME [222,347]. The finding that also ME, which is a strong attractant for some *Bactrocera* species [169,353,354] but not for medfly, displays binding activity (although moderate) to the above medfly proteins is intriguing. Methyl eugenol is known to induce an electrophysiological response in medfly [300]. In *B. scutellata*, ME elicits significant electrophysiological responses too, but it is not behaviourally active [355]. In medfly, ME has been shown to induce poor behavioural responses in binary choice bioassays, while *o*-eugenol was instead strongly attractive [300]. Thus, it appears that the presence of substituents on the aromatic ring can be essential to confer attraction to chemical compounds. It will be interesting to further explore the chemistry of candidate molecules able to bind tephritid OBPs to shed new light on structures that can be exploited as novel attractants. Both OBPs showed the highest binding affinity to (E,E)-α-farnesene, which is one of the major components of medfly male pheromone bouquet, and is known to attract females [319], suggesting its role as a natural ligand for these OBPs. The verification of the behavioural responses to the presence of ligands *in vivo* is essential to identify volatile semiochemicals with active roles in fruit fly behaviour.

Ligand-binding assays have also been performed in *B. dorsalis* using 13 chemicals, including pheromone components and attractant molecules, and six proteins with high expression in the antennae (five OBPs and one CSP) [316]. Authors showed that OBPs displayed the highest affinity to the attractants, and, in the case of BdorOBP83a-2, RNAi led to a decrease in neuronal responses to tested molecules, as shown by EAG recordings and behavioural responses.

Computational reverse chemical ecology (CRCE) is another method applied to the discovery of behaviourally active chemicals [356,357]. OBP sequences can be exploited for 3D model prediction, producing 3D structure for docking studies using specific tools. Molecular docking is commonly employed in pharmaceutical research for structure-based drug design [358,359]. It implies the use of programs based on different algorithms applied to model the interaction between a small molecule and a protein at the atomic level. This allows the exploration of the behaviour of small molecules in the binding pocket of target proteins [360]. CRCE has been implemented in *B. dorsalis* to screen 25 chemicals for their binding potential to a general OBP (GOBP) showing that this approach may be extremely useful to quickly predict behaviourally-active semiochemicals, for example selecting chemicals belonging to specific classes [356]. The described approach is particularly beneficial especially in tephritids given the wide absence of direct crystallographic data for OBP binding modes.

Identification and Functional Analysis of OR Genes

Although OBPs are excellent study targets to either understanding the molecular and biochemical mechanisms of odour perception in insects, and to explore the development of pest control agents, they have broad binding specificity, are also distributed in non-olfactory tissues and have different functions [361]. Conversely, ORs are transmembrane proteins showing high specificity and sensitivity. Thus, genes encoding for chemosensory receptors are also becoming to be identified and characterised in tephritids, such as *B. dorsalis* [336,362–364], *B. minax* [339], *B. oleae* [365], *B. latifrons* and *Z. cucurbitae* [366], and *C. capitata* [342]. Olfactory receptors have been described as heteromeric ligand-gated ion channels consisting of a specific OR and the highly-conserved co-receptor Orco [367]. Olfactory receptors are transmembrane proteins for which no 3D structure is available yet, and they are a more difficult target than OBPs to be expressed and purified in heterologous systems. So far, the only three-dimensional structure currently available, obtained using a cryo-electron microscopy-based approach, is for a tetramer of Orco, described in the parasitic fig wasp *Apocrypta bakeri* (Joseph) [368]. Thus, only limited data on the functional activity of tephritid ORs are available. In a recent study, ten *B. dorsalis* ORs were co-expressed with their essential co-receptor BdorORCO in *Xenopus laevis* Daudin (Anura: Pipidae) oocytes. Two-electrode voltage clamp was then used to record currents from injected oocytes when ligands (i.e., 1-octen-3-ol, geranyl acetate, farnesenes, and linalyl acetate) were diluted in the assay buffer [363]. Some of the identified ORs have been shown to respond to plant volatiles [363] or ME [369]. Further research efforts oriented to clarify the structure of tephritid ORs are essential to understand the molecular recognition mechanisms they are involved in, as well as their interactions with OBPs, and thus their functional roles. Insect ORs display a different topology from those of other animal G protein-coupled receptors (GPCRs) [370], with a C-terminal faced to the extracellular section and the N-terminal to the intracellular section. This feature makes insect ORs ideal targets to be explored for the development of insect-specific pest control strategies. These may include the inhibition of either Orco or the ORx/Orco complex by antagonists able to, for example, disrupt mating behaviour through the manipulation of pheromone receptivity. Interestingly, in *B. oleae*, transient knockdown via RNAi gene silencing in adult individuals showed that knockdown of Orco expression reduces the mating ability in both sexes and completely inhibits oviposition [365].

Tephritid CSPs have been identified [334,336,371], but their variable pattern of tissue distribution, the different potential functions, the still unproven binding ability [316] and unavailability of structural information [361] is locating them in a less attractive field of investigation.

4. Host-Marking Pheromones

4.1. Chemical Identity, Production and Analytical Approaches to Their Characterisation

In tephritids, host-marking behaviour was first described in *R. pomonella* [372]. Later, Hafliger speculated that the biological role of this behaviour was to equally distribute the offspring in available host fruits [373]. Following these earlier observations, Prokopy and Cirio were the first that experimentally described in 1972 the HMP deposition in *R. pomonella* [111] and *R. completa* (Cresson) [134]. Since then, the host marking behaviour has been reported in 25 tephritid species, particularly frugivorous species, belonging to the *Anastrepha*, *Ceratitis* and *Rhagoletis* genera [113] (Table 3; Table S3).

Table 3. List of tephritid species for which host-marking behaviour has been identified.

Genus	Species	Chemical Identity	References
Anastrepha	*A. suspensa*	-	[374]
	A. sororcula	-	[375]
	A. fraterculus	-	[376]
	A. pseudoparallela	-	[377]
	A. bistrigata	-	[378]
	A. grandis	-	[379]
	A. ludens	2-(2,14-Dimethylpentadecanoylamino)pentane-dioic acid	[122,380]
	A. striata	-	[96]
	A. obliqua	-	[131]
	A. serpentina	-	[131]
Bactrocera	-	-	-
Ceratitis	*C. capitata*	-	[110,125]
	C. cosyra	Glutathione	[33]
	C. rosa	Glutamic acid	[34]
Rhagoletis	*R. pomonella*	-	[111,372,381,382]
	R. cerasi	N-[15(β-Glucopyranosyl)-oxy-8-hydroxypalmitoyl]-taurine	[30,383,384]
	R. completa	-	[134]
	R. fausta	-	[385]
	R. cingulata	-	[386]
	R. cornivora	-	[386]
	R. indifferens	-	[386]
	R. mendax	-	[386]
	R. tabellaria	-	[386]
	R. basiola	-	[129]
	R. zephyria	-	[130]
	R. alternata	-	[387]

However, it is not a general feature of the family; it seems to be common in *Rhagoletis* spp., sporadic in *Anastrepha* and *Ceratitis* spp. and rather absent in others (e.g., *Bactrocera* spp.) [29,35,128,388,389].

HMPs are low-volatility and highly polar molecules [390]. They are also soluble in water and methanol [30,130,380,391,392]. HMPs can persist on the surfaces either when they are directly deposited by the fruit flies or as extracts [107]. For instance, the HMP half-life has been estimated to 10.7 days with persisting activity for three weeks in *R. pomonella* [381], 9 days in *R. fausta* [385], 12 days in *R. cerasi* [384], 6 days in *A. suspensa* [374], 6 days in *C. capitata* [110] and 4 days in *R. indifferens* [390].

To the extent of our knowledge, HMPs are produced and stored in the posterior half of the midgut and, thus, the faecal matter contains a huge quantity of these pheromones, suggesting the existence of two main routes for HMP deposition: through ovipositor dragging after egg-lying and through defaecation [131,393,394]. In *R. pomonella*, HMP accumulates in the midgut, Malpighian tubules, hindgut and faeces of mature females [393].

HMPs have been isolated from faecal matter extracts using approaches based on liquid chromatography (LC) and MS, in all four species (Table 3), namely HPLC-FAB-MS (Fast Atom Bombardment Mass Spectrometry) for *A. ludens* [380] and *R. cerasi* [30], and LC-quadrupole time-of-flight-mass spectrometry (LC-QTOF-MS) in the case of *Ceratitis* species [388,395]. Therefore, all the HMPs that have been chemically identified were isolated from the aqueous or methanol extract of adult female faecal matter. To date, it remains to be determined whether HMPs are produced by specific glands [393].

HMP chemical identity has been so far determined in a few tephritid species (Table 3). The first chemical characterisation of an HMP was achieved in *R. cerasi*. The pheromone was a complex molecule, i.e., N-[15(β-glucopyranosyl)-oxy-8-hydroxypalmitoyl]-taurine, with four stereoisomers [30], showing two chiral centres at the C-8 and C-15 positions. After the synthesis of the four different stereoisomers [396], it has been demonstrated that a racemic mixture of two isomers (8*R*, 15*S* and 8*S*, 15*R* isomers) is able to deter oviposition [397].

Later, the HMP [2-(2,14-dimethylpentadecanoylamino)pentanedioic acid (or *N*-[2,14-dimethyl-1-oxopentadecyl)glutamic acid)] of *A. ludens* has been chemically characterised and synthesised [380,398]. It presented a relatively simple structure containing an isopalmitic fatty acid chain substituted by methyl at the C-2 position and coupled to glutamic acid (GA) as a single diastereomer [380,398]. The HMP of *A. ludens* exhibited not only intraspecific but also interspecific oviposition deterring activity to *A. obliqua* and *A. serpentina* [131]. Of note, the HMPs of *R. cerasi* and that of *A. ludens* display similarities in structure (i.e., both contain a long fatty acid residue attached to an amino acid).

Recently, the HMP of *C. cosyra* and *C. rosa* have been isolated and both chemical structures have been determined [33,34]. The *C. cosyra* and *C. rosa* HMP identified are the tripeptide glutathione (GSH) (consisting of glycine, cysteine, and GA [33]), and GA [34], respectively. Interestingly, GSH and GA levels were 5–10 and 10–20 times higher in the faecal matter than in the ovipositor or haemolymph extracts of the respective females. These results suggest that the HMPs may be transferred from the gut into the ovipositor through the haemolymph and the excess amount may be expelled with the faecal matter. GSH was shown to express pheromone and allomone action respectively, reducing the oviposition in individuals of the same species and in those of different species, such as *C. rosa*, *C. fasciventris*, *C. capitata*, *Z. cucurbitae*. Interestingly, the GSH acts as kairomone inducing arrestment behaviour in the egg parasitoid *Fopius arisanus* (Sonan) (Hymenoptera: Braconidae). On the other hand, GA perception resulted in oviposition reduction in *C. rosa* and *C. fasciventris*, but not in *C. cosyra*. It is noteworthy that the HMPs of the two *Ceratitis* species are highly distinct from the HMPs identified in the other fruit flies. However, they appear to be more closely related to the HMP of the Mexican fruit fly, which contains GA, than of *R. cerasi* that is a fatty acid glucoside. Taken together, these findings may indicate that closely related species may utilise a similar pathway for HMP synthesis.

Host marking through HMP deposition is noticeably absent in the *Bactrocera* genus, even if contrasting results have been reported [35,138]. To prevent other females from ovipositing on the same fruit, *B. oleae* females do not depose an HMP but use their labella to spread olive juice leaking from the oviposition wound, with the main compounds responsible for this repulsion being (*E*)-hex-2-enal and oleuropein derivatives, such as the hydroxytyrosol [142,399].

The molecular machinery underlying the perception of these substances by other females remains to be determined. However, early studies in *R. pomonella* suggested that D-hairs on specific segments of the ventral tarsal surface and short marginal hairs on the labellum carry the receptors for HMP detection [393].

4.2. Behavioural Assays

Behavioural studies to assess the ecological role of a potential HMP rely on dual choice oviposition assays conducted under a completely randomised design [110], where tephritid females can choose to oviposit on a fruit marked by conspecifics over a control fruit [33,400].

5. Cuticular Hydrocarbons

Cuticular hydrocarbons act as pheromones in a variety of orders, including Diptera [25,401–403]. Their behavioural function in flies was first described in the housefly, *Musca domestica* L. (Diptera: Muscidae), where (*Z*)-tricos-9-ene was identified as the main compound on the female cuticle acting as a sex pheromone for males [401,404–406]. Extensive evidence of the importance of hydrocarbons (7-monoenes) as short-range signals and contact pheromones comes from *Drosophila* spp. [23]. These semiochemicals are perceived by antennae and maxillary palps and/or by contact with the taste organs that are mostly located on the tarsi and proboscis [407–409]. *Drosophila* spp. show relatively stable CH profiles, although their production can vary as the flies age and even after reproductive maturation [410,411]. In tephritids, sex- and species-specificity of CHs have been described in species of the genera *Anastrepha*, *Bactrocera* and *Ceratitis* (Table 4; Table S4).

Table 4. List of tephritid species for which CHs have been characterised.

Genus	Species	Developmental Stage	References
Anastrepha	*A. ludens*	larvae and adults	[412,413]
	A. suspensa	larvae and adults	[412,414–416]
	A. fraterculus	larvae and adults	[89,92,287,414,415,417]
	A. acris	larvae	[414,415]
	A. obliqua	larvae and adults	[415]
	A. serpentina	larvae	[415]
	A. pickeli	larvae	[415]
	A. striata	larvae	[415]
Bactrocera	*B. dorsalis*	larvae and adults	[412,418,419]
	B. carambolae	adults	[419]
	B. oleae	adults	[420]
	B. tryoni	adults	[421]
	B. zonata	adults	[420]
Zeugodacus	*Z. cucurbitae*	larvae and adults	[204,412]
Ceratitis	*C. capitata*	larvae and adults	[272,287,412,416]
	C. fasciventris		[272,287]
	C. anonae		[272,287]
	C. rosa	larvae and adults	[91,272,287,412]
Rhagoletis	-	-	-

Characteristic CHs profiles have been successfully applied for the chemotaxonomic clarification of fruit fly species complexes of *A. fraterculus*, *B. dorsalis* and the so-called African *Ceratitis* FAR complex [89,272,412,414–416,418,419]. Nevertheless, studies focused on the elucidation of CHs behavioural function in fruit fly mating system are still missing, except for those performed on *B. oleae* [79,84,286].

5.1. CHs in Tephritid Species and Their Described Roles

Earlier comparative studies on adults of *A. ludens*, *A. suspensa*, *C. capitata*, *C. rosa*, *Z. cucurbitae* and *B. dorsalis* failed to identify substantial sex-specific differences in CH profiles [412,418]. It seems that the only difference detected is the much higher amount of *n*-alkanes in males compared to females in both *B. dorsalis* and *Z. cucurbitae* [418]. However, further research reported sex- and age-dependent differences in CH production for a laboratory population of *A. fraterculus* (Brazilian-1 morphotype). Sexually mature males have specific unsaturated hydrocarbons (7-monoenes) on their cuticles that lack in females [89,287]. In follow-up studies, sexual dimorphism has been evaluated in the Brazilian-1 (Argentina), Brazilian-3, Andean, Peruvian and Mexican morphotypes of the *A. fraterculus* species complex [91,92]. The Brazilian-1 morphotype expresses the highest sexual dimorphism (29.46%), followed by the Mexican (15.42%) and Peruvian (13.79%) ones [92]. Males and females from the five abovementioned morphotypes diverge in alkene and alkadiene content. In *A. ludens*, the age-dependent CHs production in males originated from a standard mass-reared colony, a genetic-sexing strain, a hybrid strain and a wild population has been recently described [413]. Wild males of *A. ludens* differ from the mass-reared strains in the amount of nonacosane, while genetic sexing strain expressed higher values of 2-methylhexacosane. It has been suggested that the observed differences in CHs profiles may be due to environmental pressures [413], but additional research efforts are still needed to clarify these issue.

Further research also focused on the CH composition of males and females of the African fruit fly cryptic species FAR complex, demonstrating that sex-specific differences in the CH composition do exist [213,272]. The CH sex-specificity was proved by multivariate statistical analyses of the GC×GC-MS data of 59 CHs identified in the epicuticular extracts of *C. capitata*, *C. fasciventris*, and *C. rosa*. In contrast, the cuticular profiles of *C. anonae* display no sex-specificity [272].

In species of the *B. dorsalis* complex, abundant complex mixtures of sex-specific oxygenated lipids (i.e., three fatty acids and 22 fatty acid esters) with so far unknown biological function were identified in epicuticle extracts from females [419]. Such sex-specificity may be driven by sexual selection if the chemical composition of the cuticle is used as a pheromone signal in mate choice. Although early studies suggested that *B. dorsalis* males are able to recognise females at short distance and that physical contact may play key roles in courtship and *copula* [422], only limited functional information are currently available. These include the data describing the strong attractiveness to males exerted by the female-specific CH 4-allyl-2,6-dimethoxyphenol (4-DMP), which has been regarded as close-range sex pheromone [423]. This compound elicits electrophysiological responses in the mid legs of *B. dorsalis* males [424]. Moreover, after stimulation with 4-DMP, five OBP genes are found upregulated in males, with one of them, BdorOBP2, being a promising candidate for the binding and transport of 4-DMP [424], in addition to its proposed role in the perception of ME [343]. Recently, the cuticular components of *B. tryoni* have been described [421]. The spiroacetals and esters were found to be female-specific, while amides were presented in both sexes. Nevertheless, the role of these and other CHs in short-range chemical communication of *Bactrocera* spp. and other fruit flies needs to be further elucidated through expansion of the molecular machinery underlying their production and perception and with proper behavioural assays.

5.2. Analytical Approaches to Trace CH Profiles

Preparation of samples for gas chromatographic analysis of tephritid CHs is usually made by solvent extraction of whole insect bodies with solvents such as pentane, hexane, dichlormethane, and chloroform [89,425]. The most widely used method for recovering tephritid CHs is hexane washing [272,419,426]. This process may contaminate the cuticular sample with other materials, such as those from the exocrine glands. To collect only the CH fraction of the extract, the hexane solution is placed on a short chromatographic column and the CHs eluted with a small volume of hexane. For pre-cleaning of the CHs, thin layer chromatography (TLC) has proven useful in identifying novel lipid pheromones. TLC plates consist of glass or aluminium coated with an adsorbent layer of silica gel. Using TLC, components of the chemical extract can be separated into discrete fractions according to hydrophobicity [427,428]. After the hydrocarbon fraction has been collected, the solution must be concentrated to a suitable volume before the analysis. For the determination of the final concentration of the sample, the number of insects extracted must also be known. Another reported method involves continuous extraction of the insect in a refluxing solvent [429]. This procedure also requires column chromatography and re-concentration of solutes. Solvent extraction procedure involves several steps, in which more volatile compounds may be lost during the process, and moreover the use of high-purity solvents is required. Besides, the extraction and sample preparation account for most of the analysis time [426].

The solvent-free SPME technique has been used for the analysis of cuticular components of *Trupanea vicina* (Wulp), an Asteraceae-feeding tephritid [430]. SPME polydimethylsiloxane coated fibre (PDMS) was used to wipe samples from various body parts of male *T. vicina* and the subsequent GC-MS analyses showed that 1-nonanol, the male-specific compound, was concentrated on the abdomens of males exhibiting pleural distension [430]. In *A. ludens*, a PDMS fibre was used to rub down the male wing CHs, which were subsequently analysed by GC-MS [413]. Recently, the direct immersion-SPME (DI-SPME) coupled with GC-MS analyses was employed for characterisation of *C. capitata* semiochemicals in three different mating stages. This study demonstrated that medfly compound compositions were not significantly different before and during mating. However, new chemical compounds were generated after mating, such as (Z)-tricos-9-ene and hentriacontane among ten other components [258]. Female medflies seem to discriminate mated from virgin conspecifics and express higher rates of aggressive behaviour against the later [431]. Considering the recent findings, it is plausible to argue that changes in CHs of mated females drive the differential aggressive interactions between virgin and mated female medflies.

The method of choice for the analysis of the CH profile in tephritids has been GC-MS or GC×GC-MS using two different ionization techniques: electron (EI) and chemical ionization (CI) [92,287,425,432,433]. EI-MS is the primary tool for assessing the location of methyl branch points in long-chain alkanes, but it is often difficult to identify the molecular ion. The EI mass spectra interpretation allows complete identification of a compound, but often microscale reactions or derivatizations are necessary to provide additional structural information [434]. A mass spectrum also provides a 'finger print' of a compound, which can be compared with the libraries and mass spectra registries such as the NIST library, the Wiley/NBS registry of mass spectral data, and the published retention indices [435,436]. Mass spectra of an CI-MS yields the $(M-1)^+$ ion as an intense peak (sometimes as the only peak), but when the peak consists of mixture of isomeric methyl or dimethyl alkanes, placement of methyl group position becomes more difficult [433]. Another problematic aspect is the determination of the double bound position using EI-MS, because of the lack of cleavage between carbon-carbon double bonds or extensive and facile hydrogen rearrangement along the chain after molecular ion fragmentation [437]. Nevertheless, several works documented the CI-MS/MS with acetonitrile ionization gas to be a suitable method for double bound position determination [89,432,438].

There have also been studies using MALDI-TOFMS and ultraviolet laser desorption ionization orthogonal time-of-flight mass spectrometry (UV-LDI o-TOFMS) for tephritid CHs identification [89,427,439]. Cvačka and colleagues [427] applied MALDI-MS for the identification of insect CHs, using lithium 2,5-dihydroxybenzoate as a matrix [440]. This work demonstrates that MALDI-TOFMS is a convenient analytical method for the identification of high molecular weight hydrocarbons from insect cuticles, including saturated hydrocarbons and highly unsaturated and/or cyclic compounds. In *A. fraterculus*, application of MALDI-TOFMS method allowed for characterisation of high molecular weight saturated and unsaturated hydrocarbons, up to C37 in length [89]. Mass spectrometric imaging (MSI) of male and female *A. fraterculus* was performed using a previously developed protocol [441]. The preliminary MALDI-MSI experiments indicate differences in the CH distribution on the wings of males and females [287]. Nevertheless, additional detailed analyses using MALDI-MSI techniques are necessary for further conclusions concerning the CHs of *A. fraterculus*. The imaging data will show if some of the CHs have unique locations on *A. fraterculus* body surface and can also indicate if the compounds direct male/female sexual contacts.

5.3. Behavioural Assays

Behavioural tools currently used for the evaluation of CHs do not substantially differ from the set up described above for volatile pheromones. Indeed, observation chambers, as well as two- or multiple-choice systems (i.e., Y-tube/T-maze and multiple-arm olfactometers, respectively) are widely used to evaluate behavioural responses triggered by CHs with potential pheromone activity [69,74]. Further validation of the observed behavioural response can be documented in flight tunnel as well as in field and semi-field assays [170,317,442].

As a final remark, semiochemical candidates, such as potential volatile pheromones and CHs, could be evaluated for their biological functions relying on the mixed society approach [443]. Indeed, mixed societies composed of living insects and small-sized robots mimicking their conspecifics can represent a valid approach to shed light on factors guiding insect behaviour, including mating approaches. This ethorobotics-based approach has been validated on several insect species, such as cockroaches, beetles and blowflies [444–446]. However, no studies have been conducted relying on ethorobotics in tephritid research. In our opinion, this represents a challenge for future studies.

6. Tephritid Sexual Chemoecology: Real-World Applications and Challenges

Semiochemical-based interactions have been extensively studied in tephritid fruit flies and several aspects of the generated knowledge had already been exploited for practical purposes including (a) trapping and population monitoring, (b) direct population control approaches (lure and kill methodologies, and push and pull strategies), and (c) support and improvement of other methods, such as the SIT. More recent progress in the characterisation of (a) the olfactory molecular machinery, including OBPs and ORs, and (b) CHs may open new venues in developing inspiring approaches for artificial olfaction and hence generation of novel long- and medium-range attractants. These tools can be used to address issues regarding species complexes that may be of vast importance for regulatory and control aspects. Nonetheless, a deeper knowledge of semiochemical-based communication is essential to further understand how tephritid species adapt to complex ecosystems, also with respect to the invasive potential of several pests belonging to this family, further advancing applied research from several perspectives.

6.1. Population Monitoring and Early Detection of Tephritid Outbreaks

Given the influence semiochemicals play on insect behaviour, a better understanding of their identity, specificity and biological role(s) may benefit tephritid research by developing novel specific and environmentally-friendly attractants. Effective attractants are essential for adult trapping, which is a key tool used to (a) monitor density and seasonal patterns of established tephritid pest populations, (b) detect new infestations of exotic species, (c) delimit the detected populations, and (d) confirm the results of eradication campaigns. Early detection of small populations is particularly important to delimit the outbreak and thus implement control and eradication measures while the pest population is still present at low densities [447,448].

The list of odour attractants for tephritid fruit flies is quite long including food-based (protein-based, ammonia releasing compounds) and mating-related chemicals. The later can either be purely synthetic or derived from plants. The history of their discovery and development goes beyond the purposes of this review, and their identity, efficacy, and the related practical aspects of their use in monitoring and detection are widely reviewed in [170]. Attractants that are related to the mating behaviour or physiology of fruit flies have been more thoroughly studied in the case of *Bactrocera* and *Ceratitis* genera and they exclusively concern male lures. For example, both the plant derived α-copaene and the synthetic trimedlure and Ceralure are highly attractive to male but not female medflies [170,449,450]. Interestingly, both plant-derived and synthetic attractants are related to male reproductive success and may enhance lekking behaviour and mating competitiveness [451–456]. It should be stressed here that trimedlure is rather the most commonly used attractant for medfly regular population monitoring, control programs and in detection and eradication campaigns [457,458].

Some *Bactrocera* species respond to Cue Lure (CL)/RK, others to ME, the most powerful attractant so far identified, others appear to be non-responsive to both. Eugenol analogues (isoeugenol, methyl-isoeugenol and dihydroeugenol) are also proving successful in attracting *Bactrocera* species [459,460]. Similar to species of the genus *Ceratitis*, the above compounds attract only males of the *Bactrocera* species and again are related to their mating success [461–466]. Methyl eugenol is considered one of the most powerful attractants for male fruit flies and it is extensively used in detection, population monitoring, delimitation and eradication campaigns worldwide [169,467,468].

Conversely, no male lures are currently available for *Anastrepha* and *Rhagoletis* spp. [170]. Exploring the potential of tephritid pheromones to develop novel specific attractants is thus important, not only for *Anastrepha* and *Rhagoletis*, but also for non-responsive species belonging to the *Bactrocera* and *Ceratitis* genera. Moreover, ME has been suggested to be carcinogenic [469] and alternatives are required that also do not exert effects on non-target species. Finally, the available lures are generally effective in attracting one sex, and do not have a species-specific action.

Although food-based attractants [470] are still dominant in population monitoring of the olive fruit fly, the use of the female pheromone can provide additional information on both the population density of wild populations and the age structure of wild populations. More recently, Sarles and colleagues identified two lactones released exclusively by males of *R. completa* and used lactone-baited traps in walnut orchards [471]. These traps have been found particularly effective for *R. completa* monitoring and allowed its earlier detection in the season, supporting the idea that the analysis of pheromone components may be particularly promising for trapping tephritids in the field.

6.2. Eradication and Suppression of Fruit Fly Populations Employing Semiochemicals

Male lures such as trimedlure and especially methyl-eugenol (ME), besides being employed for population monitoring, have been used for more direct control purposes. Male annihilation technique (MAT) (i.e., elimination of males, mating and oviposition of fertile eggs, based on strong male-specific lures that are deployed in a mass trapping or lure and kill approach) is considered as a very successful option to eliminate low populations of *B. dorsalis* and a tool that may drive invaded population to extirpation or even eradication. For example, male annihilation against *B. dorsalis* has been used as a main tool in attempts to eradicate incursions or isolated established populations in California, Hawaii and Florida [472–476], South Africa [477,478], the Marianas Islands in Micronesia [479], the Okinawa Islands in Japan [480], and Mauritius [481]. Eradication efforts using the male attractant CL against other *Bactrocera* species such as *B. frauenfeldi* were not successful [482]. Male attraction to CL remained consistent until advanced age in *B. tryoni*, although it sharply declined after 12 weeks of age, with potential implications for pest management [483]. Despite its broad use as a population monitoring tool, trimedlure is not considered as an eradication tool against medfly.

Male lures and the MAT have been considered in suppression programs often in combination with other methods and preferably in the frame of an Area Wide application strategy. As it was demonstrated in Hawaii, combination of MAT with field sanitation, protein bait, sterile male releases and biological control resulted in satisfactory reduction of the fruit fly population [484]. Methyl eugenol and CL for *Bactrocera* species, as well as trimedlure for the medfly, have been considered. Interestingly, the use of ME for the suppression of *B. dorsalis* in Southern Ethiopia proved to be successful [485].

The use of the female pheromone alone or in combination with other baits has been evaluated for the control of the olive fruit fly [486,487]. These efforts include a male annihilation component but also tools against females.

Apparently, classical mating disruption approaches involving saturation of target area with species-specific pheromones are not effective against fruit flies, and the MAT is prevailing. Indeed, mating disruption approaches against *B. oleae* in Spain and Greece led to inconclusive results [488,489]. However, a more recent study, in which authors observed a decrease in fly catches in the presence of high pheromone (i.e., 1,7-dioxaspiro[5.5]undecane) concentrations, supports the applicability of a mating disruption approach against this species [490]. For the medfly, for both trimedlure and the iodinated trimedlure analogue Ceralure, no mating disruption effects have been so far described [490]. Such an absence of disruption effect in this species was explained by the lack of saturation in response to higher trimedlure concentrations [490].

Alternative strategies aimed at interfering with mating deserve to be explored, including the potential applications of CHs as disruptors. CHs are known to play a role in the mating behaviour of different *Drosophila* species [491–493]. In *D. suzukii*, alteration of C23 alkanes ratios results in disrupting mate recognition and, as a consequence, courtship and mating behaviour [494]. The identification of OBP candidates potentially able to transport the *B. dorsalis* female-biased CH 4-DMP [423,424] is a promising step in the clarification of the functional role of CHs in the mating behaviour of fruit flies.

Moreover, in order to improve the efficacy of any semiochemical-based approach for fruit fly eradication and suppression, additional studies should be devoted to better understand the role of abiotic factors on trapping. Indeed, although temperature, humidity, rainfall and other exogenous abiotic factors do affect the temporal and spatial activity of fruit flies, data about the role of such factors in tephritid captures are still limited and mostly available for liquid protein-baited traps (see [267] for an overview). In the case of semiochemicals, it is known that medfly attractiveness to trimedlure is related to the release rate of this compound, which is, in turn, dependent on temperature [495,496]. Recently, Cameron and colleagues examined the vapour pressures and thermodynamic properties of seven attractants (i.e., RK, CL, raspberry ketone trifluoroacetate-RKTA, ME, methyl isoeugenol, dihydroeugenol, and zingerone) currently used for trapping *Bactrocera*, *Dacus* and *Zeugodacus* species [497]. The authors provide valuable data regarding the volatility of these attractants. In particular, they found that (Z)-methyl isoeugenol is the most volatile of the ME-type compounds, while RKTA is the most volatile among the RK-type compounds. Interestingly though, the field life of RKTA is not long due to its susceptibility to humidity [498]. Expanding our understanding of the features of these chemicals as well as of the identity/impact of multiple abiotic factors that may affect their activity is essential to determine the design of the eradication and suppression programs, as well as the location and density of the traps to be used.

6.3. Push and Pull Approaches Based on Repellent Semiochemicals

HMPs have been regarded as attractive tools for tephritid pest control since the '70s. In 1976, Katsoyannos and Boller proposed to use HMPs to prevent fruit fly oviposition into the fruit. They performed the first field experiment spraying raw HMP extract obtained from the faecal matter of *R. cerasi*. In this way, they achieved over 90% reduction of *R. cerasi* infestation in cherries orchards [499,500]. Later, once the chemical structure and the synthesis of *R. cerasi* HMP have been obtained [29], Aluja and Boller [31] tested the synthetic *R. cerasi* HMP in the field and, interestingly, this was the first application of a "push-pull strategy" in fruit flies. A push-pull strategy exploits a combination of behaviour-modifying stimuli to manipulate the distribution and abundance of the insect targets. Pests are repelled from their resource (push) by using stimuli that mask the host or that acts as repellents. Simultaneously, they are driven away from the resource (pull), by using highly attractive stimuli such as traps, facilitating their elimination. A reduction of the infestation of about 90% in cherry plants was achieved by treating one half of tree canopies with a synthetic HMP. The repelled females were then trapped with visual traps placed on the other half of the canopy [31,501]. The efficacy of the synthetic HMP was further supported by another field trial in which the infestation by cherry fruit flies was eliminated [502].

Similarly, field tests were performed using raw pheromone extract from the medfly's faecal matter achieving 84% decrease in infestation in sprayed coffee plants [388]. Exploiting the cross-recognition observed in the *Anastrepha* genus, Aluja and co-workers tested three potential oviposition deterrents for *A. obliqua* in tropical plum and mango orchards. In their experiments, they used *A. ludens* faecal extracts and two fully synthetic simplified analogues of the naturally occurring pheromone, namely desmethyl *A. ludens* HMP (DM-HMP) and anastrephamide. They obtained a significant reduction in fruit damage rates with all substances tested, and interestingly, the simplified analogues displayed comparable levels of efficacy to the natural HMP [32].

The good efficacy of the synthetic HMP analogues found in both *R. cerasi* and *A. obliqua* field tests is promising in view of their potential use as pest control strategy. However, there is evidence that, after prolonged exposure to the HMP, flies can lay eggs in the treated fruits. This behaviour could be associated with the sensorial adaptation by the insect [122,501].

Although the use of HMP in the management of fruit flies was initially suggested as a push-pull system [32,111,391,398,503], the push-pull strategy is not suitable for the species with high population growth rates [504]. A recent work used anastrephamide in combination with a protein bait to reduce grapefruit infestation by *A. ludens* [505]. The authors found that anastrephamide can push flies out of the treated tree, but the push-pull system requires a more effective attractant.

Lastly, an alternative use of HMP has been proposed, which implies HMP application in commercial crops in which the fruit fly populations are not resident. This option allows achieving pest suppression because of the small population and lower risk for the occurrence of adaptation [113].

6.4. Implications for Biological Control

The recent findings that fruit fly parasitoids such as *F. arisanus* responded to HMP of fruit fly species [113,506] may pave the way for the development of methods and approaches to enhance the biological control efforts against fruit flies. Intensifying the research towards understanding the intraspecific interactions among parasitoids and fruit fly semiochemicals, in particular, oviposition-induced volatiles, is required to further increase parasitoid ability to localise and parasitise tephritid pests in the field [507]. In particular, expanding our understanding of the identity of HMPs will facilitate approaches based on the manipulation of parasitoid behaviour to benefit fruit fly control. Indeed, the incorporation of the identified HMP chemicals in the mass rearing process of the target parasitoid has the potential to facilitate the associate learning process that allows the parasitoid to distinguish HMPs from plant-emitted volatiles [113,508]. Moreover, treatments of fields and orchards with synthetic HMPs may not only deter fruit fly oviposition, but also attract parasitoids [509,510]. Among the factors mediating semiochemicals' production and perception by tephritid and/or parasitoids, the insect microbiota likely plays a key role. Interestingly, a recent study showed that the production of β-caryophyllene emitted by host plants and mediating oviposition avoidance in egg-infested fruit by *B. dorsalis* is induced by egg-surface bacteria such as *Providencia* sp. and *Klebsiella* sp. [511]. Achieving a deeper understanding of the interplay between tephritid, their microbiota and host plants will shed new light on the multifaceted field of trans-kingdom communication [512] and will provide novel targets to be exploited for pest management.

6.5. Semiochemical-Based Tools to Enhance the Sterile Insect Technique

The identification and functional characterisation of tephritid semiochemicals may be beneficial for SIT applications and can also favour the integration of SIT and MAT to maximise the efficacy of pest management campaigns.

In SIT programs, sterilised males are released in the field to mate with wild females and induce sterility and hence no viable egg production [513]. Typically, sterile males are release as immature adults, which have to survive until sexual maturation, localise females and achieve copulation for SIT programs to be successful [514]. Thus, survival, dispersal capacity and mating competitiveness, for which male signalling is essential [327,515], are key factors that have to be fulfilled. Mating-enhancing semiochemicals have been widely described in several tephritid species and include both plant-derived compounds and synthetic chemicals, such as ME, Cue-Lure, RK, ginger root oil (GRO), citrus and guava fruit volatiles, and manuka oil (see [516,517] for reviews, and [456]).

SIT campaigns are more effective when pest population density is reduced before the release of sterile males. MAT applications, as well as inundative releases of biological control agents, have the potential to reduce the density of feral population size and hence to precede the implementation of the SIT. In *Bactrocera* species, the pre-release exposure of males to plant-derived semiochemicals and synthetic lures has been shown to reduce their subsequent response to attractants used in MAT [35,518]. Recently, the use of a diet containing RK fed to immature sterile *B. tryoni* prior to release resulted in increased subsequent survival and reduced response to MAT [519]. These results are particularly promising for the implementation of SIT-MAT simultaneous used in the field, to both increase control effectiveness and reduce operational costs.

6.6. Artificial Olfaction and Pheromone-Based Nanosensors

The ability of insect to detect olfactory stimuli at low concentrations over long distances [520] stimulated researchers to exploit these phenomena to develop biosensors based on insect behaviour [521], using isolated antennae [522] to detect explosives [523,524], food toxins [525], and for disease control and diagnosis [526,527]. Biosensors require a biological and artificial component, able to make the signals readable. OBPs can be expressed and purified easily and are stable to perturbations in temperature, pH and proteases [528]. Thus, they are considered ideal candidates to be exploited in biosensors development and were used to engineer systems able to detect floral odorants, alcohols and explosives in *Drosophila* and *Apis mellifera* L. (Hymenoptera: Apidae) [529–532]. More recently, also a member of the Tephritidae family was target of this type of research: an OBP from *B. dorsalis*, BdorOBP2, was expressed, purified and immobilised on an interdigitated electrode and it was shown to work as an efficient biosensor for chemicals emitted by host plants (e.g., isoamyl acetate, β-ionone, benzaldehyde) [533]. As previously described, given their higher specificity and sensitivity ORs are ideal candidates to be explored for the development of biosensors. Research in the field of OR-based bioelectronic sensors is indeed recently emerging as a simpler strategy with respect to the use of the mammalian ORs to detect environmentally significant volatile organic compounds (VOCs), as shown in mosquitoes (Diptera: Culicidae) and *Drosophila* [534–536]. Although interesting, this type of application is still far from being used in the field for tephritid management, also because only a few insect ORs have been deorphanised. Conversely, the exploitation of pheromone components for the development of innovative strategies to monitor early infestations is emerging as a powerful alternative to currently adopted strategies. Indeed, by targeting the major olive fruit fly volatile pheromone component, 1,7-dioxaspiro[5.5]undecane [317], Moitra and colleagues developed a β-cyclodextrinylated nanosensor specific to the female volatile pheromone of *B. oleae* [537]. This device is currently being tested in open field conditions and may be important not only for the control of *B. oleae* but also for the development of similar sensitive microelectromechanical system (MEMS) devices for other tephritid pests.

6.7. Cuticular Hydrocarbons as a Tool for Chemical Taxonomy

Cuticular lipid profiles, which are species-specific both in solitary and social insects, serve as fingerprints, making it possible to discriminate species taxonomically or to recognise sibling species [441]. The first successful use of CHs for the taxonomic discrimination of tephritid fruit fly species was reported 20 years ago in five articles on the hydrocarbon profile identification of adult Malaysian *B. dorsalis* complex flies [418] and larvae of *Anastrepha* (*A. acris, A. fraterculus, A. suspensa, A. ludens, A. obliqua*), *Ceratitis* (*C. capitata, C. rosa*), *Bactrocera* (*B. dorsalis*), and *Z. cucurbitae* species [412]. Recently, Vaníčková and co-workers reported CH profiling is an efficient tool for the resolution of entities in the African fruit fly cryptic species FAR complex [272,287,412,414–416]. In these studies, twelve potential chemotaxonomic markers were identified for the distinction of adult male and female flies of *C. fasciventris, C. rosa, C. anonae* and *C. capitata*. Some of the geographically distinct subspecies hidden in the *A. fraterculus* complex can also be identified using their specific CH profiles. For example, Peruvian and Brazilian-1 morphotypes have unique CH profiles, suggesting CHs could be used to distinguish between these two subspecies [92]. In the *B. dorsalis* complex, clear segregation of complex cuticle profiles of both *B. carambolae* sexes from *B. dorsalis* was documented, supporting both taxonomic synonymisation of *B. invadens, B. papayae,* and *B. philippinensis* with *B. dorsalis,* as well as the exclusion of *B. carambolae* from *B. dorsalis* [419].

7. Conclusions and Challenges

Research in the last decades amazingly expanded our knowledge in the field of tephritid fruit fly semiochemical-based communication at the genomic, molecular, physiological and behavioural level, as outlined by the huge amount of available literature. On the other side, it is clear that semiochemical-based communication has not been characterised to an equal extent in all relevant tephritid pests. Moreover, most studies have been focused on volatile pheromones, with far less information available for CHs and HMPs. Thus, a major research need is to expand our knowledge to achieve an exhaustive understanding of all the semiochemical-based communication modalities in the target species. In addition, many specific questions remain to be answered. In the case of volatile pheromones, the application of different techniques and conditions for sampling, as well as the chosen source (i.e., rectal gland content or headspace), often provided different results in the detection and quantification of volatiles, making comparisons among identified sets of chemicals challenging. This urges for the parallel adoption of more than one method to ensure a comprehensive analysis of volatile pheromones. Moreover, future research on volatile pheromones will be essential to clarify (i) the identity of the molecular/genomic machinery underlying rectal (and salivary) gland physiology leading to volatile pheromone production, (ii) the relative impact of genomic background, evolutionary history and feeding preference in shaping the volatile pheromone bouquet. With respect to HMP research, it remains to be determined: (i) whether HMP is produced by specific glands, and how it is produced, (ii) which genes and pathways are involved in its synthesis, (iii) which is the chemical composition of most HMPs; although their widespread presence in tephritids, these pheromones were so far characterised in very few species; thus, expanding their characterisation represents a major challenge; (iv) whether HMPs are really absent in the *Bactrocera* genus. Although CHs have been widely described and novel studies are continuously being published, we still do not know which is their exact function in true fruit fly mating. Providing an answer to these and to several other questions that may arise when diving into this multifaceted field will be essential for the implementation of novel/improved approaches to tephritid pest control. Indeed, a series of knowledge gaps do exist, which limit the toolbox in the field. First of all, strategies to improve the formulation of semiochemical-based lures currently used in field applications are needed. These include the optimisation of liquid and solid dispensers containing individual compounds or mixtures (including isomers). Moreover, no male lures are currently available for *Anastrepha* and *Rhagoletis* spp., and some *Bactrocera* and *Ceratitis* species do not respond to the existing substances. Alternative and species-specific attractants are needed and the example of the lactone-baited traps used to capture *R. completa* is stimulating further studies aiming at exploring volatile pheromones for trapping. In addition to the MAT-based approaches, other strategies able to interfere with mating need to be explored, including the potential applications of CHs as disruptors. Overall, expanding our understanding of the identity of HMPs will facilitate the integration of multiple approaches for fruit fly control, including biological control, with special reference to the programs relying on the use of parasitoids. Finally, novel technologies allowing the production of nanosensors able to specifically detect pheromone components may open new routes for tephritid pest control.

Supplementary Materials: The following are available online at https://www.mdpi.com/article/10.3390/insects12050408/s1, Table S1: Chemicals detected in the volatile pheromone (headspace) of 18 tephritid species of high agricultural relevance. Table S2: List of unique and overlapping chemicals characterised in the volatile pheromone of tephritid species. Table S3: Chemicals with a HMP role detected in the adult female faecal matter of four tephritid species of high agricultural relevance. Table S4: Chemicals detected in the cuticular hexane body washes of 19 tephritid species of high agricultural relevance.

Author Contributions: Conceptualization: F.S., L.V. and F.V. Literature analysis: F.S., L.V., F.V., G.B. and N.T.P. Writing of the original draft preparation: F.S., L.V. and F.V. Writing, review and editing: F.S., L.V., F.V., G.B. and N.T.P. All authors have read and agreed to the published version of the manuscript.

Funding: This research received no external funding.

Institutional Review Board Statement: Not applicable.

Acknowledgments: This study was performed within the Food and Agriculture Organization/International Atomic Energy Agency (FAO/IAEA) Coordinated Research Project "Assessment of Simultaneous Application of SIT and MAT to Enhance *Bactrocera* Fruit Fly Management" (D41027), contract no. 23126/R0 (to L.V.).

Conflicts of Interest: The authors declare no conflict of interest.

Abbreviations

The following abbreviations and acronyms are used in this manuscript:

AW-IPM	Area-Wide Integrated Pest Management
CCE	Conventional Chemical Ecology
CHs	Cuticular Hydrocarbons
CI	Chemical Ionization
CL	Cue Lure
CRCE	Computational Reverse Chemical Ecology
CSD	Current Source Density
CSPs	Chemosensory Proteins
DESI	Desorption Electrospray Ionization
DM-HMP	Desmethyl Host-Marking Pheromone
DVB/CAR/PDMS	Divinylbenzene/Carboxen/Polydimethylsiloxane
EAG	Electroantennogram
EI	Electron Ionization
ELSD	Evaporative Light Scattering Detectors
FAR	*Ceratitis* complex including *C. fasciventris*, *C. anonae* and *C. rosa*
GA	Glutamic Acid
GC-EAD	Gas Chromatography-Electroantennographic Detection
GC×GC-TOFMS	Two-dimensional Gas Chromatography with Time-Of-Flight Mass Spectrometry
GC-FID	Gas Chromatography-Flame Ionization Detection
GC-FTIR	Gas Chromatography-Fourier Transformed Infrared Spectroscopy
GC-MS	Gas Chromatography-Mass Spectrometry
GOBP	General OBP
GPCRs	Protein-Coupled Receptors
GRO	Ginger Root Oil
GRs	Gustatory Receptors
GSH	Tripeptide Glutathione
HMP	Host-Marking Pheromone
HPLC	High Performance Liquid Chromatography
IPM	Integrated Pest Management
IRs	Ionotrophic Receptors
LC-MS	Liquid Chromatography-Mass Spectrometry
LC-QTOF-MS	LC-Quadrupole Time-Of-Flight-Mass Spectrometry
MALDI-TOFMS	Matrix-Assisted Laser Desorption/Ionization-Time-Of-Flight Mass Spectrometry
MAT	Male Annihilation Technique
ME	Methyl Eugenol
MEMS	Microelectromechanical System
MSI	Mass Spectrometric Imaging
NMR	Nuclear Magnetic Resonance
OBPs	Odorant Binding Proteins
Orco	Olfactory receptor co-receptor
ORs	Olfactory Receptors
OSNs	Olfactory Sensory Neurons
PDMS/DVB	Polydimethylsiloxane/Divinylbenzene
RCE	Reverse Chemical Ecology
RK	Raspberry Ketone
RKTA	Raspberry Ketone Trifluoroacetate
RNAi	RNA Interference
SIT	Sterile Insect Technique
SPME	Solid Phase Microextraction
SSR	Single Sensillum Recording
TLC	Thin Layer Chromatography
UV-LDI o-TOFMS	Ultraviolet Laser Desorption Ionization orthogonal Time-Of-Flight Mass Spectrometry
VOCs	Volatile Organic Compounds
1-NPN	*N*-phenylnaphthalen-1-amine
4-DMP	4-allyl-2,6-dimethoxyphenol

References

1. Brezolin, A.N.; Martinazzo, J.; Muenchen, D.K.; de Cezaro, A.M.; Rigo, A.A.; Steffens, C.; Steffens, J.; Blassioli-Moraes, M.C.; Borges, M. Tools for detecting insect semiochemicals: A review. *Anal. Bioanal. Chem.* **2018**, *410*, 4091–4108. [CrossRef] [PubMed]
2. Tinsworth, E.F. Regulation of pheromones and other semiochemicals in the United States. In *Behavior-Modifying Chemicals for Insect Management: Applications of Pheromones and Other Attractants*; Ridgway, R.L., Silverstein, R.M., Inscoe, M.N., Eds.; Marcel Dekker Inc.: New York, NY, USA, 1990; pp. 569–603.
3. Rodriguez, L.C.; Niemeyer, H.M. Integrated pest management, semiochemicals and microbial pest-control agents in Latin American agriculture. *Crop Prot.* **2005**, *24*, 615–623. [CrossRef]
4. McNeil, J.N.; Millar, J.G.; Chapman, R.F. Chemical communication: Pheromones and allelochemicals. In *The Insects: Structure and Function*; Simpson, S.J., Douglas, A.E., Eds.; Cambridge University Press: Cambridge, UK, 2012; pp. 857–900.
5. Abd El-Ghany, N.M. Semiochemicals for controlling insect pests. *J. Plant Prot. Res.* **2019**, *59*, 1–11. [CrossRef]
6. Ruther, J.; Meiners, T.; Steidle, J.L.M. Rich in phenomena-lacking in terms. A classification of kairomones. *Chemoecology* **2002**, *12*, 161–167. [CrossRef]
7. Schulz, S. *The Chemistry of Pheromones and Other Semiochemicals I*; Springer: Berlin/Heidelberg, Germany, 2004.
8. Hansson, B.S.; Stensmyr, M.C. Evolution of insect olfaction. *Neuron* **2011**, *72*, 698–711. [CrossRef]
9. Joseph, R.M.; Carlson, J.R. *Drosophila* chemoreceptors: A molecular interface between the chemical world and the brain. *Trends Genet.* **2015**, *31*, 683–695. [CrossRef]
10. Tumlinson, J.H.; Teal, P.E.A. Relationship of structure and function to biochemistry in insect pheromone systems. In *Pheromone Biochemistry*; Prestwich, G.D., Blomquist, G.J., Eds.; Academic Press: New York, NY, USA, 1987; pp. 3–26.
11. Gut, L.J.; Stelinski, L.L.; Thomson, D.R.; Miller, J.R. Behaviour-modifying chemicals: Prospects and constraints in IPM. In *Integrated Pest Management: Potential, Constraints, and Challenges*; Koul, O., Dhaliwal, G.S., Cuperus, G.W., Eds.; CABI Publishing: Cambridge, MA, USA, 2004; pp. 73–120.
12. Wyatt, T.D. *Pheromones and Animal Behavior*; Cambridge University Press: Cambridge, UK, 2014.
13. Karlson, P.; Luscher, M. "Pheromones": A new term for a class of biologically active substances. *Nature* **1959**, *183*, 55–56. [CrossRef]
14. Butenandt, V.A.; Beckmann, R.; Stamm, D.; Hecker, E. Über den sexual-lockstoff des seidenspinners *Bombyx mori*—Reindarstellung und konstitution. *Z. Naturforsch.* **1959**, *14*, 283–284.
15. Levine, J.D.; Millar, J.G. Chemical signalling: Laser on the fly reveals a new male-specific pheromone. *Curr. Biol.* **2009**, *19*, 653–655. [CrossRef]
16. Roelofs, W.L. Chemistry of sex attraction. *Proc. Natl. Acad. Sci. USA* **1995**. [CrossRef]
17. Kaissling, K.E.; Kasang, G.; Bestmann, H.J.; Stransky, W.; Vostrowsky, O. A new pheromone of the silkworm moth *Bombyx mori*. *Sci. Nat.* **1978**, *65*, 382–384. [CrossRef]
18. Kaissling, K.-E. Pheromone reception in insects. In *Neurobiology of Chemical Communication*; Mucignat-Caretta, C., Ed.; CRC Press/Taylor & Francis: Boca Raton, FL, USA, 2014.
19. Lebreton, S.; Borrero-Echeverry, F.; Gonzalez, F.; Solum, M.; Wallin, E.A.; Hedenstrom, E.; Hansson, B.S.; Gustavsson, A.L.; Bengtsson, M.; Birgersson, G.; et al. A *Drosophila* female pheromone elicits species-specific long-range attraction via an olfactory channel with dual specificity for sex and food. *BMC Biol.* **2017**, *15*, 88. [CrossRef]
20. Pankiw, T. Cued in: Honey bee pheromones as information flow and collective decision-making. *Apidologie* **2004**, *35*, 217–226. [CrossRef]
21. Yew, J.Y.; Chung, H. Insect pheromones: An overview of function, form, and discovery. *Prog. Lipid Res.* **2015**, *59*, 88–105. [CrossRef]
22. Greenfield, M.D. *Signalers and Receivers: Mechanisms and Evolution of Arthropod Communication*; Oxford University Press: Oxford, UK, 2002.
23. Ferveur, J.F. Cuticular hydrocarbons: Their evolution and roles in *Drosophila* pheromonal communication. *Behav. Genet.* **2005**, *35*, 279–295. [CrossRef]
24. Blomquist, G.J.; Jackson, L.L. Chemistry and biochemistry of insect waxes. *Prog. Lipid Res.* **1979**, *17*, 319–345. [CrossRef]
25. Blomquist, G.J.; Bagnères, A.G. *Insect Hydrocarbons Biology, Biochemistry, and Chemical Ecology*; Cambridge University Press: Cambridge, UK, 2010; ISBN 9780511711909.
26. Vaníčková, L.; Canale, A.; Benelli, G. Sexual chemoecology of mosquitoes (Diptera, Culicidae): Current knowledge and implications for vector control programs. *Parasitol. Int.* **2017**, *66*, 190–195. [CrossRef]
27. Blomquist, G.J.; Howard, R.W. Pheromone biosynthesis in social insects. In *Insect Pheromone Biochemistry and Molecular Biology*; Blomquist, G.J., Vogt, R.G., Eds.; Elsevier: New York, NY, USA, 2003; pp. 323–340.
28. Howard, R.W.; Blomquist, G.J. Ecological, behavioral, and biochemical aspects of insect hydrocarbons. *Annu. Rev. Entomol.* **2005**, *50*, 371–393. [CrossRef]
29. Nufio, C.R.; Papaj, D.R. Host marking behavior in phytophagous insects and parasitoids. *Entomol. Exp. Appl.* **2001**, *99*, 273–293. [CrossRef]
30. Hurter, J.; Boller, E.F.; Städler, E.; Blattmann, B.; Buser, H.R.; Bosshard, N.U.; Damm, L.; Kozlowski, M.W.; Schöni, R.; Raschdorf, F.; et al. Oviposition-deterring pheromone in *Rhagoletis cerasi* L.: Purification and determination of the chemical constitution. *Experientia* **1987**, *43*, 157–164. [CrossRef]

31. Aluja, M.; Boller, E.F. Host marking pheromone of *Rhagoletis cerasi*: Field deployment of synthetic pheromone as a novel cherry fruit fly management strategy. *Entomol. Exp. Appl.* **1992**, *65*, 141–147. [CrossRef]

32. Aluja, M.; Díaz-Fleischer, F.; Boller, E.F.; Hurter, J.; Edmunds, A.J.F.; Hagmann, L.; Patrian, B.; Reyes, J. Application of feces extracts and synthetic analogues of the host marking pheromone of *Anastrepha ludens* significantly reduces fruit infestation by *A. obliqua* in Tropical plum and mango backyard orchards. *J. Econ. Entomol.* **2009**, *102*, 2268–2278. [CrossRef] [PubMed]

33. Cheseto, X.; Kachigamba, D.L.; Ekesi, S.; Ndung'u, M.; Teal, P.E.A.; Beck, J.J.; Torto, B. Identification of the ubiquitous antioxidant tripeptide glutathione as a fruit fly semiochemical. *J. Agric. Food Chem.* **2017**, *65*, 8560–8568. [CrossRef] [PubMed]

34. Cheseto, X.; Kachigamba, D.L.; Benderap, M.; Ekesi, S.; Ndung'u, M.; Beck, J.J.; Torto, B. Identification of glutamic acid as a host marking pheromone of the African fruit fly species *Ceratitis rosa* (Diptera: Tephritidae). *J. Agric. Food Chem.* **2018**, *66*, 9933–9941. [CrossRef]

35. Benelli, G.; Daane, K.M.; Canale, A.; Niu, C.Y.; Messing, R.H.; Vargas, R.I. Sexual communication and related behaviours in Tephritidae: Current knowledge and potential applications for Integrated Pest Management. *J. Pest Sci.* **2014**, *87*, 385–405. [CrossRef]

36. Roitberg, B.D.; Prokopy, R.J. Insects that mark host plants. *Bioscience* **1987**, *37*, 400–406. [CrossRef]

37. Hoffmeister, T.S.; Roitberg, B.D. Counterespionage in an insect herbivore-parasitoid system. *Naturwissenschaften* **1997**, *84*, 117–119. [CrossRef]

38. White, I.M.; Elson-Harris, M.M. *Fruit Flies of Economic Significance: Their Identification and Bionomics*; CAB International: Wallingford, UK, 1992.

39. Roskov, Y.; Ower, G.; Orrell, T.; Nicolson, D.; Bailly, N.; Kirk, P.M.; Bourgoin, T.; DeWalt, R.E.; Decock, W.; van Nieukerken, E.; et al. *Species 2000 & ITIS Catalogue of Life, 2019 Annual Checklist 2019*; Naturalis: Leiden, The Netherlands, 2019.

40. Malavasi, A. Introductory remarks. In *Trapping and the Detection, Control, and Regulation of Tephritid Fruit Flies*; Shelly, T.E., Epsky, N., Jang, E.B., Reyes-Flores, J., Vargas, R.I., Eds.; Springer: Dordrecht, The Netherlands, 2014.

41. Clarke, A.R. *Biology and Management of Bactrocera and Related Fruit Flies*; CABI: Wallingford, UK, 2019.

42. Nugnes, F.; Russo, E.; Viggiani, G.; Bernardo, U. First record of an invasive fruit fly belonging to *Bactrocera dorsalis* complex (Diptera: Tephritidae) in Europe. *Insects* **2018**, *9*, 182. [CrossRef]

43. Malacrida, A.R.; Gomulski, L.M.; Bonizzoni, M.; Bertin, S.; Gasperi, G.; Guglielmino, C.R. Globalization and fruitfly invasion and expansion: The medfly paradigm. *Genetica* **2007**, *131*, 1–9. [CrossRef]

44. Papadopoulos, N.T. Fruit Fly Invasion: Historical, Biological, Economic Aspects and Management. In *Trapping and the Detection, Control, and Regulation of Tephritid Fruit Flies*; Springer: Dordrecht, The Netherlands, 2014; pp. 219–252. ISBN 9789401791939.

45. Vreysen, M.J.B.; Robinson, A.S.; Hendrichs, J. *Area-Wide Control of Insect Pests from Research to Field Implementation*; Springer: Dordrecht, The Netherlands, 2007.

46. Sivinski, J.; Aluja, M.; Dodson, G.N.; Freidberg, A.; Headrick, D.H.; Kaneshiro, K.Y.; Landolt, P.J. Topics in the evolution of fruit fly mating behavior. In *Fruit Flies (Tephritidae): Phylogeny and Evolution of Behavior*; Aluja, M., Norrbom, A., Eds.; CRC Press: Boca Raton, FL, USA, 2000; pp. 751–792.

47. Feron, M. L'instinct de reproduction chez la mouche méditerranéenne des fruits *Ceratitis capitata* Wied. (Dipt. Trypetidae). Comportement sexuel. Comportement de ponte. *Rev. Pathol. Végétale d'Entomol. Agric. Fr.* **1962**, *41*, 1–129.

48. Ekanayake, W.M.T.D.; Clarke, A.R.; Schutze, M.K. Close-distance courtship of laboratory reared *Bactrocera tryoni* (Diptera: Tephritidae). *Austral. Entomol.* **2019**, *58*, 578–588. [CrossRef]

49. Benelli, G.; Canale, A.; Bonsignori, G.; Ragni, G.; Stefanini, C.; Raspi, A. Male wing vibration in the mating behavior of the olive fruit fly *Bactrocera oleae* (Rossi) (Diptera: Tephritidae). *J. Insect Behav.* **2012**, *25*, 590–603. [CrossRef]

50. Poramarcom, R. Sexual communication in the Oriental fruit fly, *Dacus dorsalis* Hendel (Diptera: Tephritidae). Ph.D. Thesis, University of Hawaii, Honolulu, HI, USA, 1988.

51. Benelli, G.; Giunti, G.; Canale, A.; Messing, R.H. Lek dynamics and cues evoking mating behavior in tephritid flies infesting soft fruits: Implications for behavior-based control tools. *Appl. Entomol. Zool.* **2014**, *49*, 363–373. [CrossRef]

52. Bradbury, J.W. The evolution of leks. In *Natural Selection and Social Behaviour*; Alexander, R.D., Tinkle, D.W., Eds.; Chiron Press: New York, NY, USA, 1981; pp. 138–169.

53. Shelly, T.E. Sexual selection on leks: A fruit fly primer. *J. Insect Sci.* **2018**, *18*. [CrossRef]

54. Iwahashi, O.; Majima, T. Lek formation and male-male competition in the melon fly, *Dacus cucurbitae* Coquillett (Diptera, Tephritidae). *Appl. Entomol. Zool.* **1986**, *21*, 70–75. [CrossRef]

55. Mir, S.H.; Mir, G.M. Lekking behaviour and male-male rivalry in the melon fly *Bactrocera cucurbitae* (Coquillett) (Diptera: Tephritidae). *J. Insect Behav.* **2016**, *29*, 379–384. [CrossRef]

56. Raghu, S.; Clarke, A.R. Spatial and temporal partitioning of behaviour by adult dacines: Direct evidence for methyl eugenol as a mate rendezvous cue for *Bactrocera cacuminata*. *Physiol. Entomol.* **2003**, *28*, 175–184. [CrossRef]

57. Messina, F.J.; Subler, J.K. Conspecific and heterospecific interactions of male *Rhagoletis* flies (Diptera, Tephritidae) on a shared host. *J. Kansas Entomol. Soc.* **1995**, *68*, 206–213.

58. Prokopy, R.J.; Bennett, E.W.; Bush, G.L. Mating behavior in *Rhagoletis pomonella* (Diptera: Tephritidae): II. Temporal organization. *Can. Entomol.* **1972**, *104*, 97–104. [CrossRef]

59. Smith, D.C.; Prokopy, R.J. Mating behavior of *Rhagoletis mendax* (Diptera: Tephritidae) flies in nature. *Ann. Entomol. Soc. Am.* **1982**, *75*, 388–392. [CrossRef]

60. Buda, V.; Blazyte-Cereskiene, L.; Radziute, S.; Apsegaite, V.; Stamm, P.; Schulz, S.; Aleknavicius, D.; Mozuraitis, R. Male-produced (-)-delta-heptalactone, pheromone of fruit fly *Rhagoletis batava* (Diptera: Tephritidae), a sea buckthorn berries pest. *Insects* **2020**, *11*, 138. [CrossRef] [PubMed]
61. Nishida, R.; Shelly, T.E.; Whittier, T.S.; Kaneshiro, K.Y. Alpha-copaene, a potential rendezvous cue for the mediterranean fruit fly, *Ceratitis capitata*? *J. Chem. Ecol.* **2000**, *26*, 87–100. [CrossRef]
62. Shelly, T.; Dang, C.; Kennelly, S. Exposure to orange (*Citrus sinensis* L.) trees, fruit, and oil enhances mating success of male Mediterranean fruit flies (*Ceratitis capitata* [Wiedemann]). *J. Insect Behav.* **2004**, *17*, 303–315. [CrossRef]
63. Warthen, J.D.; McInnis, D.O. Isolation and identification of male medfly attractive components in *Litchi chinensis* stems and *Ficus* spp. stem exudates. *J. Chem. Ecol.* **1989**, *15*, 1931–1946. [CrossRef] [PubMed]
64. Flath, R.A.; Cunningham, R.T.; Mon, T.R.; John, J.O. Additional male Mediterranean fruit fly (*Ceratitis capitata* Wied) attractants from Angelica seed oil (*Angelica archangelica* L). *J. Chem. Ecol.* **1994**, *20*, 1969–1984. [CrossRef]
65. Flath, R.A.; Cunningham, R.T.; Mon, T.R.; John, J.O. Male lures for Mediterranean fruit fly (*Ceratitis capitata* Wied.): Structural analogs of α-copaene. *J. Chem. Ecol.* **1994**, *20*, 2595–2609. [CrossRef]
66. Niogret, J.; Gill, M.A.; Espinoza, H.R.; Kendra, P.E. Attraction and electroantennogram responses of male Mediterranean fruit fly (Diptera: Tephritidae) to six plant essential oils. *J. Entomol. Zool. Stud.* **2017**, *5*, 958–964.
67. Nishida, R.; Tan, K.H.; Serit, M.; Lajis, N.H.; Sukari, A.M.; Takahashi, S.; Fukami, H. Accumulation of phenylpropanoids in the rectal glands of males of the Oriental fruit fly, *Dacus dorsalis*. *Experientia* **1988**, *44*, 534–536. [CrossRef]
68. Tan, K.H.; Nishida, R. Sex pheromone and mating competition after methyl eugenol consumption in *Bactrocera dorsalis* complex. In *Fruit Fly Pests—A World Assessment of Their Biology and Management*; McPheron, B.A., Steck, G.J., Eds.; St. Lucie Press: Delray Beach, FL, USA, 1996; pp. 147–153.
69. Hee, A.K.W.; Tan, K.H. Attraction of female and male *Bactrocera papayae* to conspecific males fed with methyl eugenol and attraction of females to male sex pheromone components. *J. Chem. Ecol.* **1998**, *24*, 753–764. [CrossRef]
70. Khoo, C.C.-H.; Tan, K.-H. Attraction of both sexes of melon fly, *Bactrocera cucurbitae* to conspecific males—A comparison after pharmacophagy of cue-lure and a new attractant—zingerone. *Entomol. Exp. Appl.* **2000**, *97*, 317–320. [CrossRef]
71. Wee, S.L.; Tan, K.H. Female sexual response to male rectal volatile constituents in the fruit fly, *Bactrocera carambolae* (Diptera: Tephritidae). *Appl. Entomol. Zool.* **2005**, *40*, 365–372. [CrossRef]
72. Wee, S.L.; Tan, K.H.; Nishida, R. Pharmacophagy of methyl eugenol by males enhances sexual selection of *Bactrocera carambolae* (Diptera: Tephritidae). *J. Chem. Ecol.* **2007**, *33*, 1272–1282. [CrossRef]
73. Kobayashi, R.M.; Ohinata, K.; Chambers, D.L.; Fujimoto, M.S. Sex pheromones of the Oriental fruit fly and the melon fly: Mating behavior, bioassay method, and attraction of females by live males and by suspected pheromone glands of males. *Environ. Entomol.* **1978**, *7*, 107–112. [CrossRef]
74. Drew, R.A.I.; Hooper, G.H.S.; Bateman, M.A. *Economic fruit flies of the South Pacific Region*, 2nd ed.; Queensland Department of Primary Industries: Brisbane, Australia, 1982.
75. Noushini, S.; Perez, J.; Park, S.J.; Holgate, D.; Jamie, I.; Jamie, J.; Taylor, P. Rectal gland chemistry, volatile emissions, and antennal responses of male and female banana fruit fly, *Bactrocera musae*. *Insects* **2020**, *11*, 32. [CrossRef]
76. Haniotakis, G.E. Sexual attraction in the olive fruit fly *Dacus oleae* (Gmelin). *Environ. Entomol.* **1974**, *3*, 82–86. [CrossRef]
77. Haniotakis, G.E. Male olive fly attraction to virgin females in the field. *Ann. Zool. Ecol. Anim.* **1977**, *9*, 273–276.
78. Baker, R.; Herbert, R.; Howse, P.E.; Jones, O.T.; Francke, W.; Reith, W. Identification and synthesis of the major sex pheromone of the olive fly (*Dacus oleae*). *J. Chem. Soc. Chem. Commun.* **1980**, *1*, 52. [CrossRef]
79. Carpita, A.; Canale, A.; Raffaelli, A.; Saba, A.; Benelli, G.; Raspi, A. (Z)-9-tricosene identified in rectal gland extracts of *Bactrocera oleae* males: First evidence of a male-produced female attractant in olive fruit fly. *Naturwissenschaften* **2012**, *99*, 77–81. [CrossRef]
80. De Marzo, L.; Nuzzaci, L.; Solinas, M. Studio anatomico, istologico, ultrastrutturale e fisiologico del retto ed osservazioni etologiche in relazione alla possibile produzione di feromoni sessuali nel maschio di *Dacus oleae* (Gmelin). *Entomologica* **1978**, *XIV*, 203–266.
81. Mavraganis, V.G.; Papadopoulos, N.T.; Kouloussis, N.A. Extract of olive fruit fly males (Diptera: Tephritidae) attract virgin females. *Entomol. Hell.* **2017**, *19*, 14. [CrossRef]
82. Levi-Zada, A.; Nestel, D.; Fefer, D.; Nemni-Lavy, E.; Deloya-Kahane, I.; David, M. Analyzing diurnal and age-related pheromone emission of the olive fruit fly, *Bactrocera oleae* by sequential SPME-GCMS analysis. *J. Chem. Ecol.* **2012**, *38*, 1036–1041. [CrossRef]
83. Benelli, G. Aggressive behavior and territoriality in the olive fruit fly, *Bactrocera oleae* (Rossi) (Diptera: Tephritidae): Role of residence and time of day. *J. Insect Behav.* **2014**, *27*, 145–161. [CrossRef]
84. Canale, A.; Benelli, G.; Germinara, G.S.; Fusini, G.; Romano, D.; Rapalini, F.; Desneux, N.; Rotundo, G.; Raspi, A.; Carpita, A. Behavioural and electrophysiological responses to overlooked female pheromone components in the olive fruit fly, *Bactrocera oleae* (Diptera: Tephritidae). *Chemoecology* **2015**, *25*, 147–157. [CrossRef]
85. Eberhard, W.G. Sexual behavior and sexual selection in the Mediterranean fruit fly, *Ceratitis capitata* (Dacinae: Ceratitidini). In *Fruit flies (Tephritidae): Phylogeny and Evolution of Behavior*; Aluja, M., Norrbom, A.L., Eds.; CRC Press: Boca Raton, FL, USA, 2000; pp. 459–489.
86. Robacker, D.C.; Hart, W.G. Courtship and territoriality of laboratory-reared Mexican fruit-flies, *Anastrepha ludens* (Diptera, Tephritidae), in cages containing host and nonhost trees. *Ann. Entomol. Soc. Am.* **1985**, *78*, 488–494. [CrossRef]

87. Silva, M.T.; Polloni, Y.J.; Bressan, S. Mating behavior of some fruit flies of the genus *Anastrepha* Schiner, 1868 (Diptera: Tephritidae) in the laboratory. *Rev. Bras. Entomol.* **1985**, *29*, 155–164.

88. Tychsen, P.H. Mating behavior of the Queensland fruit fly, *Dacus tryoni* (Diptera: Tephritidae), in field cages. *J. Aust. Entomol. Soc.* **1977**, *16*, 459–465. [CrossRef]

89. Vaníčková, L.; Svatos, A.; Kroiss, J.; Kaltenpoth, M.; Do Nascimento, R.R.; Hoskovec, M.; Břízová, R.; Kalinova, B. Cuticular hydrocarbons of the South American fruit fly *Anastrepha fraterculus*: Variability with sex and age. *J. Chem. Ecol.* **2012**, *38*, 1133–1142. [CrossRef] [PubMed]

90. Aluja, M.; Norrbom, A. *Fruit flies (Tephritidae) Phylogeny and Evolution of Behavior*; CRC Press: Boca Raton, FL, USA, 2000.

91. Vaníčková, L.; Břízová, R.; Pompeiano, A.; Ekesi, S.; De Meyer, M. Cuticular hydrocarbons corroborate the distinction between lowland and highland Natal fruit fly (Tephritidae, *Ceratitis rosa*) populations. *Zookeys* **2015**, 507–524. [CrossRef]

92. Vaníčková, L.; Břízová, R.; Mendonca, A.L.; Pompeiano, A.; Do Nascimento, R.R. Intraspecific variation of cuticular hydrocarbon profiles in the *Anastrepha fraterculus* (Diptera: Tephritidae) species complex. *J. Appl. Entomol.* **2015**, *139*, 679–689. [CrossRef]

93. Lewis, S.M.; Vahed, K.; Koene, J.M.; Engqvist, L.; Bussière, L.F.; Perry, J.C.; Gwynne, D.; Lehmann, G.U.C. Emerging issues in the evolution of animal nuptial gifts. *Biol. Lett.* **2014**, *10*, 20140336. [CrossRef]

94. Vahed, K. The function of nuptial feeding in insects: A review of empirical studies. *Biol. Rev.* **2007**, *73*, 43–78. [CrossRef]

95. Benelli, G.; Romano, D. Does indirect mating trophallaxis boost male mating success and female egg load in Mediterranean fruit flies? *J. Pest Sci.* **2018**, *91*, 181–188. [CrossRef]

96. Aluja, M.; Jácome, I.; Birke, A.; Lozada, N.; Quintero, G. Basic patterns of behavior in wild *Anastrepha striata* (Diptera: Tephritidae) flies under field-cage conditions. *Ann. Entomol. Soc. Am.* **1993**, *86*, 776–793. [CrossRef]

97. Perez-Staples, D.; Aluja, M. *Anastrepha striata* (Diptera: Tephritidae) females that mate with virgin males live longer. *Ann. Entomol. Soc. Am.* **2004**, *97*, 1336–1341. [CrossRef]

98. Guillen, L.; Pascacio-Villafan, C.; Stoffolano, J.G.; Lopez-Sanchez, L.; Velazquez, O.; Rosas-Saito, G.; Altuzar-Molina, A.; Ramirez, M.; Aluja, M. Structural differences in the digestive tract between females and males could modulate regurgitation behavior in *Anastrepha ludens* (Diptera: Tephritidae). *J. Insect Sci.* **2019**, *19*. [CrossRef]

99. Sivinski, J.M.; Epsky, N.; Heath, R.R. Pheromone deposition on leaf territories by male Caribbean fruit flies, *Anastrepha suspensa* (loew) (Diptera: Tephritidae). *J. Insect Behav.* **1994**, *7*, 43–51. [CrossRef]

100. Papaj, D.R.; Garcia, J.M.; Alonso-Pimentel, H. Marking of host fruit by male *Rhagoletis boycei* Cresson flies (Diptera: Tephritidae) and its effect on egg-laying. *J. Insect Behav.* **1996**, *9*, 585–598. [CrossRef]

101. Jang, E.B. Effects of mating and accessory-gland injections on olfactory-mediated behavior in the female Mediterranean fruit fly, *Ceratitis capitata*. *J. Insect Physiol.* **1995**, *41*, 705–710. [CrossRef]

102. Gomulski, L.M.; Dimopoulos, G.; Xi, Z.Y.; Scolari, F.; Gabrieli, P.; Siciliano, P.; Clarke, A.R.; Malacrida, A.R.; Gasperi, G. Transcriptome profiling of sexual maturation and mating in the Mediterranean fruit fly, *Ceratitis capitata*. *PLoS ONE* **2012**, *7*. [CrossRef]

103. Zheng, W.W.; Luo, D.Y.; Wu, F.Y.; Wang, J.L.; Zhang, H.Y. RNA sequencing to characterize transcriptional changes of sexual maturation and mating in the female oriental fruit fly *Bactrocera dorsalis*. *BMC Genomics* **2016**, *17*. [CrossRef]

104. Kumaran, N.; van der Burg, C.A.; Qin, Y.J.; Cameron, S.L.; Clarke, A.R.; Prentis, P.J. Plant-mediated female transcriptomic changes post-mating in a tephritid fruit fly, *Bactrocera tryoni*. *Genome Biol. Evol.* **2018**, *10*, 94–107. [CrossRef]

105. Campanini, E.B.; Congrains, C.; Torres, F.R.; de Brito, R.A. Odorant-binding proteins expression patterns in recently diverged species of *Anastrepha* fruit flies. *Sci. Rep.* **2017**, *7*, 2194. [CrossRef]

106. Devescovi, F.; Hurtado, J.; Taylor, P.W. Mating-induced changes in responses of female Queensland fruit fly to male pheromones and fruit: A mechanism for mating-induced sexual inhibition. *J. Insect Physiol.* **2021**, *129*, 104195. [CrossRef]

107. Averill, A.L.; Prokopy, R.J. Intraspecific competition in the tephritid fruit fly *Rhagoletis pomonella*. *Ecology* **1987**, *68*, 878–886. [CrossRef]

108. Prokopy, R.J.; Roitberg, B.D. Fruit fly foraging behavior. In *Fruits Flies, Their Biology, Natural Enemies and Control*; Robinson, A.S., Hooper, G., Eds.; Elsevier: Amsterdam, The Netherlands, 1989; pp. 293–306.

109. Fletcher, B.S.; Prokopy, R.J. Host location and oviposition in tephritid fruit flies. In *Reproductive Behaviour of Insects: Individuals and Populations*; Bailey, W.J., Ridsdill-Smith, J., Eds.; Chapman & Hall: London, UK, 1991; pp. 141–171.

110. Prokopy, R.J.; Ziegler, J.R.; Wong, T.T.Y. Deterrence of repeated oviposition by fruit-marking pheromone in *Ceratitis capitata* (Diptera: Tephritidae). *J. Chem. Ecol.* **1978**, *4*, 55–63. [CrossRef]

111. Prokopy, R.J. Evidence for a marking pheromone deterring repeated oviposition in apple maggot flies. *Environ. Entomol.* **1972**, *1*, 326–332. [CrossRef]

112. Díaz-Fleischer, F.; Papaj, D.R.; Prokopy, R.J.; Norrbom, A.L.; Aluja, M. Evolution of fruit fly oviposition behavior. In *Fruit Flies (Tephritidae): Phylogeny and Evolution of Behavior*; Aluja, M., Norrbom, A.L., Eds.; CRC Press: Boca Raton, FL, USA, 2000; pp. 811–841.

113. Silva, M.A.; Bezerra-Silva, G.C.D.; Mastrangelo, T. The host marking pheromone application on the management of fruit flies—A Review. *Braz. Arch. Biol. Technol.* **2012**, *55*, 835–842. [CrossRef]

114. Leal, T.A.B.S.; Zucoloto, F.S. Selection of artificial hosts for oviposition by wild *Anastrepha obliqua* (Macquart) (Diptera, Tephritidae): Influence of adult food and effect of experience. *Rev. Bras. Entomol.* **2008**, *52*, 467–471. [CrossRef]

115. Crnjar, R.M.; Prokopy, R.J. Morphological and electrophysiological mapping of tarsal chemoreptors of oviposition deterring pheromone in *Rhagoletis pomonella* flies. *J. Insect Physiol.* **1982**, *28*, 393–400. [CrossRef]

116. Zhang, G.N.; Hu, F.; Dou, W.; Wang, J.J. Morphology and distribution of sensilla on tarsi and ovipositors of six fruit flies (Diptera: Tephritidae). *Ann. Entomol. Soc. Am.* **2012**, *105*, 319–327. [CrossRef]

117. Eisemann, C.H.; Rice, M.J. Behavioural evidence for hygro- and mechanoreception by ovipositor sensilla of *Dacus tryoni* (Diptera: Tephritidae). *Physiol. Entomol.* **1989**, *14*, 273–277. [CrossRef]

118. Liscia, A.; Crnjar, R.; Angioy, A.M.; Pietra, P.; Stoffolano, J.G.J. I chemosensilli dell'ovopositore in *Tabanus nigrovittatus* (Macq.), *Chrysops fuliginosus* (Wied.), e *Rhagoletis pomonella* (Walsh). *Boll. Soc. Ital. Biol. Sper.* **1982**, *58*, 1325–1329.

119. Girolami, V.; Crnjar, R.; Angioy, A.M.; Strapazzon, A.; Pietra, P.; Stoffolano, J.G.J.; Prokopy, R.J. Behavior and sensory physiology of *Rhagoletis pomonella* in relation to oviposition stimulants and deterrents in fruit. In *Fruitflies of Economic Importance*; Cavalloro, R., Ed.; Balkema Pub: Rotterdam, The Netherlands; Boston, MA, USA, 1986; pp. 183–190.

120. Stoffolano, J.G.; Yin, L.R.S. Structure and function of the ovipositor and associated sensilla of the apple maggot, *Rhagoletis pomonella* (Walsh) (Diptera, Tephritidae). *Int. J. Insect Morphol. Embryol.* **1987**, *16*, 41–69. [CrossRef]

121. Norrbom, A.L.; Kim, K.C. Revision of the schausi group of *Anastrepha* Schiner (Diptera: Tephritidae), with a discussion of the terminology of the female terminalia in the Tephritoidea. *Ann. Entomol. Soc. Am.* **1988**, *81*, 164–173. [CrossRef]

122. Papaj, D.R.; Aluja, M. Temporal dynamics of host-marking in the tropical tephritid fly, *Anastrepha ludens*. *Physiol. Entomol.* **1993**, *18*, 279–284. [CrossRef]

123. Papaj, D.R.; Roitberg, B.D.; Opp, S.B.; Aluja, M.; Prokopy, R.J.; Wong, T.T.Y. Effect of marking pheromone on clutch size in the Mediterranean fruit-fly. *Physiol. Entomol.* **1990**, *15*, 463–468. [CrossRef]

124. Papaj, D.R.; Roitberg, B.D.; Opp, S.B. Serial effects of host infestation on egg allocation by the Mediterranean fruit-fly—A rule of thumb and its functional-significance. *J. Anim. Ecol.* **1989**, *58*, 955–970. [CrossRef]

125. Papaj, D.R.; Averill, A.L.; Prokopy, R.J.; Wong, T.T.Y. Host-marking pheromone and use of previously established oviposition sites by the Mediterranean fruit-fly (Diptera, Tephritidae). *J. Insect Behav.* **1992**, *5*, 583–598. [CrossRef]

126. Roitberg, B.D.; van Lenteren, J.C.; van Alphen, J.M.M.; Galis, F.; Prokopy, R.J. Foraging of *Rhagoletis pomonella*, a parasite of hawthorn (Crataegus), in nature. *J. Anim. Ecol.* **1982**, *48*, 307–325. [CrossRef]

127. Roitberg, B.D.; Cairl, R.S.; Prokopy, R.J. Oviposition deterring pheromone influences dispersal distance in tephritid fruit flies. *Entomol. Exp. Appl.* **1984**, *35*, 217–220. [CrossRef]

128. Nufio, C.R.; Papaj, D.R. Host-marking behaviour as a quantitative signal of competition in the walnut fly *Rhagoletis juglandis*. *Ecol. Entomol.* **2004**, *29*, 336–344. [CrossRef]

129. Averill, A.L.; Prokopy, R.J. Oviposition-deterring fruit marking pheromone in *Rhagoletis basiola*. *Florida Entomol.* **1981**, *64*, 221–226. [CrossRef]

130. Averill, A.L.; Prokopy, R.J. Oviposition-deterring fruit marking pheromone in *Rhagoletis zephyria*. *J. Georg. Entomol. Soc.* **1982**, *17*, 315–319.

131. Aluja, M.; Diaz-Fleischer, F. Foraging behavior of *Anastrepha ludens*, *A. obliqua*, and *A. serpentina* in response to feces extracts containing host marking pheromone. *J. Chem. Ecol.* **2006**, *32*, 367–389. [CrossRef]

132. Prokopy, R.J.; Papaj, D.R. Behavior of flies of the genera *Rhagoletis*, *Zonosemata*, and *Carpomya*. In *Fruit Flies (Tephritidae): Phylogeny and Evolution of Behavior*; Aluja, M., Norrbom, A.L., Eds.; CRC Press: Boca Raton, FL, USA, 2000; pp. 219–252.

133. Papaj, D.R. Oviposition site guarding by male walnut flies and its possible consequences for mating success. *Behav. Ecol. Sociobiol.* **1994**, *34*, 187–195. [CrossRef]

134. Cirio, U.; Italiana, S.E. Osservazioni sul comportamento di ovideposizione della *Rhagoletis completa* Cresson (Diptera, Trypetidae) in laboratorio. In Proceedings of the Atti del IX Congresso Nazionale Italiano di Entomologia, Tipografia Bertelly & Picardi, Siena, Italy, 21–25 June 1972; Volume 1, pp. 99–117.

135. Papaj, D.R. Use and avoidance of occupied hosts as a dynamic process in tephritid flies. In *Insect-Plant Interactions*; Bernays, E.A., Ed.; CRC Press: Boca Raton, FL, USA, 1993; Volume 5, pp. 25–46.

136. Lalonde, R.G.; Mangel, M. Seasonal effects on superparasitism by *Rhagoletis completa*. *J. Anim. Ecol.* **1994**, *63*, 583–588. [CrossRef]

137. Prokopy, R.J.; Green, T.A.; Olson, W.A.; Vargas, R.F.; Kaneshia, D.; Wong, T.Y. Discrimination by *Dacus dorsalis* females (Diptera: Tephritidae) against larval-infested fruit. *Florida Entomol.* **1989**, *72*, 319–323. [CrossRef]

138. Fitt, G.P. Oviposition behaviour of two tephritid fruit flies, *Dacus tryoni* and *Dacus jarvisi*, as influenced by the presence of larvae in the host fruit. *Oecologia* **1984**, *62*, 37–46. [CrossRef]

139. Prokopy, R.J.; Koyama, J. Oviposition site partitioning in *Dacus cucurbitae*. *Entomol. Exp. Appl.* **1982**, *31*, 428–432. [CrossRef]

140. Cirio, U. Reperti sul meccanismo stimolo-risposta nell'ovideposizione del *Dacus oleae* (Gmelin) (Diptera: Trypetidae). *Redia* **1971**, *52*, 10.

141. Girolami, V.; Vianello, A.; Strapazzon, A.; Ragazzi, E.; Veronese, G. Ovipositional deterrents in *Dacus oleae*. *Entomol. Exp. Appl.* **1981**, *29*, 177–188. [CrossRef]

142. Lo Scalzo, R.; Scarpati, M.L.; Verzegnassi, B.; Vita, G. *Olea europaea* chemicals repellent to *Dacus oleae* females. *J. Chem. Ecol.* **1994**, *20*, 1813–1823. [CrossRef]

143. Arita, L.H.; Kaneshiro, K.Y. Structure and function of the rectal epithelium and anal glands during mating behavior in the Mediterranean fruit fly. *Proc. Hawaiian Entomol. Soc.* **1986**, *26*, 27–30.

144. Quilici, S.; Franck, A.; Peppuy, A.; Correia, E.D.R.; Mouniama, C.; Blard, F. Comparative studies of courtship behavior of *Ceratitis* spp. (Diptera: Tephritidae) in Reunion island. *Florida Entomol.* **2002**, *85*, 138–142. [CrossRef]

145. Fletcher, B.S. Storage and release of a sex pheromone by the Queensland fruit fly, *Dacus tryoni* (Diptera: Trypetidae). *Nature* **1968**, *219*, 631–632. [CrossRef]

146. Kuba, H.; Sokei, Y. The production of pheromone clouds by spraying in the melon fly, *Dacus cucurbitae* Coquillett (Diptera, Tephritidae). *J. Ethol.* **1988**, *6*, 105–110. [CrossRef]

147. Poramarcom, R.; Baimai, V. Sexual behavior and signals used for mating of *Bactrocera correcta*. In *Fruit Fly Pests: A World Assessment of Their Biology and Management*; McPheron, B.A., Steck, G.J., Eds.; St. Lucie Press: Delray Beach, FL, USA, 1996; pp. 51–58.

148. Nation, J.L. Biology of pheromone release by male Caribbean fruit-flies, *Anastrepha suspensa* (Diptera, Tephritidae). *J. Chem. Ecol.* **1990**, *16*, 553–572. [CrossRef]

149. Nation, J.L. The role of pheromones in the mating system of *Anastrepha* fruit flies. In *Fruit Flies: Their Biology, Natural Enemies and Control*; Robinson, A.S., Hooper, G., Eds.; Elsevier Science Publishers: Amsterdam, The Netherlands, 1989; Volume 16, pp. 189–205.

150. Nation, J.L. Courtship behavior and evidence for a sex attractant in the male Caribbean fruit fly, *Anastrepha suspensa*. *Ann. Entomol. Soc. Am.* **1972**, *65*, 1364–1367. [CrossRef]

151. Yip, E.C.; Mikó, I.; Ulmer, J.M.; Cherim, N.A.; Townley, M.A.; Poltak, S.; Helms, A.M.; De Moraes, C.M.; Mescher, M.C.; Tooker, J.F. Giant polyploid epidermal cells and male pheromone production in the tephritid fruit fly *Eurosta solidaginis* (Diptera: Tephritidae). *J. Insect Physiol.* **2021**, *130*, 104210. [CrossRef]

152. Caetano, F.H.; Solferini, V.N.; de Britto, F.B.; Lins, D.S.; Aluani, T.; de Brito, V.G.; Zara, F.J. Ultra morphology of the digestive system of *Anastrepha fraterculus* and *Ceratitis capitata* (Diptera Tephritidae). *Braz. J. Morphol. Sci.* **2006**, *23*, 455–462.

153. Barros, M.D.; Malavasi, A. Morphology of adult male rectum of seven species of *Anastrepha* from Brazil and mating behavior correlations. In *Fruit Fly Pests—A World Assessment of Their Biology and Management*; McPheron, B.A., Steck, G.J., Eds.; Taylor & Francis Inc.: Abingdon, UK, 1996.

154. Lee, L.W.Y.; Chang, T.H. Morphology of sex-pheromone gland in male oriental fruit fly and its suspected mechanism of pheromone release. In Proceedings of the Second International Symposium, 16–21 September; Economopoulos, A.P., Ed.; Symposium Organizing Committee: Kolymbari, Crete, Greece, 1986; pp. 16–21.

155. Economopoulos, A.P.; Gianakakis, A.; Tzanakakis, M.E.; Voyatzoglou, A. Reproductive behaviour and physiology of the olive fruit fly. Anatomy of the adult rectum and odours emitted by adults. *Ann. Entomol. Soc. Am.* **1971**, *64*, 1112–1116. [CrossRef]

156. Khoo, C.C.H.; Tan, K.H. Rectal gland of *Bactrocera papayae*: Ultrastructure, anatomy and sequestration of auto fluorescent compounds upon methyl eugenol consumption by the male fly. *Microsc. Res. Tech.* **2005**, *67*, 219–226. [CrossRef] [PubMed]

157. Arita, L.H. The Mating Behavior of the Mediterranean Fruit Fly, *Ceratitis capitata* (Wiedemann). Ph.D. Thesis, University of Hawaii, Honolulu, HI, USA, 1983.

158. Gomez Cendra, P.; Calcagno, G.; Belluscio, L.; Vilardi, J.C. Male courtship behavior of the South American fruit fly, *Anastrepha fraterculus*, from an Argentinean laboratory strain. *J. Insect Sci.* **2011**, *11*, 1–18. [CrossRef] [PubMed]

159. Nation, J.L. Sex-specific glands in tephritid fruit flies of the genera *Anastrepha*, *Ceratitis*, *Dacus* and *Rhagoletis* (Diptera: Tephritidae). *Int. J. Insect Morphol. Embryol.* **1981**, *4*, 27–30. [CrossRef]

160. Llosie, J.; Roche, A. Organes odoriferants des males de *Ceratitis capitata*. *Bull. Soc. Entomol. Fr.* **1960**, *65*, 206–209.

161. Fletcher, B.S. The structure and the function of the sex pheromone glands of the male Queensland fruit fly, *Dacus tryoni*. *J. Insect Physiol.* **1969**, *15*, 1309–1322. [CrossRef]

162. Teles, M.C. Structure and development of specific sex glands in males of some Brazilian fruit flies of the genus *Anastrepha* Schiner, 1868 (Diptera: Tephritidae). In Proceedings of the CEC/IOBC International Symposium on Fruit Flies of Economic Importance 87, Rome, Italy, 7–10 April 1987; pp. 179–189.

163. Lima, I.S.; Howse, P.E.; Stevens, D.R. Volatile components from the salivary glands of calling males of the south American fruit fly, *Anastrepha fraterculus*: Partial identification and behavioural activity. In *Fruit Fly Pests. A World Assessment of Their Biology and Management*; MacPheron, B.A., Steck, G.J., Eds.; St. Lucie Press: Delray Beach, FL, USA, 1996; pp. 107–113.

164. Nation, J.L. The structure and development of two sex specific glands in male Caribbean fruit flies. *Ann. Entomol. Soc. Am.* **1974**, *67*, 731–734. [CrossRef]

165. Goncalves, G.B.; Silva, C.E.; Dos Santos, J.C.G.; Dos Santos, E.S.; Do Nascimento, R.R.; Da Silva, E.L.; Mendonca, A.D.L.; De Freitas, M.D.; Sant'Ana, A.E.G. Comparison of the volatile components released by calling males of *Ceratitis capitata* (Diptera: Tephritidae) with those extractable from the salivary glands. *Florida Entomol.* **2006**, *89*, 375–379. [CrossRef]

166. Heath, R.R.; Landolt, P.J.; Robacker, D.C.; Dueben, B.D.; Epsky, N.D. Sexual pheromones of tephritid flies: Clues to unravel phylogeny and behavior. In *Fruit Flies (Tephritidae): Phylogeny and Evolution of Behavior*; Aluja, M., Norrbom, A.L., Eds.; CRC Press: Boca Raton, FL, USA, 2000; pp. 793–809.

167. Nishida, R.; Tan, K.H.; Takahashi, S.; Fukami, H. Volatile components of male rectal glands of the melon fly, *Dacus cucurbitae* Coquillett (Diptera, Tephritidae). *Appl. Entomol. Zool.* **1990**, *25*, 105–112. [CrossRef]

168. Nishida, R. Ecological significance of male fruit fly attractants. In Proceedings of the International Symposium Biology Control Fruit Flies, University of Ryukyus, Ginowan, Japan; Kawasaki, K., Iwahashi, O., Kaneshiro, K.Y., Eds.; 1991; pp. 246–254.

169. Tan, K.H.; Nishida, R. Methyl eugenol: Its occurrence, distribution, and role in nature, especially in relation to insect behavior and pollination. *J. Insect Sci.* **2012**, *12*, 1–74. [CrossRef]
170. Tan, K.H.; Nishida, R.; Jang, E.B.; Shelly, T.E. Pheromones, Male Lures, and Trapping of Tephritid Fruit Flies. In *Trapping and the Detection, Control, and Regulation of Tephritid Fruit Flies*; Shelly, T.E., Epsky, N., Jang, E.B., Reyes-Flores, J., Vargas, R., Eds.; Springer: New York, NY, USA, 2014; Volume 2, pp. 15–74.
171. Shelly, T.E.; Dewire, A.L.M. Chemically mediated mating success in male Oriental fruit-flies (Diptera, Tephritidae). *Ann. Entomol. Soc. Am.* **1994**, *87*, 375–382. [CrossRef]
172. Light, D.M.; Jang, E.B.; Binder, R.G.; Flath, R.A.; Kint, S. Minor and intermediate components enhance attraction of female Mediterranean fruit flies to natural male odor pheromone and its synthetic major components. *J. Chem. Ecol.* **1999**, *25*, 2757–2777. [CrossRef]
173. Mazomenos, B.E.; Haniotakis, G.E. A multicomponent female sex pheromone of *Dacus oleae* Gmelin: Isolation and bioassay. *J. Chem. Ecol.* **1981**, *7*, 437–444. [CrossRef]
174. Milet-Pinheiro, P.; Navarro, D.M.A.; De Aquino, N.C.; Ferreira, L.L.; Tavares, R.F.; da Silva, R.D.C.; Lima-Mendonca, A.; Vanickova, L.; Mendonca, A.L.; Do Nascimento, R.R. Identification of male-borne attractants in *Anastrepha fraterculus* (Diptera: Tephritidae). *Chemoecology* **2015**, *25*, 115–122. [CrossRef]
175. Caceres, C.; Segura, D.F.; Vera, M.T.; Wornoayporn, V.; Cladera, J.L.; Teal, P.; Sapountzis, P.; Bourtzis, K.; Zacharopoulou, A.; Robinson, A.S. Incipient speciation revealed in *Anastrepha fraterculus* (Diptera; Tephritidae) by studies on mating compatibility, sex pheromones, hybridization, and cytology. *Biol. J. Linn. Soc.* **2009**, *97*, 152–165. [CrossRef]
176. Břízová, R.; Mendonca, A.L.; Vaníčková, L.; Mendonca, A.L.; Da Silva, C.E.; Tomcala, A.; Paranhos, B.A.J.; Dias, V.S.; Joachim-Bravo, I.S.; Hoskovec, M.; et al. Pheromone analyses of the *Anastrepha fraterculus* (Diptera: Tephritidae) cryptic species complex. *Florida Entomol.* **2013**, *96*, 1107–1115. [CrossRef]
177. Lima, I.S.; House, P.E.; do Nascimento, R.R. Volatile substances from male *Anastrepha fraterculus* Wied. (Diptera: Tephritidae): Identification and behavioural activity. *J. Braz. Chem. Soc.* **2001**, *12*, 196–201. [CrossRef]
178. Juárez, M.L.; Devescovi, F.; Břízová, R.; Bachmann, G.; Segura, D.F.; Kalinová, B.; Fernández, P.; Ruiz, M.J.; Yang, J.; Teal, P.E.A.; et al. Evaluating mating compatibility within fruit fly cryptic species complexes and the potential role of sex pheromones in pre-mating isolation. *Zookeys* **2015**, *540*, 125–155. [CrossRef]
179. Rocca, J.R.; Nation, J.L.; Strekowski, L.; Battiste, M.A. Comparison of volatiles emitted by male Caribbean and Mexican fruit-flies. *J. Chem. Ecol.* **1992**, *18*, 223–244. [CrossRef]
180. Robacker, D.C. Behavioral responses of female Mexican fruit flies, *Anastrepha ludens*, to components of male-produced sex pheromone. *J. Chem. Ecol.* **1988**, *14*, 1715–1726. [CrossRef]
181. Robacker, D.C.; Hart, W.G. (Z)-3-nonenol, (Z,Z)-3,6-nonandienol and (S,S)-(-)-epianastrephin: Male produced pheromone of the Mexican fruit fly. *Entomol. Exp. Appl.* **1985**, *39*, 103–108. [CrossRef]
182. Esponda-Gaxiola, R.E. Contribución al Estudio Quimico Del Atrayente Sexual de la Mosca Mexicana de la Fruta, *Anastrepha ludens* (Loew). Bachelor's Thesis, Monterrey Institute of Technology and Higher Studies, Monterrey, Mexico, 1977.
183. Stokes, J.B.; Uebel, E.C.; Warthen, J.D., Jr.; Jacobson, M.; Flippen-Anderson, J.L.; Gilardi, R.; Spishakoff, L.M.; Wilzer, K.R. Isolation and identification of novel lactones from male Mexican fruit flies. *J. Agric. Food Chem.* **1983**, *31*, 1162–1167. [CrossRef]
184. Battiste, M.A.; Strekowski, L.; Vanderbilt, D.P.; Visnick, M.; King, R.W.; Nation, J.L. Anastrephin and epianastrephin, novel lactone components isolated from the sex pheromone blend of male Caribbean and Mexican fruit flies. *Tetrahedron Lett.* **1983**, *24*, 2611–2614. [CrossRef]
185. Bosa, C.F.; Cruz-López, L.; Zepeda-Cisneros, C.S.; Valle-Mora, J.; Guillén-Navarro, K.; Liedo, P. Sexual behavior and male volatile compounds in wild and mass-reared strains of the Mexican fruit fly *Anastrepha ludens* (Loew) (Diptera: Tephritidae) held under different colony management regimes. *Insect Sci.* **2016**, *23*, 105–116. [CrossRef]
186. Gonçalves, G.B.; Silva, C.E.; Mendonça, A.D.L.; Vaníčková, L.; Tomčala, A.; Do Nascimento, R.R. Pheromone communication in *Anastrepha obliqua* (Diptera: Tephritidae): A comparison of the volatiles and salivary gland extracts of two wild populations. *Florida Entomol.* **2013**, *96*, 1365–1374. [CrossRef]
187. Lopez-Guillen, G.; Lopez, L.C.; Malo, E.A.; Rojas, J.C. Olfactory responses of *Anastrepha obliqua* (Diptera: Tephritidae) to volatiles emitted by calling males. *Florida Entomol.* **2011**, *94*, 874–881. [CrossRef]
188. Dos Silva, C.S.; dos Melo, R.S.; Tavares, R.A.N.; de Aquino, N.C.; de Tavares, R.F.; Vanícková, L.; de Mendonça, A.L.; Daza, N.A.C.; Nascimento, R.R.; dos Melo, R.S.; et al. do Estudo comparativo do feromonio sexual de duas populacoes sul Americanas de *Anastrepha obliqua*. In *A Diversidade de Debates na Pesquisa em Química*; Atena Editora: Porto Alegre, Brazil, 2019.
189. López-Guillén, G.; Cruz-López, L.; Malo, E.A.; González-Hernández, H.; Cazares, C.L.; López-Collado, J.; Toledo, J.; Rojas, J.C. Factors influencing the release of volatiles in *Anastrepha obliqua* males (Diptera: Tephritidae). *Environ. Entomol.* **2008**, *37*, 876–882. [CrossRef]
190. Meza-Hernandez, J.S.; Hernandez, E.; Salvador-Figueroa, M.; Cruz-Lopez, L. Sexual compatibility, mating performance and sex pheromone release of mass-reared and wild *Anastrepha obliqua* (Diptera: Tephritidae) under field-cage conditions. In Proceedings of the International Fruit Fly Symposium, Stellenbosch, South Africa, 6–10 May 2002; pp. 99–104.
191. Ibanez-Lopez, A.; Cruz-Lopez, L. Glandulas salivales de *Anastrepha obliqua* (Macquart) (Diptera: Tephritidae): Analisis quimico y morfologico, y actividad biologica de los componentes volatiles. *Folia Entomol. Mex.* **2001**, *40*, 221–231.

192. Robacker, D.C.; Aluja, M.; Cosse, A.A.; Sacchetti, P. Sex pheromone investigation of *Anastrepha serpentina* (Diptera: Tephritidae). *Ann. Entomol. Soc. Am.* **2009**, *102*, 560–566. [CrossRef]

193. Lima-Mendonça, A.; de Mendonça, A.L.; Sant'Ana, A.E.G.; Nascimento, R.R. Do Semioquímicos de moscas das frutas do gênero *Anastrepha. Quim. Nova* **2014**, *37*. [CrossRef]

194. Nation, J.L. Sex pheromone components of Anastrepha suspensa and their role in mating behavior. In Proceedings of the International Symposium on the Biology and Control of Fruit Flies; Kawasaki, K., Iwahashi, O., Kaneshiro, K.Y., Eds.; University of Ryukyus: Okinawa, Ginowan, Japan, 1991; pp. 224–236.

195. Ponce, W.P.; Nation, J.L.; Emmel, T.C.; Smittle, B.J.; Teal, P.E.A. Quantitative analysis of pheromone production in irradiated Caribbean fruit fly males, *Anastrepha suspensa* (Loew). *J. Chem. Ecol.* **1993**, *19*, 3045–3056. [CrossRef]

196. Chuman, T.; Sivinski, J.; Heath, R.R.; Calkins, C.O.; Tumlinson, J.H.; Battiste, M.A.; Wydra, R.L.; Strekowski, L.; Nation, J.L. Suspensolide, a new macrolide component of male Caribbean fruit fly (*Anastrepha suspensa* [Loew]) volatiles. *Tetrahedron Lett.* **1988**, *29*, 6561–6564. [CrossRef]

197. Robacker, D.C.; Hart, W.G. Electroantennograms of male and female Caribbean fruit-flies (Diptera, Tephritidae) elicited by chemicals produced by males. *Ann. Entomol. Soc. Am.* **1987**, *80*, 508–512. [CrossRef]

198. Epsky, N.D.; Heath, R.R. Food availability and pheromone production by males of *Anastrepha suspensa* (Diptera, Tephritidae). *Environ. Entomol.* **1993**, *22*, 942–947. [CrossRef]

199. Tumlinson, J.H. Contemporary frontiers in insect semiochemical research. *J. Chem. Ecol.* **1988**, *14*, 2109–2130. [CrossRef]

200. Heath, R.R.; Manukian, A.; Epsky, N.D.; Sivinski, J.; Calkins, C.O.; Landolt, P.J. A bioassay system for collecting volatiles while simultaneously attracting tephritid fruit flies. *J. Chem. Ecol.* **1993**. [CrossRef]

201. Epsky, N.D.; Heath, R.R. Pheromone production by male *Anastrepha suspensa* (Diptera: Tephritidae) under natural light cycles in greenhouse studies. *Environ. Entomol.* **1993**, *22*, 464–469. [CrossRef]

202. Nation, J.L. Sex pheromone of the Caribbean fruit fly: Chemistry and field ecology. In *Natural Products*; Elsevier: Amsterdam, The Netherlands, 1983; pp. 109–110.

203. Baker, R.; Bacon, A. The identification of spiroacetals in the volatile secretions of two species of fruit fly (*Dacus dorsalis, Dacus cucurbitae*). *Cell. Mol. Life Sci.* **1985**, *41*, 1484–1485. [CrossRef]

204. Ohinata, K.; Jacobson, M.; Kobayashi, R.M.; Chambers, D.L.; Fujimoto, M.S.; Higa, H.H. Oriental fruit fly and melon fly: Biological and chemical studies of smoke produced by males. *J. Environ. Sci. Health Part A* **1982**, *17*, 197–216. [CrossRef]

205. Nishida, R.; Shelly, T.E.; Kaneshiro, K.Y.; Tan, K.-H. Roles of semiochemicals in mating systems: A comparison between oriental fruit fly and medfly. In *Area Wide Control Fruit Flies Other Insect Pests*; CRC Press: Boca Raton, FL, USA, 2000.

206. Haniotakis, G.; Francke, W.; Mori, K.; Redlich, H.; Schurig, V. Sex-specific activity of (R)-(-)- and (S)- (+)-1,7-dioxaspiro[5.5]undecane, the major pheromone of *Dacus oleae. J. Chem. Ecol.* **1986**, *12*, 1559–1568. [CrossRef]

207. Gariboldi, P.; Jommi, G.; Rossi, R.; Vita, G. Studies on the chemical constitution and sex pheromone activity of volatile substances emitted by *Dacus oleae. Experientia* **1982**, *38*, 441–444. [CrossRef]

208. Bellas, T.E.; Fletcher, B.S. Identification of the major components in the secretion from the rectal pheromone glands of the Queensland fruit flies *Dacus tryoni* and *Dacus neohumeralis* (Diptera:Tephritidae). *J. Chem. Ecol.* **1979**, *5*, 795–803. [CrossRef]

209. Booth, Y.K.; Schwartz, B.D.; Fletcher, M.T.; Lambert, L.K.; Kitching, W.; Voss, J.J. De A diverse suite of spiroacetals, including a novel branched representative, is released by female *Bactrocera tryoni* (Queensland fruit fly). *Chem. Commun.* **2006**, 3975. [CrossRef]

210. El-Sayed, A.M.; Venkatesham, U.; Unelius, C.R.; Sporle, A.; Pérez, J.; Taylor, P.W.; Suckling, D.M. Chemical composition of the rectal gland and volatiles released by female Queensland fruit fly, *Bactrocera tryoni* (Diptera: Tephritidae). *Environ. Entomol.* **2019**, *48*, 807–814. [CrossRef]

211. Tan, H.K.; Nishida, R. Incorporation of raspberry ketone in the rectal glands of males of the Queensland fruit fly, *Bactrocera tryoni* Frogatt (Diptera: Tephritidae). *Appl. Entomol. Zool.* **1995**, *30*, 494–497. [CrossRef]

212. Levi-Zada, A.; Levy, A.; Rempoulakis, P.; Fefer, D.; Steiner, S.; Gazit, Y.; Nestel, D.; Yuval, B.; Byers, J.A. Diel rhythm of volatile emissions of males and females of the peach fruit fly *Bactrocera zonata. J. Insect Physiol.* **2020**, *120*. [CrossRef] [PubMed]

213. Břízová, R.; Vaníčková, L.; Fatarova, M.; Ekesi, S.; Hoskovec, M.; Kalinova, B. Analyses of volatiles produced by the African fruit fly species complex (Diptera, Tephritidae). *Zookeys* **2015**, 385–404. [CrossRef]

214. Jacobson, M.; Ohinata, K.; Chambers, D.L.; Jones, W.A.; Fujimoto, M.S. Insect sex attractants. Isolation, identification, and synthesis of sex pheromones of male Mediterranean fruit fly. *J. Med. Chem.* **1973**, *13*, 248–251. [CrossRef] [PubMed]

215. Ohinata, K.; Nakagawa, S.; Fujimoto, M.; Higa, H.; Jacobson, M. Mediterranean fruit fly: Laboratory and field evaluations of synthetic sex pheromones. *J. Environ. Sci. Health Part A* **1977**. [CrossRef]

216. Ohinata, K.; Jacobson, M.; Nakagawa, S.; Urago, T.; Fujimoto, M.; Higa, H. Methyl (E)-6-nonenoate: A new Mediterranean fruit fly male attractant. *J. Econ. Entomol.* **1979**, *72*, 648–650. [CrossRef]

217. Jang, E.B.; Light, D.M.; Dickens, J.C.; McGovern, T.P.; Nagata, J.T. Electroantennogram responses of mediterranean fruit fly, *Ceratitis capitata* (Diptera: Tephritidae) to trimedlure and its trans isomers. *J. Chem. Ecol.* **1989**, *15*, 2219–2231. [CrossRef] [PubMed]

218. Flath, R.A.; Jang, E.B.; Light, D.M.; Mon, T.R.; Carvalho, L.; Binder, R.G.; John, J.O. Volatile pheromonal emissions from the male Mediterranean fruit fly—Effects of fly age and time of day. *J. Agric. Food Chem.* **1993**, *41*, 830–837. [CrossRef]

219. Merli, D.; Mannucci, B.; Bassetti, F.; Corana, F.; Falchetto, M.; Malacrida, A.R.; Gasperi, G.; Scolari, F. Larval diet affects male pheromone blend in a laboratory strain of the medfly, *Ceratitis capitata* (Diptera: Tephritidae). *J. Chem. Ecol.* **2018**, *44*, 339–353. [CrossRef]

220. Vaníčková, L.; do Nascimento, R.R.; Hoskovec, M.; Jezkova, Z.; Břízová, R.; Tomcala, A.; Kalinova, B. Are the wild and laboratory insect populations different in semiochemical emission? The case of the medfly sex pheromone. *J. Agric. Food Chem.* **2012**, *60*, 7168–7176. [CrossRef]

221. Alfaro, C.; Vacas, S.; Zarzo, M.; Navarro-Llopis, V.; Primo, J. Solid phase microextraction of volatile emissions of *Ceratitis capitata* (Wiedemann) (Diptera: Tephritidae): Influence of fly sex, age, and mating status. *J. Agric. Food Chem.* **2011**, *59*, 298–306. [CrossRef]

222. Siciliano, P.; He, X.L.; Woodcock, C.; Pickett, J.A.; Field, L.M.; Birkett, M.A.; Kalinova, B.; Gomulski, L.M.; Scolari, F.; Gasperi, G.; et al. Identification of pheromone components and their binding affinity to the odorant binding protein CcapOBP83a-2 of the Mediterranean fruit fly, *Ceratitis capitata*. *Insect Biochem. Mol. Biol.* **2014**, *48*, 51–62. [CrossRef]

223. Baker, R.; Herbert, R.H.; Grant, G.G. Isolation and identification of the sex pheromone of the Mediterranean fruit fly, *Ceratitis capitata* (Wied.). *J. Chem. Soc. Chem. Commun.* **1985**, *12*, 824–825. [CrossRef]

224. Heath, R.R.; Landolt, P.J.; Tumlinson, J.H.; Chambers, D.L.; Murphy, R.E.; Doolittle, R.E.; Dueben, B.D.; Sivinski, J.; Calkins, C.O. Analysis, synthesis, formulation, and field testing of three major components of male mediterranean fruit fly pheromone. *J. Chem. Ecol.* **1991**, *17*, 1925–1940. [CrossRef]

225. Raptopoulos, D.; Haniotakis, G.; Koutsaftikis, A.; Kelly, D.; Mavraganis, V. Biological activity of chemicals identified from extracts and volatiles of male *Rhagoletis cerasi*. *J. Chem. Ecol.* **1995**, *21*, 1287–1297. [CrossRef]

226. Robacker, D.C.; Garcia, J.A. Responses of laboratory-strain Mexican fruit flies, *Anastrepha ludens*, to combinations of fermenting fruit odor and male-produced pheromone in laboratory bioassays. *J. Chem. Ecol.* **1990**, *16*, 2027–2038. [CrossRef]

227. Perkins, M.; Fletcher, M.; Kitching, W.; Drew, R.; Moore, C. Chemical studies of rectal gland secretions of some species of *Bactrocera dorsalis* complex of fruit flies (Diptera: Tephritidae). *J. Chem. Ecol.* **1990**, *16*, 2475–2487. [CrossRef]

228. Fletcher, M.T.; Jacobs, M.F.; Kitching, W.; Krohn, S.; Drew, R.A.I.; Haniotakis, G.E.; Francke, W. Absolute stereochemistry of the 1,7-dioxaspiro[5.5]undecanols in fruit-fly species, including the olive-fly. *J. Chem. Soc. Chem. Commun.* **1992**, 1457. [CrossRef]

229. Kitching, W.; Lewis, J.A.; Perkins, M.V.; Drew, R.; Moore, C.J.; Schurig, V.; Konig, W.A.; Francke, W. Chemistry of fruit-flies—Composition of the rectal gland secretion of (male) *Dacus cucumis* (cucumber fly) and *Dacus halfordiae*—Characterization of (Z,Z)-2,8-Dimethyl-1,7-Dioxaspiro[5.5]Undecane. *J. Org. Chem.* **1989**, *54*, 3893–3902. [CrossRef]

230. Zhang, X.G.; Wei, C.M.; Miao, J.; Zhang, X.J.; Wei, B.; Dong, W.X.; Xiao, C. Chemical compounds from female and male rectal pheromone glands of the guava fruit fly, *Bactrocera correcta*. *Insects* **2019**, *10*, 78. [CrossRef]

231. Tokushima, I.; Orankanok, W.; Tan, K.H.; Ono, H.; Nishida, R. Accumulation of phenylpropanoid and sesquiterpenoid volatiles in male rectal pheromonal glands of the guava fruit fly, *Bactrocera correcta*. *J. Chem. Ecol.* **2010**, *36*, 1327–1334. [CrossRef]

232. Fletcher, M.T.; Wells, J.A.; Jacobs, M.F.; Krohn, S.; Kitching, W.; Drew, R.A.I.; Moore, C.J.; Francke, W. Chemistry of fruit-flies. Spiroacetal-rich secretions in several *Bactrocera* species from the South-West Pacific region. *J. Chem. Soc. Perkin Trans. 1* **1992**, 2827–2831. [CrossRef]

233. Mazomenos, B.E.; Pomonis, J.G. Male olive fruit fly pheromone: Isolation, identification and lab-bioassays. In Proceedings of the CEC/IOBC International Symposium on Fruit Flies of Economic Importance, Athens, Greece, 16 November 1982; pp. 96–103.

234. Noushini, S.; Park, S.J.; Jamie, I.; Jamie, J.; Taylor, P. Sampling technique biases in the analysis of fruit fly volatiles: A case study of Queensland fruit fly. *Sci. Rep.* **2020**, *10*, 19799. [CrossRef]

235. Perkins, M.V.; Kitching, W.; Drew, R.A.I.; Moore, C.J.; König, W.A. Chemistry of fruit flies: Composition of the male rectal gland secretions of some species of South-East Asian Dacinae. Re-examination of *Dacus cucurbitae* (melon fly). *J. Chem. Soc., Perkin Trans. 1* **1990**, 1111–1117. [CrossRef]

236. Baker, R.; Herbert, R.H.; Lomer, R.A. Chemical components of the rectal gland secretions of male *Dacus cucurbitae*, the melon fly. *Experientia* **1982**, *38*, 232–233. [CrossRef]

237. Nishida, R.; Iwahashi, O.; Tan, K.H. Accumulation of *Dendrobium superbum* (orchidaceae) fragrance in the rectal glands by males of the melon fly, *Dacus cucurbitae*. *J. Chem. Ecol.* **1993**, *19*, 713–722. [CrossRef] [PubMed]

238. Ioannou, C.S.; Papadopoulos, N.T.; Kouloussis, N.A.; Tananaki, C.I.; Katsoyannos, B.I. Essential oils of citrus fruit stimulate oviposition in the Mediterranean fruit fly *Ceratitis capitata* (Diptera: Tephritidae). *Physiol. Entomol.* **2012**, *37*, 330–339. [CrossRef]

239. Cosse, A.A.; Todd, J.L.; Millar, J.G.; Martinez, L.A.; Baker, T.C. Electroantennographic and coupled gas chromatographic-electroantennographic responses of the mediterranean fruit fly, *Ceratitis capitata*, to male-produced volatiles and mango odor. *J. Chem. Ecol.* **1995**, *21*, 1823–1836. [CrossRef] [PubMed]

240. Nation, J.L. The sex pheromone blend of Caribbean fruit fly males: Isolation biological activity, and partial chemical characterization. *J. Environ. Entomol.* **1975**, *4*, 27–30. [CrossRef]

241. De Aquino, N.C.; Ferreira, L.L.; Tavares, R.; Silva, C.S.; Mendonça, A.; Joachim-Bravo, I.S.; Milet-Pinheiro, P.; Navarro, D.; De Abreu Galdino, F.C.; Do Nascimento, R.R. Bioactive male-produced volatiles from *Anastrepha obliqua* and their role in attraction of conspecific females. *J. Chem. Ecol.* **2021**, *47*, 167–174. [CrossRef]

242. Robacker, D.C.; Hart, W.G. Behavioral-responses of male and female Mexican fruit-flies, *Anastrepha ludens*, to male-produced chemicals in laboratory experiments. *J. Chem. Ecol.* **1986**, *12*, 39–47. [CrossRef]

243. Aluja, M.; Cabagne, G.; Altuzar-Molina, A.; Pascacio-Villafan, C.; Enciso, E.; Guillen, L. Host plant and antibiotic effects on scent bouquet composition of *Anastrepha ludens* and *Anastrepha obliqua* calling males, two polyphagous tephritid pests. *Insects* **2020**, *11*, 309. [CrossRef]

244. Teal, P.E.A.; Gomez-Simuta, Y.; Proveaux, A.T. Mating experience and juvenile hormone enhance sexual signaling and mating in male Caribbean fruit flies. *Proc. Natl. Acad. Sci. USA* **2000**, *97*, 3708–3712. [CrossRef]

245. Aceves-Aparicio, E.; Pérez-Staples, D.; Arredondo, J.; Corona-Morales, A.; Morales-Mávil, J.; Díaz-Fleischer, F. Combined effects of methoprene and metformin on reproduction, longevity, and stress resistance in *Anastrepha ludens* (Diptera: Tephritidae): Implications for the Sterile Insect Technique. *J. Econ. Entomol.* **2021**, *114*, 142–151. [CrossRef]

246. Landolt, P.J.; Averill, A.L. Fruit flies. In *Pheromones of Non-Lepidopteran Insects Associated with Agricultural Plants*; Hardie, J., Minks, A.K., Eds.; CABI Publication: Wallingford, UK, 1999; pp. 3–25.

247. Farine, J.P.; Ferveur, J.F.; Everaerts, C. Volatile *Drosophila* cuticular pheromones are affected by social but not sexual experience. *PLoS ONE* **2012**, *7*, e40396. [CrossRef]

248. Droney, D.C.; Hock, M.B. Male sexual signals and female choice in *Drosophila grimshawi* (Diptera: Drosophildae). *J. Insect Behav.* **1998**, *11*, 59–71. [CrossRef]

249. Muller, M.; Buchbauer, G. Essential oil components as pheromones. A review. *Flavour Fragr. J.* **2011**, *26*, 357–377. [CrossRef]

250. Engl, T.; Kaltenpoth, M. Influence of microbial symbionts on insect pheromones. *Nat. Prod. Rep.* **2018**, *35*, 386–397. [CrossRef]

251. Henneken, J.; Goodger, J.Q.D.; Jones, T.M.; Elgar, M.A. Diet-mediated pheromones and signature mixtures can enforce signal reliability. *Front. Ecol. Evol.* **2017**, *4*. [CrossRef]

252. Ren, L.; Ma, Y.; Xie, M.; Lu, Y.; Cheng, D. Rectal bacteria produce sex pheromones in the male oriental fruit fly. *Curr. Biol.* **2021**, *31*. [CrossRef]

253. Conway, J.R.; Lex, A.; Gehlenborg, N. UpSetR: An R package for the visualization of intersecting sets and their properties. *Bioinformatics* **2017**, *33*, 2938–2940. [CrossRef]

254. Barbosa-Cornelio, R.; Cantor, F.; Coy-Barrera, E.; Rodriguez, D. Tools in the investigation of volatile semiochemicals on insects: From sampling to statistical analysis. *Insects* **2019**, *10*, 241. [CrossRef]

255. Leal, W.S. Reverse chemical ecology at the service of conservation biology. *Proc. Natl. Acad. Sci. USA* **2017**, *114*, 12094–12096. [CrossRef]

256. Brito, N.F.; Moreira, M.F.; Melo, A.C.A. A look inside odorant-binding proteins in insect chemoreception. *J. Insect Physiol.* **2016**, *95*, 51–65. [CrossRef]

257. Torto, B. Chemical signals as attractants, repellents and aggregation stimulants. In *Chemical Ecology. Encyclopedia of Life Support Systems*; EOLSS Publishers Co. Ltd: Oxford, UK, 2009.

258. Al-Khshemawee, H.; Du, X.; Agarwal, M.; Yang, J.O.; Ren, Y.L. Application of Direct Immersion Solid-Phase Microextraction (DI-SPME) for understanding biological changes of Mediterranean Fruit Fly (*Ceratitis capitata*) during mating procedures. *Molecules* **2018**, *23*, 2951. [CrossRef]

259. Arthur, C.L.; Pawliszyn, J. Solid-Phase Microextraction with thermal-desorption using fused silica optical fibers. *Anal. Chem.* **1990**, *62*, 2145–2148. [CrossRef]

260. Pawliszyn, J.; Pawliszyn, B.; Pawliszyn, M. Solid Phase Microextraction (SPME). *Chem. Educ.* **1997**, *2*, 1–7. [CrossRef]

261. Monnin, T.; Malosse, C.; Peeters, C. Solid-phase microextraction and cuticular hydrocarbon differences related to reproductive activity in queenless ant *Dinoponera quadriceps*. *J. Chem. Ecol.* **1998**, *24*, 473–490. [CrossRef]

262. Turillazzi, S.; Sledge, M.F.; Moneti, G. Use of a simple method for sampling cuticular hydrocarbons from live social wasps. *Ethol. Ecol. Evol.* **1998**, *10*, 293–297. [CrossRef]

263. Tentschert, J.; Bestmann, H.J.; Heinze, J. Cuticular compounds of workers and queens in two *Leptothorax* ant species—A comparison of results obtained by solvent extraction, solid sampling, and SPME. *Chemoecology* **2002**, *12*, 15–21. [CrossRef]

264. AL-Khshema, H.; Agarwal, M.; Ren, Y.L. Optimization and validation for determination of volatile organic compounds from Mediterranean Fruit Fly (Medfly) *Ceratitis capitata* (Diptera: Tephritidae) by using HS-SPME-GC-FID/MS. *J. Biol. Sci.* **2017**, *17*, 347–352. [CrossRef]

265. Robacker, D.C.; Warfield, W.C.; Flath, R.A. A four-component attractant for the mexican fruit fly, *Anastrepha ludens* (Diptera: Tephritidae), from host fruit. *J. Chem. Ecol.* **1992**, *18*, 1239–1254. [CrossRef]

266. Cruz-Lopez, L.; Malo, E.A.; Toledo, J.; Virgen, A.; del Mazo, A.; Rojas, J.C. A new potential attractant for *Anastrepha obliqua* from *Spondias mombin* fruits. *J. Chem. Ecol.* **2006**, *32*, 351–365. [CrossRef]

267. Diaz-Fleischer, F.; C., P.J.; Shelly, T.E. Interactions between tephritid fruit fly physiological state and stimuli from baits and traps: Looking for the pied piper of Hamelin to lure pestiferous fruit flies. In *Trapping and the Detection, Control, and Regulation of Tephritid Fruit Flies*; Shelly, T.E., Epsky, N., Jang, E.B., Reyes-Flores, J., Vargas, R., Eds.; Springer: Berlin/Heidelberg, Germany, 2014; pp. 145–172.

268. Jang, E.B.; Light, D.M.; Binder, R.G.; Flath, R.A.; Carvalho, L.A. Attraction of female mediterranean fruit flies to the five major components of male-produced pheromone in a laboratory flight tunnel. *J. Chem. Ecol.* **1994**, *20*, 9–20. [CrossRef]

269. Oldham, N.J. Chemical Studies on Exocrine Gland Secretion and Pheromones of Some Social Insects. Ph.D. Thesis, Keele University, Keele, UK, 1994.

270. Sinha, A.E.; Fraga, C.G.; Prazen, B.J.; Synovec, R.E. Trilinear chemometric analysis of two-dimensional comprehensive gas chromatography-time-of-flight mass spectrometry data. *J. Chromatogr. A* **2004**, *1027*, 269–277. [CrossRef]

271. van Deursen, M.M.; Beens, J.; Janssen, H.G.; Leclercq, P.A.; Cramers, C.A. Evaluation of time-of-flight mass spectrometric detection for fast gas chromatography. *J. Chromatogr. A* **2000**, *878*, 205–213. [CrossRef]

272. Vanickova, L.; Virgilio, M.; Tomcala, A.; Brizova, R.; Ekesi, S.; Hoskovec, M.; Kalinova, B.; Do Nascimento, R.R.; De Meyer, M. Resolution of three cryptic agricultural pests (*Ceratitis fasciventris, C. anonae, C. rosa*, Diptera: Tephritidae) using cuticular hydrocarbon profiling. *Bull. Entomol. Res.* **2014**, *104*, 631–638. [CrossRef]

273. Warthen, J.D.; Mcgovern, T.P. Gc/Ftir analyses of Trimedlure isomers and related esters. *J. Chromatogr. Sci.* **1986**, *24*, 451–457. [CrossRef]

274. Karas, M.; Bachmann, D.; Bahr, U.; Hillenkamp, F. Matrix-assisted ultraviolet laser desorption of non-volatile compounds. *Int. J. Mass Spectrom. Ion Process.* **1987**, *78*, 53–68. [CrossRef]

275. Patton, G.M.; Fasulo, J.M.; Robins, S.J. Analysis of lipids by high performance liquid chromatography: Part I. *J. Nutr. Biochem.* **1990**, *1*, 493–500. [CrossRef]

276. McHowad, J.; Jones, J.H.; Creer, H.M. Quantification of individual phospholipid molecular species by UV absorption measurements. *J. Lipid Res.* **1996**, *37*, 2450–2460. [CrossRef]

277. McHowad, J.; Jones, J.H.; Creer, H.M. Gradient elution reverse-phase chromatographic isolation of individual glycerolphospholipid molecular species. *J. Chromatogr. B* **1997**, *702*, 21–32. [CrossRef]

278. Olsson, N.U.; Harding, A.J.; Harper, C.; Salem, N.J. High-performance liquid chromatography method with light-scattering detection for measurement of lipid class composition: Analysis of brain from alcoholics. *J. Chromatogr. B* **1996**, *681*, 213–218. [CrossRef]

279. Zhou, Y.; Qin, Q.; Zhang, P.W.; Chen, X.T.; Liu, B.J.; Cheng, D.M.; Zhang, Z.X. Integrated LC-MS and GC-MS-based untargeted metabolomics studies of the effect of azadirachtin on *Bactrocera dorsalis* larvae. *Sci. Rep.* **2020**, *10*, 2306. [CrossRef]

280. Buděšínský, M.; Pelař, J. *Cyklus Organická Chemie*; Institute of Organic Chemistry and Biochemistry: Praha, Czech Republic, 2000.

281. Baker, J.D.; Heath, R.R. NMR spectral assignment of lactone pheromone components emitted by Caribbean and Mexican fruit flies. *J. Chem. Ecol.* **1993**, *19*, 1511–1519. [CrossRef]

282. Mori, K. Significance of chirality in pheromone science. *Bioorg. Med. Chem.* **2007**, *15*, 7505–7523. [CrossRef]

283. Olszewski, T.K.; Grison, C. A concise synthesis of sex pheromone of Mediterranean fruit fly, *Ceratitis capitata* via lithiated carbanion derived from enephosphoramide. *Heteroat. Chem.* **2010**, *21*, 139–147. [CrossRef]

284. Mori, K.; Uematsu, T.; Yanagi, K.; Minobe, M. Synthesis of the optically active forms of 4,10-dihydroxy-1,7-dioxaspiro[5.5]undecane and their conversion to the enantiomers of 1,7-dioxaspiro[5.5]undecane, the olive fly pheromone. *Tetrahedron* **1985**, *41*, 2751–2758. [CrossRef]

285. Fusini, G.; Barsanti, D.; Angelici, G.; Casotti, G.; Canale, A.; Benelli, G.; Lucchi, A.; Carpita, A. Identification and synthesis of new sex-specific components of olive fruit fly (*Bactrocera oleae*) female rectal gland, through original Negishi reactions on supported catalysts. *Tetrahedron* **2018**, *74*, 4381–4389. [CrossRef]

286. Canale, A.; Germinara, S.G.; Carpita, A.; Benelli, G.; Bonsignori, G.; Stefanini, C.; Raspi, A.; Rotundo, G. Behavioural and electrophysiological responses of the olive fruit fly, *Bactrocera oleae* (Rossi) (Diptera: Tephritidae), to male-and female-borne sex attractants. *Chemoecology* **2013**. [CrossRef]

287. Vaníčková, L. Chemical Ecology of Fruit Flies: Genera *Ceratitis* and *Anastrepha*. Ph.D. Thesis, Institute of Chemical Technology, Prague, Czech Republic, 2012.

288. Břízová, R. Analyses of Male Sex Pheromone of *Anastrepha fraterculus* (Diptera: Tephritidae). Master's Thesis, Institute of Chemical Technology, Prague, Czech Republic, 2011.

289. Zykova, K. Quantitative Composition Changes of Sex Pheromone in *Anastrepha fraterculus* Depending on Age. Master's Thesis, Institute of Chemical Technology, Prague, Czech Republic, 2013.

290. Cruz-Lopez, L.; Malo, E.A.; Rojas, J.C. Sex Pheromone of *Anastrepha striata*. *J. Chem. Ecol.* **2015**, *41*, 458–464. [CrossRef]

291. Noushini, S.; Perez, J.; Park, S.J.; Holgate, D.; Alvarez, V.M.; Jamie, I.; Jamie, J.; Taylor, P. Attraction and electrophysiological response to identified rectal gland volatiles in *Bactrocera frauenfeldi* (Schiner). *Molecules* **2020**, *25*, 1275. [CrossRef] [PubMed]

292. Robacker, D.C.; Chapa, B.E.; Hart, W.G. Electroantennograms of Mexican fruit flies to chemicals produced by males. *Entomol. Exp. Appl.* **1986**, *40*, 123–127. [CrossRef]

293. Van Der Pers, J.C.N.; Haniotakis, G.E.; King, B.M. Electroantennogram responses from olfactory receptors in *Dacus oleae*. *Entomol. Hell.* **1984**, *2*, 47–53. [CrossRef]

294. Siderhurst, M.S.; Jang, E.B. Cucumber volatile blend attractive to female melon fly, *Bactrocera cucurbitae* (Coquillett). *J. Chem. Ecol.* **2010**, *36*, 699–708. [CrossRef]

295. Liscia, A.; Angioni, P.; Sacchetti, P.; Poddighe, S.; Granchietti, A.; Setzu, M.D.; Belcari, A. Characterization of olfactory sensilla of the olive fly: Behavioral and electrophysiological responses to volatile organic compounds from the host plant and bacterial filtrate. *J. Insect Physiol.* **2013**, *59*, 705–716. [CrossRef]

296. Cunningham, J.P.; Carlsson, M.A.; Villa, T.F.; Dekker, T.; Clarke, A.R. Do fruit ripening volatiles enable resource specialism in polyphagous fruit flies? *J. Chem. Ecol.* **2016**, *42*, 931–940. [CrossRef]

297. Biasazin, T.D.; Karlsson, M.F.; Hillbur, Y.; Seyoum, E.; Dekker, T. Identification of host blends that attract the African invasive fruit fly, *Bactrocera invadens*. *J. Chem. Ecol.* **2014**, *40*, 966–976. [CrossRef]

298. Kendra, P.E.; Epsky, N.D.; Montgomery, W.S.; Heath, R.R. Response of *Anastrepha suspensa* (Diptera: Tephritidae) to terminal diamines in a food-based synthetic attractant. *Environ. Entomol.* **2008**, *37*, 1119–1125. [CrossRef]

299. Kendra, P.E.; Montgomery, W.S.; Mateo, D.M.; Puche, H.; Epsky, N.D.; Heath, R.R. Effect of age on EAG response and attraction of female *Anastrepha suspensa* (Diptera: Tephritidae) to ammonia and carbon dioxide. *Environ. Entomol.* **2005**, *34*, 584–590. [CrossRef]

300. Tabanca, N.; Masi, M.; Epsky, N.D.; Nocera, P.; Cimmino, A.; Kendra, P.E.; Niogret, J.; Evidente, A. Laboratory evaluation of natural and synthetic aromatic compounds as potential attractants for male Mediterranean fruit fly, *Ceratitis capitata*. *Molecules* **2019**, *24*, 2409. [CrossRef]

301. Jenkins, D.A.; Kendra, P.E.; Epsky, N.D.; Montgomery, W.S.; Heath, R.R.; Jenkins, D.M.; Goenaga, R. Antennal responses of West Indian and Caribbean fruit flies (Diptera: Tephritidae) to ammonium bicarbonate and putrescine lures. *Florida Entomol.* **2012**, *95*, 28–34. [CrossRef]

302. Epsky, N.; Kendra, P.E.; Heath, R.R. Response of *Anastrepha suspensa* to liquid protein baits and synthetic lure formulations. In Proceedings of the 7th International Symposium on Fruit Flies of Economic Importance, Salvador, Brazil, 10–15 September 2006; pp. 81–88.

303. Crnjar, R.; Scalera, G.; Liscia, A.; Angioy, A.M.; Bigiani, A.; Pietra, P.; Barbarossa, I.T. Morphology and eag mapping of the antennal olfactory receptors in Dacus oleae. *Entomol. Exp. Appl.* **1989**, *51*, 77–85. [CrossRef]

304. Nagai, T. On the relationship between the electroantennogram and simultaneously recorded single sensillum response of the European corn borer *Ostrinia nubilalis*. *Arch. Insect Biochem. Physiol.* **1983**, *1*, 85–91. [CrossRef]

305. Olsson, S.B.; Hansson, B.S. Electroantennogram and single sensillum recording in insect antennae. In *Methods in Molecular Biology*; Touhara, K., Ed.; Springer Science and Business Media LLC: Berlin, Germany, 2013; Volume 1068.

306. Bigiani, A.; Scalera, G.; Crnjar, R.; Barbarossa, I.T.; Magherini, P.C.; Pietra, P. Distribution and function of the antennal olfactory sensilla in *Ceratitis capitata* Wied (Diptera, Trypetidae). *Boll. Di Zool.* **1989**, *56*, 305–311. [CrossRef]

307. Bisotto-de-Oliveira, R.; Redaelli, L.R.; Sant'ana, J. Morphometry and distribution of sensilla on the antennae of *Anastrepha fraterculus* (Wiedemann) (Diptera: Tephritidae). *Neotrop. Entomol.* **2011**, *40*, 212–216. [CrossRef]

308. Jacob, V.; Scolari, F.; Delatte, H.; Gasperi, G.; Jacquin-Joly, E.; Malacrida, A.R.; Duyck, P.F. Current source density mapping of antennal sensory selectivity reveals conserved olfactory systems between tephritids and *Drosophila*. *Sci. Rep.* **2017**, *7*. [CrossRef]

309. Jacob, V.E.J.M. Current source density analysis of electroantennogram recordings: A tool for mapping the olfactory response in an insect antenna. *Front. Cell. Neurosci.* **2018**, *12*. [CrossRef]

310. Loy, F.; Solari, P.; Isola, M.; Crnjar, R.; Masala, C. Morphological and electrophysiological analysis of tarsal sensilla in the medfly *Ceratitis capitata* (Wiedemann, 1824) (Diptera: Tephritidae). *Ital. J. Zool.* **2016**, *83*, 456–468. [CrossRef]

311. Tait, C.; Batra, S.; Ramaswamy, S.S.; Feder, J.L.; Olsson, S.B. Sensory specificity and speciation: A potential neuronal pathway for host fruit odour discrimination in *Rhagoletis pomonella*. *Proc. Biol. Sci.* **2016**, *283*. [CrossRef] [PubMed]

312. Hare, J.D. Bioassays with terrestrial invertebrates. In *Methods in Chemical Ecology, Bioassays Methods*; Hayes, K.F., Ed.; Kluwer Academic: Norwell, MA, USA, 1998; Volume 2, pp. 212–270.

313. Howse, P.E.; Stevens, I.D.R.; Jones, O.T.; Howse, P.E.; Stevens, I.D.R.; Jones, O.T. Bioassay methods. In *Insect Pheromones and Their Use in Pest Management*; Springer: Dordrecht, The Netherlands, 1998.

314. Baker, T.C.; Cardé, R.T. Techniques for Behavioral Bioassays. In *Techniques in Pheromone Research. Springer Series in Experimental Entomology*; Hummel, H.E., Miller, T.A., Eds.; Springer New York: New York, NY, USA, 1984; pp. 45–73.

315. Knudsen, G.K.; Tasin, M.; Aak, A.; Thöming, G. A wind tunnel for odor mediated insect behavioural assays. *J. Vis. Exp.* **2018**. [CrossRef] [PubMed]

316. Wu, Z.; Lin, J.; Zhang, H.; Zeng, X. BdorOBP83a-2 mediates responses of the Oriental fruit fly to semiochemicals. *Front. Physiol.* **2016**, *7*. [CrossRef] [PubMed]

317. Mazomenos, B.E.; Haniotakis, G.E. Male olive fruit fly attraction to synthetic sex pheromone components in laboratory and field tests. *J. Chem. Ecol.* **1985**, *11*, 397–405. [CrossRef]

318. Giunti, G.; Campolo, O.; Laudani, F. Olive fruit volatiles route intraspecific interactions and chemotaxis in *Bactrocera oleae* (Rossi) (Diptera: Tephritidae) females. *Sci. Rep.* **2020**, *10*. [CrossRef]

319. Landolt, P.J.; Heath, R.R.; Chambers, D.L. Oriented flight responses of female Mediterranean fruit flies to calling males, odor of calling males, and a synthetic pheromone blend. *Entomol. Exp. Appl.* **1992**, *65*, 259–266. [CrossRef]

320. Biasazin, T.D.; Herrera, S.L.; Kimbokota, F.; Dekker, T. Translating olfactomes into attractants: Shared volatiles provide attractive bridges for polyphagy in fruit flies. *Ecol. Lett.* **2019**, *22*, 108–118. [CrossRef]

321. Baker, P.; Howse, P.; Ondarza, R.N.; Reyes, J. Field trials of synthetic sex pheromone components of the male Mediterranean fruit fly (Diptera: Tephritidae) in Southern Mexico. *J. Econ. Entomol.* **1990**, *83*, 2235–2245. [CrossRef]

322. Jones, O.T.; Lisk, J.C.; Longhurst, C.; Howse, P.E. Development of a monitoring trap for the olive fruit fly, *Dacus oleae* (Gmelin) (Diptera: Tephritidae), using a multicomponent of its sex pheromone as lure. *Bull. Entomol. Res.* **1983**, *73*, 97–106. [CrossRef]

323. Mazomenos, B.E.; Haniotakis, G.E.; Ioannou, A.; Spanakis, I.; Kozirakis, A. Field evaluation of the olive fruit fly pheromone traps with various dispensers and concentrations. In Proceedings of the International Symposium of Fruit Flies of Economic Importance, Athens, Greece, 16 November 1982; pp. 506–512.

324. Perdomo, A.J.; Baranowski, R.M.; Nation, J.L. Recapture of virgin female Caribbean fruit flies from traps baited with males. *Florida Entomol.* **1975**, *58*, 291–295. [CrossRef]

325. Perdomo, A.J.; Nation, J.L.; Baranowski, R.M. Attraction of female and male caribbean fruit flies to food-baited and male-baited traps under field conditions. *Environ. Entomol.* **1976**. [CrossRef]

326. Nakagawa, S.; Steiner, L.F.; Farias, G.J. Response of virgin female Mediterranean fruit flies to live mature normal males, sterile males, and Trimedlure in plastic traps. *J. Econ. Entomol.* **1981**, *74*, 566–567. [CrossRef]

327. Shelly, T.E. Male signalling and lek attractiveness in the Mediterranean fruit fly. *Anim. Behav.* **2000**, *60*, 245–251. [CrossRef]

328. Webb, J.C.; Burk, T.; Sivinski, J. Attraction of female Caribbean fruit flies, *Anastrepha suspensa* (Diptera: Tephritidae), to the presence of males and male-produced stimuli in field cages. *Ann. Entomol. Soc. Am.* **1983**, *76*, 996–998. [CrossRef]

329. Ono, H.; Hee, A.K.-W.; Jiang, H. Recent advancements in studies on chemosensory mechanisms underlying detection of semiochemicals in Dacini fruit flies of economic importance (Diptera: Tephritidae). *Insects* **2021**, *12*, 106. [CrossRef]

330. Pelosi, P.; Iovinella, I.; Zhu, J.; Wang, G.; Dani, F.R. Beyond chemoreception: Diverse tasks of soluble olfactory proteins in insects. *Biol. Rev. Camb. Philos. Soc.* **2018**, *93*, 184–200. [CrossRef]

331. Gomulski, L.M.; Dimopoulos, G.; Xi, Z.Y.; Soares, M.B.; Bonaldo, M.F.; Malacrida, A.R.; Gasperi, G. Gene discovery in an invasive tephritid model pest species, the Mediterranean fruit fly, *Ceratitis capitata*. *BMC Genomics* **2008**, *9*. [CrossRef]

332. Zheng, W.W.; Peng, T.; He, W.; Zhang, H.Y. High-throughput sequencing to reveal genes involved in reproduction and development in *Bactrocera dorsalis* (Diptera: Tephritidae). *PLoS ONE* **2012**, *7*. [CrossRef]

333. Zheng, W.W.; Peng, W.; Zhu, C.P.; Zhang, Q.; Saccone, G.; Zhang, H.Y. Identification and expression profile analysis of odorant binding proteins in the Oriental fruit fly *Bactrocera dorsalis*. *Int. J. Mol. Sci.* **2013**, *14*, 14936–14949. [CrossRef]

334. Ramsdell, K.M.; Lyons-Sobaski, S.A.; Robertson, H.M.; Walden, K.K.; Feder, J.L.; Wanner, K.; Berlocher, S.H. Expressed sequence tags from cephalic chemosensory organs of the northern walnut husk fly, *Rhagoletis suavis*, including a putative canonical odorant receptor. *J. Insect Sci.* **2010**, *10*, 51. [CrossRef]

335. Schwarz, D.; Robertson, H.M.; Feder, J.L.; Varala, K.; Hudson, M.E.; Ragland, G.J.; Hahn, D.A.; Berlocher, S.H. Sympatric ecological speciation meets pyrosequencing: Sampling the transcriptome of the apple maggot *Rhagoletis pomonella*. *BMC Genomics* **2009**, *10*. [CrossRef]

336. Wu, Z.Z.; Zhang, H.; Wang, Z.B.; Bin, S.Y.; He, H.L.; Lin, J.T. Discovery of chemosensory genes in the Oriental fruit fly, *Bactrocera dorsalis*. *PLoS ONE* **2015**, *10*. [CrossRef]

337. Wu, Z.Z.; Kang, C.; Qu, M.Q.; Chen, J.L.; Chen, M.S.; Bin, S.Y.; Lin, J.T. Candidates for chemosensory genes identified in the Chinese citrus fly, *Bactrocera minax*, through a transcriptomic analysis. *BMC Genomics* **2019**, *20*. [CrossRef]

338. Chen, X.-F.; Xu, L.; Zhang, Y.-X.; Wei, D.; Wang, J.-J.; Jiang, H.-B. Genome-wide identification and expression profiling of odorant-binding proteins in the oriental fruit fly, *Bactrocera dorsalis*. *Comp. Biochem. Physiol. Part D Genomics Proteomics* **2019**, *31*, 100605. [CrossRef]

339. Xu, P.H.; Wang, Y.H.; Akami, M.; Niu, C.Y. Identification of olfactory genes and functional analysis of BminCSP and BminOBP21 in *Bactrocera minax*. *PLoS ONE* **2019**, *14*. [CrossRef]

340. Campanini, E.B.; de Brito, R.A. Molecular evolution of Odorant-binding proteins gene family in two closely related *Anastrepha* fruit flies. *BMC Evol. Biol.* **2016**, *16*. [CrossRef]

341. Rezende, V.B.; Congrains, C.; Lima, A.L.; Campanini, E.B.; Nakamura, A.M.; Oliveira, J.L.; Chahad-Ehlers, S.; Junior, I.S.; Alves de Brito, R. Head transcriptomes of two closely related species of fruit flies of the *Anastrepha fraterculus* group reveals divergent genes in species with extensive gene flow. *G3* **2016**, *6*, 3283–3295. [CrossRef] [PubMed]

342. Papanicolaou, A.; Schetelig, M.F.; Arensburger, P.; Atkinson, P.W.; Benoit, J.B.; Bourtzis, K.; Castanera, P.; Cavanaugh, J.P.; Chao, H.; Childers, C.; et al. The whole genome sequence of the Mediterranean fruit fly, *Ceratitis capitata* (Wiedemann), reveals insights into the biology and adaptive evolution of a highly invasive pest species. *Genome Biol.* **2016**, *17*, 192. [CrossRef] [PubMed]

343. Liu, H.; Zhao, X.F.; Fu, L.; Han, Y.Y.; Chen, J.; Lu, Y.Y. BdorOBP2 plays an indispensable role in the perception of methyl eugenol by mature males of *Bactrocera dorsalis* (Hendel). *Sci. Rep.* **2017**, *7*, 15894. [CrossRef] [PubMed]

344. Liu, Z.; Liang, X.F.; Xu, L.; Keesey, I.W.; Lei, Z.R.; Smagghe, G.; Wang, J.J. An antennae-specific Odorant-Binding Protein is involved in *Bactrocera dorsalis* olfaction. *Front. Ecol. Evol.* **2020**, *8*. [CrossRef]

345. Nakamura, A.M.; Chahad-Ehlers, S.; Lima, A.L.A.; Taniguti, C.H.; Sobrinho, I.; Torres, F.R.; de Brito, R.A. Reference genes for accessing differential expression among developmental stages and analysis of differential expression of OBP genes in *Anastrepha obliqua*. *Sci. Rep.* **2016**, *6*. [CrossRef]

346. Pelosi, P.; Zhu, J.; Knoll, W. Odorant-binding proteins as sensing elements for odour monitoring. *Sensors* **2018**, *18*, 3248. [CrossRef]

347. Falchetto, M.; Ciossani, G.; Scolari, F.; Di Cosimo, A.; Nenci, S.; Field, L.M.; Mattevi, A.; Zhou, J.J.; Gasperi, G.; Forneris, F. Structural and biochemical evaluation of *Ceratitis capitata* odorant-binding protein 22 affinity for odorants involved in intersex communication. *Insect Mol. Biol.* **2019**, *28*, 431–443. [CrossRef]

348. Pelosi, P.; Zhu, J.; Knoll, W. From radioactive ligands to biosensors: Binding methods with olfactory proteins. *Appl. Microbiol. Biotechnol.* **2018**, *102*, 8213–8227. [CrossRef]

349. Paolini, S.; Tanfani, F.; Fini, C.; Bertoli, E.; Pelosi, P. Porcine odorant-binding protein: Structural stability and ligand affinities measured by Fourier-transform infrared spectroscopy and fluorescence spectroscopy. *Biochim. Biophys. Acta Protein Struct. Mol. Enzymol.* **1999**, *1431*, 179–188. [CrossRef]

350. Wojtasek, H.; Leal, W.S. Conformational change in the pheromone-binding protein from *Bombyx mori* induced by pH and by interaction with membranes. *J. Biol. Chem.* **1999**, *274*, 30950–30956. [CrossRef]

351. Qiao, H.L.; Tuccori, E.; He, X.L.; Gazzano, A.; Field, L.; Zhou, J.J.; Pelosi, P. Discrimination of alarm pheromone (E)-beta-farnesene by aphid odorant-binding proteins. *Insect Biochem. Mol. Biol.* **2009**, *39*, 414–419. [CrossRef]

352. Ban, L.; Scaloni, A.; D'Ambrosio, C.; Zhang, L.; Yan, Y.; Pelosi, P. Biochemical characterization and bacterial expression of an odorant-binding protein from *Locusta migratoria*. *Cell. Mol. Life Sci.* **2003**, *60*, 390–400. [CrossRef]

353. Drew, R.A.I. The responses of fruit fly species (Diptera: Tephritidae) in the South Pacific area to male attractants. *Aust. J. Entomol.* **1974**, *13*, 267–270. [CrossRef]

354. Wee, S.-L.; Peek, T.; Clarke, A.R. The responsiveness of *Bactrocera jarvisi* (Diptera: Tephritidae) to two naturally occurring phenylbutaonids, zingerone and raspberry ketone. *J. Insect Physiol.* **2018**, *109*, 41–46. [CrossRef]

355. Park, K.C.; Jeong, S.A.; Kwon, G.; Oh, H.-W. Olfactory attraction mediated by the maxillary palps in the striped fruit fly, *Bactrocera scutellata*: Electrophysiological and behavioral study. *Arch. Insect Biochem. Physiol.* **2018**, *99*, e21510. [CrossRef]

356. Jayanthi, P.D.K.; Kempraj, V.; Aurade, R.M.; Roy, T.K.; Shivashankara, K.S.; Verghese, A. Computational reverse chemical ecology: Virtual screening and predicting behaviorally active semiochemicals for *Bactrocera dorsalis*. *BMC Genomics* **2014**, *15*. [CrossRef]

357. Jayanthi, P.D.K.; Kempraj, V.; Aurade, R. Computational reverse chemical ecology: Prospecting semiochemicals for pest management using in silico approach in *Plutella xylostella* Linn. *Pest Manag. Hortic. Ecosyst.* **2016**, *22*, 20–27.

358. Kuntz, I.D.; Blaney, J.M.; Oatley, S.J.; Langridge, R.; Ferrin, T.E. A geometric approach to macromolecule-ligand interactions. *J. Mol. Biol.* **1982**, *161*, 269–288. [CrossRef]

359. Meng, X.Y.; Zhang, H.X.; Mezei, M.; Cui, M. Molecular docking: A powerful approach for structure-based drug discovery. *Curr. Comput. Aided Drug Des.* **2011**, *7*, 146–157. [CrossRef]

360. McConkey, B.J.; Sobolev, V.; Edelman, M. The performance of current methods in ligand-protein docking. *Curr. Sci.* **2002**, *83*, 845–856.

361. Venthur, H.; Zhou, J.-J. Odorant receptors and odorant-binding proteins as insect pest control targets: A comparative analysis. *Front. Physiol.* **2018**, *9*. [CrossRef]

362. Liu, Z.; Smagghe, G.; Lei, Z.R.; Wang, J.J. Identification of male- and female-specific olfaction genes in antennae of the Oriental fruit fly (*Bactrocera dorsalis*). *PLoS ONE* **2016**, *11*. [CrossRef] [PubMed]

363. Miyazaki, H.; Otake, J.; Mitsuno, H.; Ozaki, K.; Kanzaki, R.; Chieng, A.C.T.; Hee, A.K.W.; Nishida, R.; Ono, H. Functional characterization of olfactory receptors in the Oriental fruit fly *Bactrocera dorsalis* that respond to plant volatiles. *Insect Biochem. Mol. Biol.* **2018**, *101*, 32–46. [CrossRef] [PubMed]

364. Jin, S.; Zhou, X.F.; Gu, F.; Zhong, G.H.; Yi, X. Olfactory plasticity: Variation in the expression of chemosensory receptors in *Bactrocera dorsalis* in different physiological states. *Front. Physiol.* **2017**, *8*. [CrossRef] [PubMed]

365. Tsoumani, K.T.; Belavilas-Trovas, A.; Gregoriou, M.-E.; Mathiopoulos, K.D. Anosmic flies: What Orco silencing does to olive fruit flies. *BMC Genet.* **2020**, *21*, 140. [CrossRef]

366. Ono, H.; Miyazaki, H.; Mitsuno, H.; Ozaki, K.; Kanzaki, R.; Nishida, R. Functional characterization of olfactory receptors in three Dacini fruit flies (Diptera: Tephritidae) that respond to 1-nonanol analogs as components in the rectal glands. *Comp. Biochem. Physiol. B Biochem. Mol. Biol.* **2020**, *239*. [CrossRef]

367. Sato, K.; Pellegrino, M.; Nakagawa, T.; Nakagawa, T.; Vosshall, L.B.; Touhara, K. Insect olfactory receptors are heteromeric ligand-gated ion channels. *Nature* **2008**, *452*, 1002–1006. [CrossRef]

368. Butterwick, J.A.; del Mármol, J.; Kim, K.H.; Kahlson, M.A.; Rogow, J.A.; Walz, T.; Ruta, V. Cryo-EM structure of the insect olfactory receptor Orco. *Nature* **2018**, *560*, 447–452. [CrossRef]

369. Liu, H.; Chen, Z.-S.; Zhang, D.-J.; Lu, Y.-Y. BdorOR88a modulates the responsiveness to methyl eugenol in mature males of *Bactrocera dorsalis* (Hendel). *Front. Physiol.* **2018**, *9*. [CrossRef]

370. Katritch, V.; Cherezov, V.; Stevens, R.C. Structure-function of the G Protein–Coupled Receptor superfamily. *Annu. Rev. Pharmacol. Toxicol.* **2013**, *53*, 531–556. [CrossRef]

371. Cheng, J.-F.; Yu, T.; Chen, Z.-J.; Chen, S.; Chen, Y.-P.; Gao, L.; Zhang, W.-H.; Jiang, B.; Bai, X.; Walker, E.D.; et al. Comparative genomic and transcriptomic analyses of chemosensory genes in the citrus fruit fly *Bactrocera* (*Tetradacus*) *minax*. *Sci. Rep.* **2020**, *10*, 18068. [CrossRef]

372. Porter, B.A. The Apple Maggot. *U.S. Dep. Agric. Tech. Bull.* **1928**, *66*, 1–48.

373. Häfliger, E. Das Auswahlvermögen der Kirschenfliege bei der Eiablage. *Mitt. Schweiz. Entomol. Ges.* **1953**, *26*, 258–264.

374. Prokopy, R.J.; Greany, P.; Chambers, D. Oviposition-deterring pheromone in *Anastrepha suspensa*. *Environ. Entomol.* **1977**, *6*, 463–465. [CrossRef]

375. Simões, M.H.; Polloni, Y.J.; Pauludetti, M.A. Biología de algunas especies de *Anastrepha* (Diptera: Tephritidae) en laboratório. In Proceedings of the III Latin-American Entomology Congress, Ilheus, Brazil, 23 July 1978.

376. Prokopy, R.J.; Malavasi, A.; Morgante, J.S. Oviposition deterring pheromone in *Anastrepha fraterculus* flies. *J. Chem. Ecol.* **1982**, *8*, 763–771. [CrossRef]

377. Poloni, Y.J.; Silva, M.T. Considerations on the reproductive of Anastrepha pseudoparallela Loew 1873 (Diptera: Tephritidae). In *Proceedings of the Fruit Flies: II International Symposium*; Elsevier Science Publishers: Kolymbari, Greece, 1986.

378. Selivon, D. Alguns Aspectos do Comportamento de *Anastrepha striata* Schiner e *Anastrepha bistrigata* Bezzii (Diptera: Tephritidae). Master's Thesis, University of São Paulo, Sao Paulo, Brazil, 1991.

379. Silva, J.G. Biologia e Comportamento de *Anastrepha grandis* (Macquart, 1846) (Diptera: T ephritidae). Master's Thesis, University of São Paulo, Sao Paulo, Brazil, 1991.

380. Aluja-Schuneman, M.R.; Díaz-Fleischer, F.; Edmunds, A.J.F.; Hagmann, L. Isolation, Structural Determination, Synthesis, Biological Activity and Application as Control Agent of the Host Marking Pheromone (and Derivatives Thereof) of the Fruit Flies of the Type Anastrepha (Diptera: Tephritidae). US6555120B1, 29 April 2003.

381. Averill, A.L.; Prokopy, R.J. Residual activity of oviposition-deterring pheromone in *Rhagoletis pomonella* (Diptera: Tephritidae) and female response to infested fruit. *J. Chem. Ecol.* **1987**, *13*, 167–177. [CrossRef]

382. Prokopy, R.J.; Reynolds, A.H. Ovipositional enhancement through socially facilitated behavior in *Rhagoletis pomonella* flies. *Entomol. Exp. Appl.* **1998**, *86*, 281–286. [CrossRef]

383. Wiesmann, R. Die orientierung der kirschfliege *Rhagoletis cerasi* L. *Landwirtsch. Jahrb. Schweiz* **1937**, *51*, 1080–1109.

384. Katsoyannos, B.I. Oviposition-deterring, male-arresting, fruit-marking pheromone in *Rhagoletis cerasi*. *Environ. Entomol.* **1975**, *4*, 801–807. [CrossRef]

385. Prokopy, R.J. Oviposition-deterring fruit marking pheromone in *Rhagoletis fausta*. *Environ. Entomol.* **1975**, *4*, 298–300. [CrossRef]

386. Prokopy, R.J.; Reissig, W.H.; Moericke, V. Marking pheromones deterring repeated oviposition in *Rhagoletis* flies. *Entomol. Exp. Appl.* **1976**, *20*, 170–178. [CrossRef]

387. Bauer, G. Life-history strategy of *Rhagoletis alternata* (Diptera: Trypetidae), a fruit fly operating in a 'non-interactive' system. *J. Anim. Ecol.* **1986**, *55*, 785. [CrossRef]

388. Arredondo, J.; Diaz-Fleischer, F. Oviposition deterrents for the Mediterranean fruit fly, *Ceratitis capitata* (Diptera: Tephritidae) from fly faeces extracts. *Bull. Entomol. Res.* **2006**, *96*, 35–42. [CrossRef] [PubMed]

389. Roitberg, B.D.; Prokopy, R.J. Host deprivation influence on response of *Rhagoletis pomonella* to its oviposition deterring pheromone. *Physiol. Entomol.* **1983**, *8*, 69–72. [CrossRef]

390. Mumtaz, M.; AliNiazee, M. The oviposition-deterring pheromone in the western cherry fruit fly, *Rhagoletis indifferens* Curran (Dipt., Tephritidae). Biological properties. *Zeitschrift für Angew. Entomol.* **1983**, *96*, 83–93. [CrossRef]

391. Boller, E. Oviposition-deterring pheromone of the European cherry fruit fly: Status of research and potential applications. In *Management of Insect Pests with Semiochemicals*; Mitchell, E.R., Ed.; Springer: Boston, MA, USA, 1981; pp. 457–462.

392. Boller, E.F.; Hurter, J. Oviposition deterring pheromone in *Rhagoletis cerasi*: Behavioral laboratory test to measure pheromone activity. *Entomol. Exp. Appl.* **1985**, *39*, 163–169. [CrossRef]

393. Prokopy, R.J.; Averill, A.L.; Bardinelli, C.M.; Bowdan, E.S.; Cooley, S.S.; Crnjar, R.M.; Dundulis, E.A.; Roitberg, C.A.; Spatcher, P.J.; Tumlinson, J.H.; et al. Site of production of an oviposition-deterring pheromone component in *Rhagoletis pomonella* flies. *J. Insect Physiol.* **1982**, *28*, 1–7. [CrossRef]

394. Hurter, J.; Katsoyannos, B.; Boller, E.F.; Wirz, P. Beitrag zur Anreicherung und teilweisen Reinigung des eiablageverhindernden Pheromons der Kirschenfliege, *Rhagoletis cerasi* L. (Dipt., Trypetidae). *Zeitschrift für Angew. Entomol.* **1976**, *80*, 50–56. [CrossRef]

395. Kachigamba, D.L. Host-marking behaviour and pheromones in major fruit fly species (Diptera: Tephritidae) infesting mango (Mangifera indica) in Kenya. Ph.D. Thesis, Jomo Kenyatta University of Agriculture and Technology, Nairobi, Kenya, 2012.

396. Ernst, B.; Wagner, B. Synthesis of the oviposition-deterring pheromone (Odp) in *Rhagoletis cerasi* L. *Helv. Chim. Acta* **1989**, *72*, 165–171. [CrossRef]

397. Boller, E.F.; Aluja, M. Oviposition deterring pheromone in *Rhagoletis cerasi* L. Biological activity of 4 synthetic isomers and HMP discrimination of two host races as measured by an improved laboratory bioassay. *J. Appl. Entomol.* **1992**, *113*, 113–119. [CrossRef]

398. Edmunds, A.J.F.; Aluja, M.; Diaz-Fleischer, F.; Patrian, B.; Hagmann, L. Host marking pheromone (HMP) in the Mexican fruit fly *Anastrepha ludens*. *Chimia* **2010**, *64*, 37–42. [CrossRef]

399. Scarpati, M.L.; Loscalzo, R.; Vita, G. *Olea europaea* volatiles attractive and repellent to the Olive fruit fly (*Dacus oleae*, Gmelin). *J. Chem. Ecol.* **1993**, *19*, 881–891. [CrossRef]

400. Kachigamba, D.L.; Ekesi, S.; Ndung'u, M.W.; Gitonga, L.M.; Teal, P.E.A.; Torto, B. Evidence for potential of managing some African fruit fly species (Diptera: Tephritidae) using the mango fruit fly host-marking pheromone. *J. Econ. Entomol.* **2012**, *105*, 2068–2075. [CrossRef]

401. Carlson, D.A.; Mayer, M.S.; Silhacek, D.L.; James, J.D.; Beroza, M.; Bierl, B.A. Sex attractant pheromone of the house fly: Isolation, identification and synthesis. *Science* **1971**. [CrossRef]

402. Antony, C.; Jallon, J.-M. The chemical basis for sex recognition in *Drosophila melanogaster*. *J. Insect Physiol.* **1982**, *28*, 873–880. [CrossRef]

403. Blomquist, G.J.; Dillwith, J.W.; Adams, T.S. Biosynthesis and endocrine regulation of sex pheromone production in Diptera. In *Pheromone Biochemistry*; Elsevier: Amsterdam, The Netherlands, 1987; pp. 217–250.

404. Rogoff, W.M.; Beltz, A.D.; Johnsen, J.O.; Plapp, F.W. A sex pheromone in the housefly, *Musca domestica* L. *J. Insect Physiol.* **1964**, *10*, 239–246. [CrossRef]

405. Mayer, M.S.; James, J.D. Response of male *Musca domestica* to a specific olfactory attractant and its initial chemical purification. *J. Insect Physiol.* **1971**, *17*, 833–842. [CrossRef]

406. Silhacek, D.L.; Carlson, D.A.; Mayer, M.S.; James, J.D. Composition and sex attractancy of cuticular hydrocarbons from houseflies: Effects of age, sex, and mating. *J. Insect Physiol.* **1972**. [CrossRef]

407. Stocker, R.F. The organization of the chemosensory system in *Drosophila melanogaster*: A review. *Cell Tissue Res.* **1994**, *275*, 3–26. [CrossRef]

408. Boll, W.; Noll, M. The *Drosophila* Pox neuro gene: Control of male courtship behavior and fertility as revealed by a complete dissection of all enhancers. *Development* **2002**, *129*, 5667–5681. [CrossRef]

409. Bray, S.; Amrein, H. A putative *Drosophila* pheromone receptor expressed in male-specific taste neurons is required for efficient courtship. *Neuron* **2003**, *39*, 1019–1029. [CrossRef]
410. Jackson, L.L.; Bartelt, R.J. Cuticular hydrocarbons of *Drosophila virilis*. Comparison by age and sex. *Insect Biochem.* **1986**, *16*, 433–439. [CrossRef]
411. Kim, Y.-K.; Phillips, D.R.; Chao, T.; Ehrman, L. Developmental isolation and subsequent adult behavior of *Drosophila paulistorum*. VI. Quantitative variation in cuticular hydrocarbons. *Behav. Genet.* **2004**, *34*, 385–394. [CrossRef] [PubMed]
412. Carlson, D.A.; Yocom, S.R. Cuticular hydrocarbons from six species of tephritid fruit flies. *Arch. Insect Biochem. Physiol.* **1986**, *3*, 397–412. [CrossRef]
413. Bosa, C.F.; Cruz-Lopez, L.; Guillen-Navarro, K.; Zepeda-Cisneros, C.S.; Liedo, P. Variation in the cuticular hydrocarbons of the Mexican fruit fly *Anastrepha ludens* males between strains and age classes. *Arch. Insect Biochem. Physiol.* **2018**, *99*. [CrossRef] [PubMed]
414. Lavine, B.K.; Carlson, D.A.; Calkins, C.O. Classification of tephritid fruit fly larvae by gas chromatography pattern/recognition techniques. *Microchem. J.* **1992**, *45*, 50–57. [CrossRef]
415. Sutton, B.D.; Carlson, D.A. Interspecific variation in tephritid fruit fly larvae surface hydrocarbons. *Arch. Insect Biochem. Physiol.* **1993**, *23*, 53–65. [CrossRef]
416. Sutton, B.D.; Steck, G.J. Discrimination of Caribbean and Mediterranean fruit fly larvae (Diptera: Tephritidae) by cuticular hydrocarbon analysis. *Florida Entomol.* **1994**, *77*, 231. [CrossRef]
417. Vaníčková, L.; Hernández-Ortiz, V.; Joachim Bravo, I.S.; Dias, V.; Passos Roriz, A.K.; Laumann, R.A.; de Mendonça, A.L.; Jordão Paranhos, B.A.; do Nascimento, R.R. Current knowledge of the species complex *Anastrepha fraterculus* (Diptera, tephritidae) in Brazil. *Zookeys* **2015**, *540*, 211–237. [CrossRef]
418. Goh, S.H.; Ooi, K.E.; Chuah, C.H.; Yong, H.S.; Khoo, S.G.; Ong, S.H. Cuticular hydrocarbons from two species of Malaysian *Bactrocera* fruit flies. *Biochem. Syst. Ecol.* **1993**, *21*, 215–226. [CrossRef]
419. Vanickova, L.; Nagy, R.; Pompeiano, A.; Kalinova, B. Epicuticular chemistry reinforces the new taxonomic classification of the *Bactrocera dorsalis* species complex (Diptera: Tephritidae, Dacinae). *PLoS ONE* **2017**, *12*. [CrossRef]
420. Galhoum, A. Taxonomic studies on two Tephritid species (Order: Diptera), *Bactrocera oleae* and *B. zonata*, using the cuticular hydrocarbons profile. *Al Azhar Bull. Sci.* **2017**, *28*, 45–54. [CrossRef]
421. Park, S.J.; Pandey, G.; Castro-Vargas, C.; Oakeshott, J.G.; Taylor, P.W.; Mendez, V. Cuticular chemistry of the Queensland fruit fly *Bactrocera tryoni* (Froggatt). *Molecules* **2020**, *25*, 4185. [CrossRef]
422. Arakaki, N.; Kuba, H.; Soemori, H. Mating behavior of the Oriental fruit fly, *Dacus dorsalis* Hendel (Diptera: Tephritidae). *Appl. Entomol. Zool.* **1984**, *19*, 42–51. [CrossRef]
423. Shen, J.; Hu, L.; Zhou, X.; Dai, J.; Chen, B.; Li, S. Allyl-2,6-dimethoxyphenol, a female-biased compound, is robustly attractive to conspecific males of *Bactrocera dorsalis* at close range. *Entomol. Exp. Appl.* **2019**, *167*, 811–819. [CrossRef]
424. Hu, L.; Chen, B.; Liu, K.; Yu, G.; Chen, Y.; Dai, J.; Zhao, X.; Zhong, G.; Zhang, Y.; Shen, J. OBP2 in the midlegs of the male *Bactrocera dorsalis* is involved in the perception of the female-biased sex pheromone 4-allyl-2,6-dimethoxyphenol. *J. Agric. Food Chem.* **2021**, *69*, 126–134. [CrossRef]
425. Bagneres, A.G.; Morgan, E.D. A simple method for analysis of insect cuticular hydrocarbons. *J. Chem. Ecol.* **1990**, *16*, 3263–3276. [CrossRef]
426. Brill, J.H.; Bertsch, W. An investigation of sampling methods for the analysis of cuticular hydrocarbons. *J. Entomol. Sci.* **1985**, *20*, 435–443. [CrossRef]
427. Cvacka, J.; Jiroš, P.; Šobotník, J.; Hanus, R.; Svatoš, A. Analysis of insect cuticular hydrocarbons using matrix-assisted laser desorption/ionization mass spectrometry. *J. Chem. Ecol.* **2006**, *32*, 409–434. [CrossRef]
428. Chin, J.S.R.; Yew, J.Y. Pheromones in the Fruit Fly. *Methods Mol. Biol.* **2013**, *1068*, 15–27. [CrossRef]
429. Lockey, K.H. Cuticular hydrocarbons of adult *Cylindrinotus laevioctostriatus* (Goeze) and *Phylan gibbus* (Fabricius) (Coleoptera: Tenebrionidae). *Insect Biochem.* **1981**, *11*, 549–561. [CrossRef]
430. Kosi, A.Z.; Chinta, S.; Headrick, D.H.; Cokl, A.; Millar, J.G. Do chemical signals mediate reproductive behavior of Trupanea vicina, an emerging pest of ornamental marigold production in California? *Entomol. Exp. Appl.* **2013**, *149*, 44–56. [CrossRef]
431. Papadopoulos, N.T.; Carey, J.R.; Liedo, P.; Muller, H.-G.; Senturk, D. Virgin females compete for mates in the male lekking species *Ceratitis capitata*. *Physiol. Entomol.* **2009**, *34*, 238–245. [CrossRef]
432. Kroiss, J.; Svatos, A.; Kaltenpoth, M. Rapid identification of insect cuticular hydrocarbons using Gas Chromatography-Ion-Trap Mass Spectrometry. *J. Chem. Ecol.* **2011**, *37*, 420–427. [CrossRef] [PubMed]
433. Howard, R.W.; McDaniel, C.A.; Nelson, D.R.; Blomquist, G.J. Chemical ionization mass spectrometry. Application to insect-derived cuticular alkanes. *J. Chem. Ecol.* **1980**, *6*, 609–623. [CrossRef]
434. Moravcova, D.; Kahle, V.; Rehulkova, H.; Chmelik, J.; Rehulka, P. Short monolithic columns for purification and fractionation of peptide samples for matrix-assisted laser desorption/ionization time-of-flight/time-of-flight mass spectrometry analysis in proteomics. *J. Chromatogr. A* **2009**, *1216*, 3629–3636. [CrossRef]
435. McLafferty, F.W. *Registry of Mass Spectral Data*, 5th ed.; Wiley: New York, NY, USA, 1989.
436. Adams, R.P. *Identification of Essential Oil Components by Gas Chromatography/Quadrupole Mass Spectroscopy*; Allured: Carol Stream, IL, USA, 2007.

437. Carlson, D.A.; Roan, C.S.; Yost, R.A.; Hector, J. Dimethyl disulfide derivatives of long-chain alkenes, alkadienes, and alkatrienes for gas-chromatography mass-spectrometry. *Anal. Chem.* **1989**, *61*, 1564–1571. [CrossRef]

438. Oldham, N.J.; Svatos, A. Determination of the double bond position in functionalized monoenes by chemical ionization ion-trap mass spectrometry using acetonitrile as a reagent gas. *Rapid Commun. Mass Spectrom.* **1999**, *13*, 331–336. [CrossRef]

439. Yew, J.Y.; Dreisewerd, K.; Luftmann, H.; Müthing, J.; Pohlentz, G.; Kravitz, E.A. A new male sex pheromone and novel cuticular cues for chemical communication in *Drosophila. Curr. Biol.* **2009**, *19*, 1245–1254. [CrossRef]

440. Cvacka, J.; Svatos, A. Matrix-assisted laser desorption/ionization analysis of lipids and high molecular weight hydrocarbons with lithium 2,5-dihydroxybenzoate matrix. *Rapid Commun. Mass Spectrom.* **2003**, *17*, 2203–2207. [CrossRef]

441. Vrkoslav, V.; Muck, A.; Cvacka, J.; Svatos, A. MALDI imaging of neutral cuticular lipids in insects and plants. *J. Am. Soc. Mass Spectrom.* **2010**, *21*, 220–231. [CrossRef]

442. Sivinski, J.M.; Calkins, C. Pheromones and parapheromones in the control of tephritids. *Florida Entomol.* **1986**, *69*, 157. [CrossRef]

443. Romano, D.; Donati, E.; Benelli, G.; Stefanini, C. A review on animal–robot interaction: From bio-hybrid organisms to mixed societies. *Biol. Cybern.* **2019**, *113*, 201–225. [CrossRef] [PubMed]

444. Halloy, J.; Sempo, G.; Caprari, G.; Rivault, C.; Asadpour, M.; Tache, F.; Said, I.; Durier, V.; Canonge, S.; Ame, J.M.; et al. Social integration of robots into groups of cockroaches to control self-organized choices. *Science* **2007**, *318*, 1155–1158. [CrossRef]

445. Romano, D.; Benelli, G.; Kavallieratos, N.G.; Athanassiou, C.G.; Canale, A.; Stefanini, C. Beetle-robot hybrid interaction: Sex, lateralization and mating experience modulate behavioural responses to robotic cues in the larger grain borer *Prostephanus truncatus* (Horn). *Biol. Cybern.* **2020**, *114*, 473–483. [CrossRef]

446. Romano, D.; Benelli, G.; Stefanini, C. Opposite valence social information provided by bio-robotic demonstrators shapes selection processes in the green bottle fly. *J. R. Soc. Interface* **2021**, *18*, rsif.2021.0056. [CrossRef]

447. Papadopoulos, N.T.; Katsoyannos, B.I.; Kouloussis, N.A.; Hendrichs, J.; Carey, J.R.; Heath, R.R. Early detection and population monitoring of *Ceratitis capitata* (Diptera: Tephritidae) in a mixed-fruit orchard in northern Greece. *J. Econ. Entomol.* **2001**, *94*, 971–978. [CrossRef]

448. Suckling, D.M.; Kean, J.M.; Stringer, L.D.; Cáceres-Barrios, C.; Hendrichs, J.; Reyes-Flores, J.; Dominiak, B.C. Eradication of tephritid fruit fly pest populations: Outcomes and prospects. *Pest Manag. Sci.* **2016**, *72*, 456–465. [CrossRef]

449. Casaña-Giner, V.; Levi, V.; Navarro-Llopis, V.; Jang, E.B. Implication of SAR of male medfly attractants in insect olfaction. *SAR QSAR Environ. Res.* **2002**, *13*, 629–640. [CrossRef]

450. Jang, E.; Khrimian, A.; Holler, T. Field response of Mediterranean fruit flies to ceralure B1 relative to most active isomer and commercial formulation of trimedlure. *J. Econ. Entomol.* **2010**. [CrossRef]

451. Shelly, T.E.; Cowan, A.N.; Edu, J.; Pahio, E. Mating success of male Mediterranean fruit flies following exposure to two sources of α-copaene, manuka oil and mango. *Florida Entomol.* **2008**, *91*, 9–15. [CrossRef]

452. Shelly, T.E.; Whittier, T.S.; Villalobos, E.M. Trimedlure affects mating success and mate attraction in male Mediterranean fruit flies. *Entomol. Exp. Appl.* **1996**, *78*, 181–185. [CrossRef]

453. Papadopoulos, N.T.; Katsoyannos, B.I.; Kouloussis, N.A.; Hendrichs, J. Effect of orange peel substances on mating competitiveness of male *Ceratitis capitata. Entomol. Exp. Appl.* **2001**, *99*, 253–261. [CrossRef]

454. Shelly, T.E.; Epsky, N.D. Exposure to tea tree oil enhances the mating success of male Mediterranean fruit flies (Diptera: Tephritidae). *Florida Entomol.* **2015**, *98*, 1127–1133. [CrossRef]

455. Shelly, T.E. Exposure to α-copaene and α-copaene-containing oils enhances mating success of male Mediterranean fruit flies (Diptera: Tephritidae). *Ann. Entomol. Soc. Am.* **2001**, *94*, 497–502. [CrossRef]

456. Kouloussis, N.A.; Gerofotis, C.D.; Ioannou, C.S.; Iliadis, I.V.; Papadopoulos, N.T.; Koveos, D.S. Towards improving sterile insect technique: Exposure to orange oil compounds increases sexual signalling and longevity in *Ceratitis capitata* males of the Vienna 8 GSS. *PLoS ONE* **2017**, *12*, e0188092. [CrossRef]

457. Dean, D.; Pierre, H.; Mosser, L.; Kurashima, R.; Shelly, T. Field longevity and attractiveness of trimedlure plugs to male *Ceratitis capitata* in Florida and Hawaii. *Florida Entomol.* **2018**, *101*, 441–446. [CrossRef]

458. (IAEA) International Atomic Energy Agency. *Trapping Guidelines for Area-Wide Fruit Fly Programmes*; IAEA: Vienna, Austria, 2003.

459. Royer, J.E.; Agovaua, S.; Bokosou, J.; Kurika, K.; Mararuai, A.; Mayer, D.G.; Niangu, B. Responses of fruit flies (Diptera: Tephritidae) to new attractants in Papua New Guinea. *Austral. Entomol.* **2018**, *57*, 40–49. [CrossRef]

460. Royer, J.E. Responses of fruit flies (Tephritidae: Dacinae) to novel male attractants in north Queensland, Australia, and improved lures for some pest species. *Austral. Entomol.* **2015**, *54*, 411–426. [CrossRef]

461. Shelly, T.E.; Villalobos, E.M. Cue Lure and the mating behavior of male melon flies (Diptera: Tephritidae). *Florida Entomol.* **1995**, *78*, 473. [CrossRef]

462. Shelly, T.E. Effects of methyl eugenol and raspberry ketone/cue lure on the sexual behavior of *Bactrocera* species (Diptera: Tephritidae). *Appl. Entomol. Zool.* **2010**, *45*, 349–361. [CrossRef]

463. Khan, M.A.M.; Shuttleworth, L.A.; Osborne, T.; Collins, D.; Gurr, G.M.; Reynolds, O.L. Raspberry ketone accelerates sexual maturation and improves mating performance of sterile male Queensland fruit fly, *Bactrocera tryoni* (Froggatt). *Pest Manag. Sci.* **2019**, *75*, 1942–1950. [CrossRef] [PubMed]

464. Akter, H.; Mendez, V.; Morelli, R.; Pérez, J.; Taylor, P.W. Raspberry ketone supplement promotes early sexual maturation in male Queensland fruit fly, *Bactrocera tryoni* (Diptera: Tephritidae). *Pest Manag. Sci.* **2017**, *73*, 1764–1770. [CrossRef] [PubMed]

465. Wee, S.-L.; Clarke, A.R. Male-lure type, lure dosage, and fly age at feeding all influence male mating success in Jarvis' fruit fly. *Sci. Rep.* **2020**, *10*, 15004. [CrossRef] [PubMed]

466. Akter, H.; Pérez, J.; Park, S.J. Raspberry ketone supplements provided to immature male Queensland fruit fly, *Bactrocera tryoni* (Froggatt), increase the amount of volatiles in rectal glands. *Chemoecology* **2020**, *31*, 89–99. [CrossRef]

467. Cunningham, R.T.; Suda, D.Y. Male annihilation through mass-trapping of male flies with methyleugenol to reduce infestation of Oriental fruit fly (Diptera: Tephritidae) larvae in papaya. *J. Econ. Entomol.* **1986**, *79*, 1580–1582. [CrossRef]

468. Haq, I.U.; Cáceres, C.; Meza, J.S.; Hendrichs, J.; Vreysen, M.J.B. Different methods of methyl eugenol application enhance the mating success of male Oriental fruit fly (Dipera: Tephritidae). *Sci. Rep.* **2018**, *8*, 6033. [CrossRef] [PubMed]

469. De Vincenzi, M.; Silano, M.; Stacchini, P.; Scazzocchio, B. Constituents of aromatic plants: I. Methyleugenol. *Fitoterapia* **2000**, *71*, 216–221. [CrossRef]

470. Haniotakis, G.; Kozyrakis, M.; Fitsakis, T.; Antonidaki, A. An effective mass trapping method for the control of *Dacus oleae* (Diptera, tephritidae). *J. Econ. Entomol.* **1991**, *84*, 564–569. [CrossRef]

471. Sarles, L.; Fassotte, B.; Boullis, A.; Lognay, G.; Verhaeghe, A.; Marko, I.; Verheggen, F.J. Improving the monitoring of the walnut husk fly (Diptera: Tephritidae) using male-produced lactones. *J. Econ. Entomol.* **2018**, *111*, 2032–2037. [CrossRef]

472. Vargas, R.I.; Souder, S.K.; Hoffman, K.; Mercogliano, J.; Smith, T.R.; Hammond, J.; Davis, B.J.; Brodie, M.; Dripps, J.E. Attraction and mortality of *Bactrocera dorsalis* (Diptera: Tephritidae) to STATIC spinosad ME weathered under operational conditions in California and Florida: A reduced-risk male annihilation treatment. *J. Econ. Entomol.* **2014**, *107*, 1362–1369. [CrossRef]

473. (CDFA) California Department of Food and Agriculture. *Oriental Fruit Fly Fact Sheet*; CDFA: Sacramento, CA, USA, 2008.

474. Vargas, R.; Piñero, J.; Leblanc, L. An overview of pest species of *Bactrocera* fruit flies (Diptera: Tephritidae) and the integration of biopesticides with other biological approaches for their management with a focus on the Pacific region. *Insects* **2015**, *6*, 297–318. [CrossRef] [PubMed]

475. Papadopoulos, N.T.; Plant, R.E.; Carey, J.R. From trickle to flood: The large-scale, cryptic invasion of California by tropical fruit flies. *Proc. R. Soc. B Biol. Sci.* **2013**, *280*, 20131466. [CrossRef] [PubMed]

476. Manoukis, N.C.; Vargas, R.I.; Carvalho, L.; Fezza, T.; Wilson, S.; Collier, T.; Shelly, T.E. A field test on the effectiveness of male annihilation technique against *Bactrocera dorsalis* (Diptera: Tephritidae) at varying application densities. *PLoS ONE* **2019**, *14*, e0213337. [CrossRef]

477. Manrakhan, A.; Venter, J.H.; Hattingh, V. The progressive invasion of *Bactrocera dorsalis* (Diptera: Tephritidae) in South Africa. *Biol. Invasions* **2015**, *17*, 2803–2809. [CrossRef]

478. Manrakhan, A.; Hattingh, V.; Venter, J.-H.; Holtzhausen, M. Eradication of *Bactrocera invadens* (Diptera: Tephritidae) in Limpopo Province, South Africa. *Afr. Entomol.* **2011**, *19*, 650–659. [CrossRef]

479. Steiner, L.F.; Hart, W.G.; Harris, E.J.; Cunningham, R.T.; Ohinata, K.; Kamakahi, D.C. Eradication of the Oriental fruit fly from the Mariana Islands by the methods of male annihilation and sterile insect release. *J. Econ. Entomol.* **1970**, *63*, 131–135. [CrossRef]

480. Koyama, J.; Teruya, T.; Tanaka, K. Eradication of the Oriental fruit fly (Diptera: Tephritidae) from the Okinawa Islands by a male annihilation method. *J. Econ. Entomol.* **1984**, *77*, 468–472. [CrossRef]

481. Seewooruthun, S.I.; Permalloo, S.; Gungah, B.; Soonnoo, A.R.; Alleck, M. Eradication of an exotic fruit fly from Mauritius. In *Area-Wide Control of Fruit Flies and Other Insect Pests*; Penerbit Universiti Sains Malaysia: Penang, Malaysia, 2000; pp. 389–394.

482. Ekesi, S.; De Meyer, M.; Mohamed, S.A.; Virgilio, M.; Borgemeister, C. Taxonomy, ecology, and management of native and exotic fruit fly species in Africa. *Annu. Rev. Entomol.* **2016**, *61*, 219–238. [CrossRef]

483. Tasnin, M.S.; Merkel, K.; Clarke, A.R. Effects of advanced age on olfactory response of male and female Queensland fruit fly, *Bactrocera tryoni* (Froggatt) (Diptera: Tephritidae). *J. Insect Physiol.* **2020**, *122*, 104024. [CrossRef]

484. Mau, R.F.L.; Jang, E.B.; Vargas, R. The Hawaii area-wide fruit fly management programme. In *Area-Wide Control of Insect Pests from research to Field Implementation*; Vreysen, M.J.B., Robinson, A.S., Hendrichs, J., Eds.; Springer: Dordrecht, The Netherlands, 2007; pp. 671–683.

485. Ballo, S.; Demissie, G.; Tefera, T.; Mohamed, S.A.; Khamis, F.M.; Niassy, S.; Ekesi, S. Use of para-pheromone methyl eugenol for suppression of the mango fruit fly, *Bactrocera dorsalis* (Hendel) (Diptera: Tephritidae) in Southern Ethiopia. In *Sustainability in Plant and Crop Protection*; Springer: Cham, Switzerland, 2020; pp. 203–217.

486. Speranza, S.; Bellocchi, G.; Pucci, C. IPM trials on attract-and-kill mixtures against the olive fly *Bactrocera oleae* (Diptera Tephritidae). *Bull. Insectol.* **2004**, *57*, 111–115.

487. Broumas, T.; Haniotakis, G.; Liaropoulos, C.; Tomazou, T.; Ragoussis, N. The efficacy of an improved form of the mass-trapping method, for the control of the olive fruit fly, *Bactrocera oleae* (Gmelin) (Dipt., Tephritidae): Pilot-scale feasibility studies. *J. Appl. Entomol.* **2002**, *126*, 217–223. [CrossRef]

488. Montiel, A.; Ramos, P.; Jones, O.T.; Lisk, J.C.; Howse, P.E.; Baker, R. Interferencias en el apareamiento de la mosca del olivo (*Dacus oleae* Gmel.) con el principal componente de su feromona sexual. *Bol. Serv. Plagas* **1982**, *8*, 193–200.

489. Montiel, A.; Jones, O.T. Alternative methods for controlling the olive fly, *Bactrocera oleae*, involving semiochemicals. *IOBC WPRS Bull.* **2002**, *25*, 147–156.

490. Navarro-Llopis, V.; Alfaro, C.; Primo, J.; Vacas, S. Response of two tephritid species, *Bactrocera oleae* and *Ceratitis capitata*, to different emission levels of pheromone and parapheromone. *Crop Prot.* **2011**, *30*, 913–918. [CrossRef]

491. Ferveur, J.F. Genetic control of pheromones in *Drosophila simulans*. I. Ngbo, a locus on the second chromosome. *Genetics* **1991**, *128*, 293–301. [CrossRef] [PubMed]

492. Gleason, J.M.; Jallon, J.M.; Rouault, J.D.; Ritchie, M.G. Quantitative trait loci for cuticular hydrocarbons associated with sexual isolation between *Drosophila simulans* and *D. sechellia*. *Genetics* **2005**, *171*, 1789–1798. [CrossRef] [PubMed]

493. Liimatainen, J.O.; Jallon, J.M. Genetic analysis of cuticular hydrocarbons and their effect on courtship in *Drosophila virilis* and *D. lummei*. *Behav. Genet.* **2007**, *37*. [CrossRef]

494. Snellings, Y.; Herrera, B.; Wildemann, B.; Beelen, M.; Zwarts, L.; Wenseleers, T.; Callaerts, P. The role of cuticular hydrocarbons in mate recognition in *Drosophila suzukii*. *Sci Rep* **2018**, *8*, 4996. [CrossRef]

495. Leonhardt, B.A.; Rice, R.E.; Harte, E.M.; Cunningham, R.T. Evaluation of dispensers containing trimedlure, the attractant for the Mediterranean fruit fly (Diptera: Tephritidae). *J. Econ. Entomol.* **1984**, *77*, 744–749. [CrossRef]

496. Domínguez-Ruiz, J.; Sanchis, J.; Navarro-Llopis, V.; Primo, J. A new long-life trimedlure dispenser for Mediterranean fruit fly. *J. Econ. Entomol.* **2008**, *101*, 1325–1330. [CrossRef]

497. Cameron, D.N.S.; McRae, C.; Park, S.J.; Taylor, P.W.; Jamie, I.M. Vapor pressures and thermodynamic properties of phenylpropanoid and phenylbutanoid attractants of male *Bactrocera*, *Dacus*, and *Zeugodacus* fruit flies at ambient temperatures. *J. Agric. Food Chem.* **2020**, *68*, 9654–9663. [CrossRef] [PubMed]

498. Lehman, K.A.; Barahona, D.C.; Manoukis, N.C.; Carvalho, L.A.F.N.; De Faveri, S.G.; Auth, J.E.; Siderhurst, M.S. Raspberry ketone trifluoroacetate trapping of *Zeugodacus cucurbitae* (Diptera: Tephritidae)in Hawaii. *J. Econ. Entomol.* **2019**, *112*, 1306–1313. [CrossRef] [PubMed]

499. Katsoyannos, B.; Boller, E. Second field application of oviposition-deterring pheromone of the European cherry fruit fly, *Rhagoletis cerasi* L. (Diptera: Tephritidae). *Zeitschrift für Angew. Entomol.* **1980**, *89*, 278–281. [CrossRef]

500. Katsoyannos, B.I.; Boller, E.F. First field application of oviposition-deterring marking pheromone of European cherry fruit fly. *Environ. Entomol.* **1976**, *5*, 151–152. [CrossRef]

501. Aluja, M.; Boller, E.F. Host marking pheromone of *Rhagoletis cerasi*: Foraging behavior in response to synthetic pheromonal isomers. *J. Chem. Ecol.* **1992**, *18*, 1299–1311. [CrossRef]

502. Boller, E.; Hurter, J. The marking pheromone of the cherry fruit fly: A novel non-toxic and ecologically safe technique to protect cherries against cherry fruit fly infestation. In Proceedings of the Second International Symposium on Insect Pheromones; 1998.

503. Prokopy, R.J. Epideitic pheromones that influence spacing patterns of phytophagous insects. In *Semiochemicals: Their Role in Pest Control*; Nordlund, D.A., Jones, R.L., Lewis, W.J., Eds.; Wiley Press: New York, NY, USA, 1981; pp. 181–213.

504. Cook, S.M.; Khan, Z.R.; Pickett, J.A. The use of push-pull strategies in integrated pest management. *Annu. Rev. Entomol.* **2007**, *52*, 375–400. [CrossRef]

505. Birke, A.; Lopez-Ramirez, S.; Jimenez-Mendoza, R.; Acosta, E.; Ortega, R.; Edmunds, A.; Aluja, M. Host marking pheromone and GF120(TM) applied in a push-pull scheme reduce grapefruit infestation by *Anastrepha ludens* in field-cage studies. *J. Pest Sci.* **2020**, *93*, 507–518. [CrossRef]

506. Cai, P.; Song, Y.; Huo, D.; Lin, J.; Zhang, H.; Zhang, Z.; Xiao, C.; Huang, F.; Ji, Q. Chemical cues induced from fly-oviposition mediate the host-seeking behaviour of *Fopius arisanus* (Hymenoptera: Braconidae), an effective egg parasitoid of *Bactrocera dorsalis* (Diptera: Tephritidae), within a tritrophic context. *Insects* **2020**, *11*, 231. [CrossRef]

507. Benelli, G.; Revadi, S.; Carpita, A.; Giunti, G.; Raspi, A.; Anfora, G.; Canale, A. Behavioral and electrophysiological responses of the parasitic wasp *Psyttalia concolor* (Szépligeti) (Hymenoptera: Braconidae) to *Ceratitis capitata*-induced fruit volatiles. *Biol. Control* **2013**, *64*, 116–124. [CrossRef]

508. Giunti, G.; Canale, A.; Messing, R.H.; Donati, E.; Stefanini, C.; Michaud, J.P.; Benelli, G. Parasitoid learning: Current knowledge and implications for biological control. *Biol. Control* **2015**, *90*, 208–219. [CrossRef]

509. Stelinski, L.L.; Rodriguez-Saona, C.; Meyer, W.L. Recognition of foreign oviposition-marking pheromone in a multi-trophic context. *Naturwissenschaften* **2009**, *96*, 585–592. [CrossRef]

510. Faraone, N.; Hillier, N.K.; Cutler, G.C. Collection of host-marking pheromone from *Rhagoletis mendax* (Diptera: Tephritidae). *Can. Entomol.* **2016**, *148*, 552–555. [CrossRef]

511. Li, H.; Ren, L.; Xie, M.; Gao, Y.; He, M.; Hassan, B.; Lu, Y.; Cheng, D. Egg-surface bacteria are indirectly associated with oviposition aversion in *Bactrocera dorsalis*. *Curr. Biol.* **2020**, *30*, 4432–4440. [CrossRef]

512. Calcagnile, M.; Tredici, S.M.; Talà, A.; Alifano, P. Bacterial semiochemicals and transkingdom interactions with insects and plants. *Insects* **2019**, *10*, 441. [CrossRef]

513. Knipling, E.F. Possibilities of insect control or eradication through the use of sexually sterile males. *J. Econ. Entomol.* **1955**, *48*, 459–462. [CrossRef]

514. Gurr, G.M.; Kvedaras, O.L. Synergizing biological control: Scope for sterile insect technique, induced plant defences and cultural techniques to enhance natural enemy impact. *Biol. Control* **2010**, *52*, 198–207. [CrossRef]

515. Whittier, T.S.; Nam, F.Y.; Shelly, T.E.; Kaneshiro, K.Y. Male courtship success and female discrimination in the mediterranean fruit fly (Diptera: Tephritidae). *J. Insect Behav.* **1994**, *7*, 159–170. [CrossRef]

516. Pereira, R.; Yuval, B.; Liedo, P.; Teal, P.E.A.; Shelly, T.E.; McInnis, D.O.; Hendrichs, J. Improving sterile male performance in support of programmes integrating the sterile insect technique against fruit flies. *J. Appl. Entomol.* **2013**, *137*, 178–190. [CrossRef]

517. Segura, D.F.; Belliard, S.A.; Vera, M.T.; Bachmann, G.E.; Ruiz, M.J.; Jofre-Barud, F.; Fernández, P.C.; López, M.L.; Shelly, T.E. Plant chemicals and the sexual behavior of male tephritid fruit flies. *Ann. Entomol. Soc. Am.* **2018**, *111*, 239–264. [CrossRef]

518. Vargas, R.I.; Shelly, T.E.; Leblanc, L.; Piñero, J.C. Recent advances in methyl eugenol and Cue-Lure technologies for fruit fly detection, monitoring, and control in Hawaii. In *Vitamins and Hormones*; Elsevier: Amsterdam, The Netherlands, 2010; pp. 575–595.

519. Khan, M.A.M.; Manoukis, N.C.; Osborne, T.; Barchia, I.M.; Gurr, G.M.; Reynolds, O.L. Semiochemical mediated enhancement of males to complement sterile insect technique in management of the tephritid pest *Bactrocera tryoni* (Froggatt). *Sci. Rep.* **2017**, *7*, 13366. [CrossRef]

520. Wehrenfennig, C.; Schott, M.; Gasch, T.; During, R.A.; Vilcinskas, A.; Kohl, C.D. On-site airborne pheromone sensing. *Anal. Bioanal. Chem.* **2013**, *405*, 6389–6403. [CrossRef]

521. Fernandez-Grandon, G.M.; Girling, R.D.; Poppy, G.M. Utilizing insect behavior in chemical detection by a behavioral biosensor. *J. Plant Interact.* **2011**, *6*, 109–112. [CrossRef]

522. Schroth, P.; Schoning, M.J.; Schutz, S.; Malkoc, U.; Steffen, A.; Marso, M.; Hummel, H.E.; Kordos, P.; Luth, H. Coupling of insect antennae to field-effect transistors for biochemical sensing. *Electrochim. Acta* **1999**, *44*, 3821–3826. [CrossRef]

523. Repasky, K.S.; Shaw, J.A.; Scheppele, R.; Melton, C.; Carsten, J.L.; Spangler, L.H. Optical detection of honeybees by use of wing-beat modulation of scattered laser light for locating explosives and land mines. *Appl. Opt.* **2006**, *45*, 1839–1843. [CrossRef]

524. Tomberlin, J.K.; Tertuliano, M.; Rains, G.; Lewis, W.J. Conditioned *Microplitis croceipes* Cresson (Hymenoptera: Braconidae) detect and respond to 2,4-DNT: Development of a biological sensor. *J. Forensic. Sci.* **2005**, *50*, 1187–1190. [CrossRef]

525. Tertuliano, M.; Tomberlin, J.K.; Jurjevic, Z.; Wilson, D.; Rains, G.C.; Lewis, W.J. The ability of conditioned *Microplitis croceipes* (Hymenoptera: Braconidae) to distinguish between odors of aflatoxigenic and non-aflatoxigenic fungal strains. *Chemoecology* **2005**, *15*, 89–95. [CrossRef]

526. Carey, A.F.; Carlson, J.R. Insect olfaction from model systems to disease control. *Proc. Natl. Acad. Sci. USA* **2011**, *108*, 12987–12995. [CrossRef]

527. Strauch, M.; Lüdke, A.; Münch, D.; Laudes, T.; Galizia, C.G.; Martinelli, E.; Lavra, L.; Paolesse, R.; Ulivieri, A.; Catini, A.; et al. More than apples and oranges—Detecting cancer with a fruit fly's antenna. *Sci. Rep.* **2015**, *4*, 3576. [CrossRef]

528. Pelosi, P.; Mastrogiacomo, R.; Iovinella, I.; Tuccori, E.; Persaud, K.C. Structure and biotechnological applications of odorant-binding proteins. *Appl. Microbiol. Biotechnol.* **2014**, *98*, 61–70. [CrossRef]

529. Lu, Y.L.; Li, H.L.; Zhuang, S.L.; Zhang, D.M.; Zhang, Q.; Zhou, J.; Dong, S.Y.; Liu, Q.J.; Wang, P. Olfactory biosensor using odorant-binding proteins from honeybee: Ligands of floral odors and pheromones detection by electrochemical impedance. *Sensors Actuators B Chem.* **2014**, *193*, 420–427. [CrossRef]

530. Sankaran, S.; Panigrahi, S.; Mallik, S. Odorant binding protein based biomimetic sensors for detection of alcohols associated with Salmonella contamination in packaged beef. *Biosens. Bioelectron.* **2011**, *26*, 3103–3109. [CrossRef] [PubMed]

531. Kuang, Z.F.; Kim, S.N.; Crookes-Goodson, W.J.; Farmer, B.L.; Naik, R.R. Biomimetic chemosensor: Designing peptide recognition elements for surface functionalization of carbon nanotube field effect transistors. *ACS Nano* **2010**, *4*, 452–458. [CrossRef] [PubMed]

532. Kotlowski, C.; Larisika, M.; Guerin, P.M.; Kleber, C.; Krober, T.; Mastrogiacomo, R.; Nowak, C.; Pelosi, P.; Schutz, S.; Schwaighofer, A.; et al. Fine discrimination of volatile compounds by graphene-immobilized odorant-binding proteins. *Sensors Actuators B Chem.* **2018**, *256*, 564–572. [CrossRef]

533. Lu, Y.L.; Yao, Y.; Li, S.; Zhang, Q.; Liu, Q.J. Olfactory biosensor based on odorant-binding proteins of *Bactrocera dorsalis* with electrochemical impedance sensing for pest management. *Sens. Rev.* **2017**, *37*, 396–403. [CrossRef]

534. Bohbot, J.D.; Dickens, J.C. Characterization of an enantioselective odorant receptor in the yellow fever mosquito *Aedes aegypti*. *PLoS ONE* **2009**, *4*, e7032. [CrossRef]

535. Stensmyr, M.C.; Dweck, H.K.M.; Farhan, A.; Ibba, I.; Strutz, A.; Mukunda, L.; Linz, J.; Grabe, V.; Steck, K.; Lavista-Llanos, S.; et al. A conserved dedicated olfactory circuit for detecting harmful microbes in *Drosophila*. *Cell* **2012**, *151*, 1345–1357. [CrossRef]

536. Bohbot, J.D.; Vernick, S. The emergence of insect Odorant Receptor-based biosensors. *Biosensors* **2020**, *10*, 26. [CrossRef]

537. Moitra, P.; Bhagat, D.; Kamble, V.B.; Umarji, A.M.; Pratap, R.; Bhattacharya, S. First example of engineered β-cyclodextrinylated MEMS devices for volatile pheromone sensing of olive fruit pests. *Biosens. Bioelectron.* **2021**, *173*, 112728. [CrossRef]

Review

Latest Developments in Insect Sex Pheromone Research and Its Application in Agricultural Pest Management

Syed Arif Hussain Rizvi [1], Justin George [2], Gadi V. P. Reddy [2], Xinnian Zeng [3,*] and Angel Guerrero [4,*]

[1] National Agricultural Research Center (NARC), Islamabad 44000, Pakistan; arifrizvi@aup.edu.pk
[2] Southern Insect Management Research Unit, USDA-ARS, Stoneville, MS 38776, USA; justin.george@usda.gov (J.G.); Gadi.Reddy@usda.gov (G.V.P.R.)
[3] College of Plant Protection, South China Agricultural University, Guangzhou 510642, China
[4] Department of Biological Chemistry, Institute of Advanced Chemistry of Catalonia-CSIC, 08034 Barcelona, Spain
* Correspondence: zengxn@scau.edu.cn (X.Z.); anggue56@outlook.com (A.G.)

Simple Summary: Insect pheromones are specific natural compounds that meet modern pest control requirements, i.e., species-specificity, lack of toxicity to mammals, environmentally benign, and a component for the Integrated Pest Management of agricultural pests. Therefore, the practical application of insect pheromones, particularly sex pheromones, have had a tremendous success in controlling low density pest populations, and long-term reduction in pest populations with minimal impact on their natural enemies. Mass trapping and mating disruption strategies using sex pheromones have significantly reduced the use of conventional insecticides, thereby providing sustainable and ecofriendly pest management in agricultural crops. In this review, we summarize the latest developments in sex pheromone research, mechanisms of sex pheromone perception, and its practical application in agricultural pest management.

Abstract: Since the first identification of the silkworm moth sex pheromone in 1959, significant research has been reported on identifying and unravelling the sex pheromone mechanisms of hundreds of insect species. In the past two decades, the number of research studies on new insect pheromones, pheromone biosynthesis, mode of action, peripheral olfactory and neural mechanisms, and their practical applications in Integrated Pest Management has increased dramatically. An interdisciplinary approach that uses the advances and new techniques in analytical chemistry, chemical ecology, neurophysiology, genetics, and evolutionary and molecular biology has helped us to better understand the pheromone perception mechanisms and its practical application in agricultural pest management. In this review, we present the most recent developments in pheromone research and its application in the past two decades.

Keywords: sex pheromones; integrated pest management; biosynthesis; pheromone perception; resistance; review

Citation: Rizvi, S.A.H.; George, J.; Reddy, G.V.P.; Zeng, X.; Guerrero, A. Latest Developments in Insect Sex Pheromone Research and Its Application in Agricultural Pest Management. *Insects* **2021**, *12*, 484. https://doi.org/10.3390/insects12060484

Academic Editors: Giovanni Benelli and Andrea Lucchi

Received: 14 April 2021
Accepted: 20 May 2021
Published: 23 May 2021

Publisher's Note: MDPI stays neutral with regard to jurisdictional claims in published maps and institutional affiliations.

1. Introduction

Sex pheromones are chemical signals emitted by an organism that elicit a sexual response in a member of the opposite sex of the same species [1,2]. Since the structural characterization of the first sex pheromone of the silkworm moth *Bombyx mori* in 1959 [3,4], more than 600 species [5] of lepidopteran pheromones have been identified. Their main features, e.g., species-specificity, non-toxicity to mammals and other beneficial organisms, their activity in minute amounts, and rapid degradation in the environment were soon envisioned to be promising tools for controlling insect pests, estimating pest populations, detecting the entry and progress of invasive pests, and preserving endangered species [6–8]. In fact, in recent years the most successful practical applications of sex pheromones in integrated pest management (IPM) have been the monitoring of pest populations, mass trapping, mating disruption, and push-pull strategies [7,9–11]. The number of articles appearing in the literature on these subjects has dramatically increased over the years, particularly in the past two decades (Figure 1), with the sex pheromones of the genera *Helicoverpa*, *Spodoptera*, *Grapholita* and *Cydia* most frequently cited (Table 1). In this review, a literature search, including patents, was conducted by the following search tools and databases: Web of Science, CABI, Agricola, SciFinder, Google Scholar, The Pherobase, and Espacenet. The following keywords were chosen: sex pheromone, sex pheromone autodetection, sex pheromone perception mechanism, sex pheromone biosynthesis, resistance to pheromone, pheromone and biological control agents, the application of pheromones in IPM, and push-pull strategy.

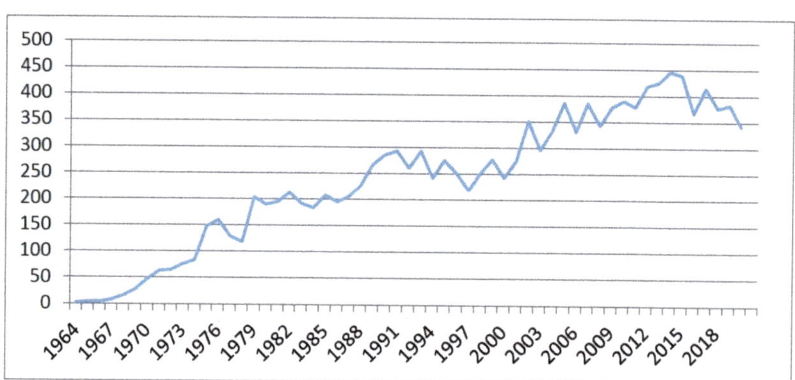

Figure 1. Number of publications on insect sex pheromones appeared in the literature (1964–2020).

Table 1. Highly concerned genera on sex pheromones since 2000.

Order	Family	Genus	Publications
Lepidoptera	Noctuidae	*Helicoverpa*	194
		Spodoptera	173
	Tortricidae	*Grapholita*	137
		Cydia	128
	Plutellidae	*Plutella*	72
	Crambidae	*Ostrinia*	63
	Bombycidae	*Bombyx*	61
Diptera	Drosophilidae	*Drosophila*	49
Lepidoptera	Pyralidae	*Chilo*	42
	Tortricidae	*Lobesia*	41
	Noctuidae	*Agrotis*	40
Diptera	Tephritidae	*Bactrocera*	37
Hemiptera	Pseudococcidae	*Planococcus*	36
Lepidoptera	Lymantriidae	*Lymantria*	36
Coleoptera	Scarabaeidae	*Holotrichia*	35
Lepidoptera	Gelechiidae	*Tuta*	33
	Geometridae	*Ectropis*	30
	Lasiocampidae	*Dendrolimus*	29

Table 1. *Cont.*

Order	Family	Genus	Publications
Diptera	Psychodidae	*Lutzomyia*	26
Lepidoptera	Sesiidae	*Synanthedon*	25
Blattodea	Blatteidae	*Blattella*	25
Coleoptera	Buprestidae	*Agrilus*	25
Lepidoptera	Crambidae	*Pyrausta*	24
	Gracillariidae	*Phyllocnistis*	23
	Erebidae	*Hyphantria*	22
Hemiptera	Pseudococcidae	*Pseudococcus*	21
Lepidoptera	Pyralidae	*Plodia*	21
	Crambidae	*Cnaphalocrocis*	18
	Tortricidae	*Choristoneura*	18
Hemiptera	Miridae	*Apolygus*	18

Sex pheromones are mainly produced by females and used as attractant compounds to show the presence of potential mating partners and their reproductive status [8,11]. Sex pheromones comprise sex attractant pheromones, which induce upwind oriented movements to the conspecific individual, and courtship pheromones, which elicit a variety of close-range responses in the insect partner [12,13]. Since the first pheromone discovery, the rapid progress of methodologies developed to identify new pheromones, mainly GC, GC-MS, NMR, electrophysiological techniques [electroantennography (EAG), gas chromatography coupled to electroantennography (GC-EAD), single sensillum recordings (SSR), and coupled GC-SSR], have allowed the identification of thousands of compounds as insect sex pheromones [14] (Table 2 contains new sex pheromones and sex pheromone components recently identified from insect pests in the period 2010–2020 and the corresponding references [15–32]). In addition, an interdisciplinary approach involving advances in analytical chemistry, neurophysiology, genetics, and molecular biology have improved our understanding of insect chemical communication and behavior to the level of discrete neural circuits [8].

Insects' sex pheromones are generally blends of two or more compounds and only in few cases, one chemical, usually the primary component, is efficient to attract conspecifics and mate [33,34]. Löfstedt et al. [35] have reported different types of moth pheromones based on their production, chemical structure, and biosynthetic origin. Type I pheromones are C_{10}–C_{18} monounsaturated or diunsaturated acetates, alcohols, and aldehydes [36] and constitute ca. 75% of all known moth sex pheromone components. Type II pheromones are polyunsaturated straight-chain hydrocarbons and the corresponding epoxide derivatives with C_{17}–C_{25} carbon atoms [36] and comprise about 15% of the moth pheromones reported. In addition to Types I and II, Löfstedt et al. [35] proposed to extent this classification by defining two more pheromone Types 0 and III. Type III pheromones include saturated and unsaturated hydrocarbons, which may be functionalized, containing one or more methyl branches. Type 0 pheromones, in turn, consist of short chain methylcarbinols and methylketones and have been found more recently in Eriocraniidae and caddisflies (Trichoptera). Another group of pheromones comprising propionate esters of secondary alcohols, methyl-branched secondary alcohols, methyl-branched methylketones, and straight-chain (Z)-7-alken-11-ones cannot be categorized in any of the Types 0-III pheromone groups because their structures are not clearly biosynthetically related to any of those classifications [35].

Table 2. New sex pheromones and sex pheromone components recently identified from insect pests (2010–2020).

Pheromone/Pheromone Components	Insect	References
(*E*)-11-hexadecenal, (*E,E*)-10,12-hexadecadienal	*Diaphania glauculalis* (Lepidoptera: Crambidae)	Ma et al. [15]
(*E*)-10-hexadecenal, (*Z*)-10-hexadecenal, (*E*)-10-hexadecenol, (*E,E*)-10,12-hexadecadienal, (*Z,Z,Z*)-3,6,9-tricosatriene	*Conogethes pluto* (Lepidoptera: Crambidae)	El Sayed et al. [16]
(*Z*)-11-hexadecenyl acetate, (*Z*)-11-hexadecenal, (*Z*)-11-hexadecenol	*Trichophysetis cretacea* (Lepidoptera: Crambidae)	Pong et al. [17]
(4a*S*,7*S*,7a*R*)-nepetalactone, (1*R*,4a*S*,7*S*,7a*R*)-nepetalactol	*Hyalopterus pruni, Brachycaudus helichrysi* (Hemiptera: Aphididae)	Symmes et al. [18]
(*E,Z*)-3,13-octadecadienyl acetate, (*Z,Z*)-3,13-octadecadienyl acetate	*Synanthedon vespiformis* (Lepidoptera: Sesiidae)	Levi-Zada et al. [19]
(*E*)-11-tetradecenyl acetate, (*E,E*)-9,11-tetradecadienyl acetate, (*E*)-11-tetradecenol, (*E*)-11-hexadecenyl acetate	*Epiphyas postvittana* (Lepidoptera: Tortricidae)	El Sayed et al. [20]
(*Z,E*)-9,12-tetradecadienyl acetate, (*Z*)-9-tetradecenyl acetate, (*Z*)-11-hexadecenyl acetate, (*Z,E*)-9,12-tetradecadienol, (*Z*)-9-tetradecenol, (*Z*)-11-hexadecenol	*Spodoptera exigua* (Lepidoptera: Noctuidae)	Acín et al. [21]
(*Z,Z*)-3,13-dodecadienolide	*Parcoblatta lata* (broad wood cochroach)	Eliyahu et al. [22]
(*R,R*)-(*Z*)-3,7,11,15-tetramethyl hexadec-2-enal, (*R,R*)-(*E*)-3,7,11,15-tetramethyl hexadec-2-enal	*Dociostaurus maroccanus* (Moroccan locust)	Guerrero et al. [23]
(4,5,5)-(trimethyl-3-methylenecyclopent-1-en-1-yl)methyl acetate	*Delottococcus aberiae* (Hemiptera: Pseudococcidae)	Vacas et al. [24]
(*Z*)-9-tetradecenyl acetate, (*Z*)-9-tetradecenol, tetradecyl acetate	*Coryphodema tristis* (Lepidoptera: Cossidae)	Bouwer et al. [25]
(-)-δ-heptalactone	*Rhagoletis batava* (Diptera: Tephritidae)	Büda et al. [26]
(*E,E,Z,Z*)-4,6,11,13-hexadecatetraenal	*Callosamia promethea* (Lepidoptera: Saturniidae)	Gago et al. [27]
(-)-iridomyrmecin	*Leptopilina heterotoma* (Hymenoptera: Figitidae)	Weiss et al. [28]
(*Z,E*)-5,7-dodecadienyl acetate, (*Z,E*)-5,7-dodecadienol, (*Z,E*)-5,7-dodecadienyl propionate	*Dendrolimus tabulaeformis* (Lepidoptera: Lasiocampidae)	Kong et al. [29]
(*E*)-7,9-decadienol, (*E*)-8-decenol	*Monema flavescens* (Lepidoptera: Limacodidae)	Shibasaki et al. [30]
(1*S*,4*R*,1′*S*)-4-(1′,5′-dimethylhex-4′-enyl)-1-methylcyclohex-2-en-1-ol	*Oebalus poecilus* (Heteroptera: Pentatomidae)	Oliveira et al. [31]
(3*S*,6*S*,7*R*)-1,10-bisaboladien-3-ol, (3*R*,6*S*,7*R*)-1,10-bisaboladien-3-ol	*Tibraca limbativentris* (Hemiptera: Pentatomidae)	Blassioli-Moraes et al. [32]

The chemical structure of pheromones is widely diverse. Thus, many of them are hydrocarbons, alcohols, esters, epoxides, aldehydes, ketones, lactones, carboxylic acids, isoprenoids, and triacylglycerides [8,37]. This structural diversity is the key for the pheromone specificity but, in a number of species in Lepidoptera, an appropriate combination of the pheromone components in specific ratio renders the pheromone species-specific. In this review, we present an update of the latest developments of insect sex pheromones reported as useful, environmentally benign, tools for the management of agricultural pests in IPM programs including the mechanisms of pheromone perception, interactions with biological control agents, autodetection, and resistance. An excellent book covering the pheromone communication in moths up to 2015 has recently appeared in the literature [38].

2. Sex Pheromone Biosynthesis

Sex pheromone components are C_{10}–C_{18} straight chain unsaturated compounds with an oxygenated functional group [39]. In many Lepidoptera species, sex pheromone production is regulated by the pheromone biosynthesis-activating neuropeptide (PBAN), a neurohormone originated in the subesophageal ganglion and released into the hemolymph to operate directly on the pheromone gland (PG) [40], which in turn activates functional group modification enzymes [41] or acetyl-coenzyme A carboxylase (ACC) [42]. In the first step of pheromone biosynthesis, the carboxylation of acetyl-CoA to malonyl-CoA is catalyzed by ACC, followed by fatty acid synthase action which leads to production of saturated fatty acids (C18:0 and C16:0) [42].

Through a series of enzymatic reactions, i.e., desaturation, chain-shortening reaction, reduction, acetylation, and oxidation, the palmitic or stearic acids are then converted to the final pheromone components in a stepwise manner [41,43]. Therefore, different enzymes are likely to be involved in different reactions, and to date, the genes encoding the essential enzymes—fatty acid desaturases (FADs), fatty acid reductases (FARs), chain shortening via peroxisomal β-oxidation, and chain elongation—have been functionally identified. The specificity of the enzymes and their combinations allow formation of an immense array of pheromone components with different chain lengths, unsaturations, and functional groups [5,35]. The discovery of the biosynthetic enzymes is generally accomplished by comparative analysis of gene expression in the female PGs relative to control tissues. The genes that are overexpressed in the pheromone glands are candidate genes to be involved in pheromone biosynthesis, such as desaturases, reductases, etc. The enzymes are expressed in heterologous hosts, such as the yeast *Saccharomyces cerevisiae* [44], insect cells [45], or plants [46], and functionally assayed. Transcriptome data from pheromone glands have been reported for many Lepidopterans, such as *Agrotis segetum*, *Ephestia cautella*, *Pectinophora gossypiella*, *Plutella xylostella*, *Spodoptera exigua*, and *Spodoptera litura* [5]. Transcriptomes from many other insects can be found in the database http://www.insect-genome.com (accessed on 8 March 2021) [47].

FADs are the most intensively reported class of enzymes involved in moth sex pheromone biosynthesis and, thus, more than 50 FADs have been functionally characterized [48]. They are able to display different specificities and introduce double bonds in different positions and geometry. The most common desaturation processes occur in positions $\Delta 9$, f.i. in *Mamestra brassicae* [43], *Trichoplusia ni* [49], and *S. litura* [50]; and $\Delta 11$, f.i. *Ostrinia* spp. [51], but they can produce also unsaturations at $\Delta 5$ in *Ctenopseustis obliquana* and *C. herana* [52], $\Delta 6$ in *Antheraea pernyi* [53], $\Delta 8$ in *Dendrolimus punctatus* [54], $\Delta 10$ in *Planotortrix octo* [55], $\Delta 14$ in *Ostrinia furnacalis* [56], and $\Delta 1$ in *Operophtera brumata* [57]. A multi-functional $\Delta 10$–$\Delta 12$ desaturase was found in *B. mori* pheromone biosynthesis [45], $\Delta 11$-$\Delta 12$ desaturases were noticed in *S. exigua* and *S. litura* [58], and a $\Delta 11$–$\Delta 13$ desaturase was characterized in *Thaumetopoea pityocampa* [59]. In the case of the homologous $\Delta 5$ FAD of *S. litura* (SlitDes5) and *S. exigua* (SexiDes5), the full-length sequences of SexiDes5 (1017 bp) and SlitDes5 (1017 bp), were obtained from the pheromone gland cDNA library transcripts [60], and both encoded proteins with 339 amino acids. In the heterologous expression system, both desaturases inserted a *cis* double bond at $\Delta 11$ position in palmitic, myristic and stearic acids. However, while both enzymes introduce only a *cis* double bond in saturated C16 and C18 substrates, in C14 substrates both *cis* and *trans* unsaturations are created. The differences between substrates and the geometry of the substrate-binding tunnel might influence the regioselectivity and stereospecificity of the desaturation reaction [61].

FARs catalyze the reduction of fatty acylCoA precursors into fatty alcohols in a two-step reaction without releasing the intermediate aldehyde forms [62]. They play an essential role in regulating the final steps of sex pheromone biosynthesis in some moths, and have been functionally identified in species from the genera *Agrotis*, *Bicyclus*, *Bombyx*, *Helicoverpa*, *Heliothis*, *Ostrinia*, *Spodoptera*, and *Yponomeuta* [48]. Reductases from *Helicoverpa* spp. and *Heliothis* spp. can act on a broad range of C8 to C16 fatty acids, preferentially on C14 substrates [63]. In the European and Asian corn borers, *Ostrinia nubilalis* and *O. furnacalis*, in vivo labeling studies demonstrated that the selectivity of the reductase system could modulate ratios among final pheromone components by exclusive conversion of specific acid moieties into their corresponding alcohols [64]. Four reductases from *Spodoptera* spp. showed different selectivity for C14 and C16 fatty acids, whereas SexpgFAR I and SlitpgFAR I selectively act on C16 fatty acids, SexpgFAR II, and SlitpgFAR II preferred C14 fatty acids as substrates [65]. The small ermine moths, namely *Yponomeuta evonymellus*, *Yponomeuta padellus*, and *Yponomeuta rorellus* use pheromone blends made of structurally related C14 and C16 fatty alcohols and their derivatives [66]. In *Y. evonymellus* and *Y. padellus*, the two primary pheromone components are Z11–14:OAc and E11–14:OAc, and complete reproductive isolation is ensured by the use of additional pheromone components. In contrast, *Y. rorellus* uses only the saturated acetate 14:OAc as a pheromone. By screening the PG FAR genes, Liénard and coworkers [66] demonstrated that the reduction step of long-chain C14- and C16-acyl pheromone precursors found in the three insects is accounted for by a single PG-specific FAR.

There are other important enzymes that have not been functionally confirmed but are postulated to be involved in the sex pheromone biosynthesis pathway. For example, biochemical studies have suggested that acetyltransferases (ACTs) and alcohol dehydrogenases (ADHs) play a crucial role by converting fatty alcohols into the corresponding fatty acetates and aldehydes, respectively, since they constitute the last step of the pheromone biosynthetic pathway in many moths. Currently, no insect ACTs have been characterized to esterify fatty alcohols, in contrast to plants [67,68] and yeast [69] from which a number of ACTs (EC: 2.3.1.84) were cloned. In a recent study on the pheromone biosynthetic pathway of *A. segetum*, Ding and Löfstedt [44] expressed 34 genes potentially encoding for ACTs but none of them successfully converted fatty alcohols into the corresponding acetates. The sex pheromone of *P. xylostella* is a mixture of (Z)-11-hexadecenal, (Z)-11-hexadecenyl acetate, and (Z)-11-hexadecenol in 8:100:18 ratio [70]. In its PG transcriptome two transcripts encoding proteins homologous to acetyltransferases ACT1 and ACT2 were found [71]. ACT1 was homologous to acetyl-CoA ACT from *Amyelois transitella* and ACT2 had similar amino acid sequence to the ACT gene from *S. litura*. However, neither of them was homologue to the genes belonging to the group of ACTs EC 2.3.1.84.

With regard to pheromone aldehydes, Chen et al. [71] found that the amino acid sequences encoded by five transcripts in *P. xylostella* PG transcriptome resemble alcohol dehydrogenase (ADH, EC 1.1.1.1) genes, but none of them were cloned and characterized. ADHs are a group of enzymes that facilitate interconversion between alcohols and aldehydes with the reduction of NAD^+ to NADH in the biosynthetic pathway of pheromone aldehydes. β-Oxidation enzymes are supposedly involved in the biosynthesis of pheromones of shorter chain lengths and may play an important role in regulating the ratio of sex pheromone compounds with different carbon lengths. The genes involved in β-oxidation have been identified in some moth PG tissues [72] but not yet characterized.

3. Mechanisms of Insect Sex Pheromone Perception

Insects detect volatile odorants/pheromone molecules using olfactory receptor (OR) sensilla present on their antennae and maxillary palps [73]. Pheromones and other odorant molecules that are absorbed on the cuticular surface of an olfactory sensillum diffuse inside through olfactory pores and the pore tubule [74]. The sensillum lymph contains the pheromone binding proteins (PBPs) that bind with volatile pheromones and solubilize them to pass across the sensillum lymph to activate pheromone receptors [75]. The structure and arrangement of olfactory sensilla and olfactory sensory neurons (OSNs) on the antennae and palps of insects are very specialized and optimized to detect odorants, especially sex pheromones in the case of male antennae. OSNs carry the olfactory information from the periphery to the antennal lobe in the insect brain. Many studies have reported the structure and peripheral olfactory mechanisms of odorant perception in insects [73,76,77]. Other works that have investigated these different olfactory elements, and the cellular and molecular mechanisms of volatile pheromone signal detection in olfactory sensilla have been published in recent review publications [75,78,79]. Recent studies have also reported on the evolution of olfactory circuits and processing of these information in the higher olfactory centers in the insect brain [80,81].

The peripheral olfactory hairs or olfactory sensilla that house the OSNs play an essential role in odorants perception, especially sex pheromones. Different morphological types and distribution of olfactory sensilla have been reported in multiple insect species based on their ecological niches. For example, the long trichoid sensilla house the OSNs tuned to sex pheromones in many moth species, whereas the antennae in flies and wasps contain basiconic and placoid sensilla, respectively, which are the most common peripheral olfactory hairs [82]. In the last decade, significant improvements have been made in understanding the ultrastructural features of peripheral olfactory sensilla and its correlation with neuronal mechanisms using electron microscopy and genetic labelling, which has enabled the integration of morphological and molecular information in different insects, such as *Drosophila* sp. and *B. mori* [83,84]. In this context, a "Cryochem method" has been developed and used in *Drosophila melanogaster*. This method rehydrates cryofixed and high-pressure frozen samples for creating three-dimensional reconstructions of genetically marked OSNs in different sensilla by serial block-face scanning electron microscopy. It is found useful in providing insights into the relationship between OSN anatomy and olfactory physiology [83,84]. A recent review of these new molecular tools and mechanisms of olfactory detection in insects has been published [85].

A significant amount of research has been done in the last decade to understand the odorant receptors (ORs) [86,87] involved in chemo-electric signal transduction and processing of odorant signals in the insect brain. Sato et al. [88] and Wicher et al. [89] reported the role of OR co-receptor (Orco) that is characteristic for each OR-expressing sensory neuron. Orco acts as heteromeric ligand-gated and cyclic-nucleotide-activated cation channel, and is highly conserved across insect species and orders [90]. Sakurai et al. [91] published a detailed review on the neural and molecular mechanisms involved in sex pheromone perception and processing in the silkworm moth, *B. mori*. Rapid progress in the sequencing technologies, genomics, and transcriptomics of insect olfactory tissues has helped identify odorant receptor (OR) gene families for multiple insect species in recent years [79]. These new technologies have helped improve our knowledge of insect sensory systems and their practical application in product development for pest management.

Many studies have been published to understand the role of odorant binding proteins (OBPs), pheromone binding proteins (PBPs), sensory neuron membrane proteins (SNMPs), and sex pheromone receptors that are involved in the transport and central processing of pheromone molecules in *Drosophila* and different moth species [75,92,93]. Genomic and transcriptomic analysis have allowed the identification of new sensory neuron membrane proteins (SNMPs) in Lepidoptera, Diptera, Coleoptera, and Hemiptera (see f.i. the excellent review by Cassau and Krieger [94]), and their role in pheromone signaling has been noticed [95]. Very recently, it has been disclosed that a new sensory membrane protein, HarmSNMP1, plays an essential role in the detection of long-chain sex pheromones in *Helicoverpa armigera*, but this protein was not required for detecting shorter chain sex pheromones of the same species [96]. Zhao et al. [97] reported the identification of 49 OBPs and 5 chemosensory proteins (CSPs) in the chive gnat *Bradysia odoriphaga*, and Zhang et al. [98] identified 64 ORs, 24 OBPs and 19 CSPs after analyzing transcripts expressed in chemosensory organs of the beet armyworm *S. exigua*. Initial studies were reported in *Drosophila*, where the pheromone binding protein LUSH carries the hydrophobic pheromone, *cis*-11-vaccenyl acetate (cVA), through the aqueous sensillum lymph to the olfactory neuron dendrites [99]. In recent years, RNAi techniques have been widely used to identify OBPs and PBPs in insects [100]. Oliveira et al. [101] has reported the use of RNAi technique to identify the PBP RproOBP27, involved in sex pheromone detection of *Rhodnius prolixus*. Identification and molecular characterization of PBPs from multiple insect species, such as *Cydia pomonella* [100], *Chilo suppressalis* [102], *Loxostege sticticalis* [103], and *Conogethes pinicolalis* [104] have been reported in recent years. PBP1 from *S. exigua* [105] and *Cyrtotrachelus buqueti* [106] can also bind to plant volatile compounds, such as benzaldehyde, linalool, indole, and other carboxylic acids, in addition to pheromone molecules. Also, a dual role of the GmolOBP7 from the oriental fruit moth *G. molesta* in the detection of both sex pheromone components and host plant volatiles has been disclosed [107].

An insect's age and the physiological state also affect its responsiveness to sex pheromones and other host plant volatiles. A very good example is the mating-dependent olfactory plasticity exhibited by *Agrotis ipsilon* males. The male copulates only once in a scotophase and results in a temporary inhibition of attraction to the sex pheromone [108], but in turn has no effect in their response to plant odors. This olfactory plasticity allows mated *A. ipsilon* males to transiently block their central responses to pheromones after mating. This leads to an increased non-pheromonal odor detection allowing more efficient finding of food sources in a natural environment [108]. Kromann et al. [109] studied this olfactory plasticity in *Spodoptera littoralis* and showed that newly mated males stopped responding to pheromones and host odors but not to food odors. In addition, Hatano et al. [110] showed that (*E*)-4,8-dimethyl-1,3,7-nonatriene (DMNT), a key cotton volatile compound used by natural enemies in finding prey, is used by females and males of *S. littoralis* to avoid induced plant sites and calling females, respectively. This chemical suppressed responses to the main pheromone component, (*Z,E*)-9,11-tetradecadienyl acetate, and to (*Z*)-3-hexenyl acetate, a host plant attractant. In neurophysiological experiments, the compound interfered with host plant and mate location through suppression of olfactory signaling pathways. Using Ca^{2+} imaging, the authors demonstrated that the major component of the pheromone elicited calcium responses in the cumulus, the largest glomerulus of the MGC, but addition of DMNT suppressed them [110]. As DMNT attracts natural enemies and deters herbivores, it may be useful in the development of push-pull strategies.

Jarriault et al. [111] found that the sensitivity of antennal lobe neurons to pheromones increases with age and is dependent on the level of octopamine and juvenile hormone (JH) in *A. ipsilon*, but posterior studies by these authors indicated that it is not mediated by octopamine or serotonin [112]. Follow-up studies by Deisig et al. [113] revealed that in *A. ipsilon* plant odors, such as heptanal, reduce pheromone sensitivity at the macroglomerular complex (MGC) level, resulting in an improved temporal resolution of pheromone pulses by antennal lobe (AL) output neurons [114]. In addition, heptanal activated the specialist olfactory receptor neuron (ORN) for (*Z*)-7-dodecenyl acetate, one of the pheromone components, and altered the ratio responses of pheromone-sensitive neurons [115]. In another noctuid moth, *H. armigera*, Ian et al. [116] used calcium imaging to reveal a reduced increase of intracellular calcium levels when stimulated with a blend of sex pheromone and complex plant odors as compared to individual odor application. Recent studies by Borrero-Echeverry et al. [117] reported that host plant volatiles enhance the selectivity for conspecific pheromone blends in the noctuid moth *S. littoralis*, and provided evidence for the evolution of pheromone specificity within a host plant odor environment, although the neural mechanisms are unknown. Other studies have reported that interactions between sex pheromone compounds and some plant-derived signals occur in olfactory receptor neurons (ORNs) [118,119].

Recent studies have shown a significant interest in understanding the neurophysiological mechanisms that regulate the interaction between female sex pheromone and behaviorally active host plant odorants by using functional imaging of the antennal lobe (AL) and intracellular recordings (IRs) of projection neurons (PNs) that transmit olfactory signals to higher centers of insect brain [120]. Galizia et al. [121] reported the specific role of the macroglomerular complex (MGC) in pheromone coding, and how the sexually isomorphic, ordinary glomeruli codes for plant volatile information. The olfactory coding of sex pheromones and general plant odors are supposed to occur in these different pathways of the olfactory system of insects. However, studies by Varela et al. [122] and Trona et al. [123] revealed some interesting observations on the processing of sex pheromones and general odorants by moths in the Tortricid family. Varela et al. reported that in *Grapholita molesta* the processing of sex pheromones occurs in olfactory glomeruli (OG) rather than in the macroglomerular complex (MGC), whereas in *C. pomonella* no clear segregation between the pheromone and the general odor were observed. Both odor classes were represented in the MGC and in OG [123] and were correlated with behavioral responses [120]. In the codling moth *C. pomonella*, it was noticed that the macroglomerular complex (MGC) in the antennal lobe (AL), involved in pheromone perception, showed an enhanced response to blends of pheromone and plant signals, whereas the response in glomeruli surrounding the MGC was suppressed [123]. This effect implied a higher attraction of males to blends of female sex pheromone and plant odor compared with single compounds. These findings show that, in nature, sex pheromone and plant odors are perceived as an ensemble, and mating and habitat cues are coded as blends in the MGC of the AL highlighting the dual role of plant signals in habitat selection and in premating sexual communication [120]. Very recently, reviews on the plasticity and modulation of olfactory circuits of insects in a complex environment with different odorants and pheromones, and new neuroecological studies directed to understand the evolution of insect sensory systems, have been published [124,125]. These studies could eventually provide us with potential tools to protect endangered species and reduce the risk for the invasion of alien species.

Research studies have also reported the effects of sex-pheromone exposure on non-sexual behaviors, such as gustatory perception and habituation (a non-associative learning) in male *A. ipsilon* moths. Hostachy et al. [12] used proboscis extension response (PER) assay to investigate the links between reproduction and gustation in *A. ipsilon* by assessing whether their sex-pheromone can modulate sucrose responsiveness and gustatory habituation. Experiments showed that the conspecific sex-pheromone (blend of Z7-dodecenyl-acetate, Z9-tetradecenyl-acetate, and Z11-hexadecenyl-acetate in 4:1:4 ratio) and the hetero-specific sex pheromone (Z5-decenyl acetate) had time-dependent effects on gustatory habituation of *A. ipsilon* moths. This study showed that the sex pheromones can play modulatory roles in gustatory perception in non-social insects, such as moths, and may affect their behavioral plasticity. Follow up studies by Murmu et al. [126] showed that pheromones facilitated both appetitive and aversive olfactory learning in *A. ipsilon* moths. The exposure to the *A. ipsilon* conspecific sex-pheromone before conditioning enhanced appetitive but not aversive learning, while exposure to a heterospecific sex-pheromone component facilitated aversive but not appetitive learning. These modulatory effects of sex pheromones on insects' learning and memory may have practical applications in developing specific traps that uses attractants or deterrents for pest control.

4. Evolutionary Aspects of Olfactory Receptors

Insects detect odorants through olfactory sensory neurons (OSNs), housed within the olfactory sensilla located in the surface of the antennae and maxillary pals [73] The olfactory capacities of an insect depend essentially on the repertoire of expressed olfactory receptor (OR) genes and the functional properties of OR proteins (OBPs), their sensitivity and their variety of responses. These two large gene families, the OBPs and the ORs, are presumably exclusive to insects but when they first appeared in the insect lineage remains to be determined [127]. The ORs form a large and highly divergent gene family, which shows no homology to the OR families of vertebrates. In contrast to the ORs of vertebrates, insect ORs form heteromeric complexes that are typically composed of a single ligand OR and the OR corecepotor ORCO [128]. Many OR repertoires have been identified and unique lineage-specific expansions of OR clades have been observed in different insect orders. This suggests that in each order ORs have followed different evolutionary trajectories as insects have adapted to new ecological niches [127]. This adaptation has been studied through the ORs repertoires from the Diptera *D. melanogaster* [129] and the malaria vector mosquito *Anopheles gambiae* [86], which have demonstrated that the ORs repertoires have been specialized in the detection of ecologically relevant natural products. Apart from Diptera, no other OR repertoire has been functionally characterized. However, De Fouchier et al. [127] have reported a functional analysis of a large array of ORs from the cotton leafworm *S. littoralis*, from which the antennal transcriptome had been previously sequenced [130]. A total of 35 candidate *S. littoralis* receptors (SlitORs) were expressed in *Drosophila* OSNs and identified SlitORs tuned to a variety of odorant molecules, which had been previously shown to be physiologically or behaviorally active in this species [127]. These include host plant and herbivore-induced volatiles, oviposition cues, larval frass volatiles, and pheromone components. The authors reported that receptors have now been deorphanized in 13 different clades of the lepidopteran OR phylogeny, including previous results obtained from *B. mori*, *Epiphyas postvittana*, and *S. exigua*, among other moths. A number of deorphanized receptors had aromatic compounds as their best ligand whereas others were best activated by terpenes and in a lower extent by aliphatic chemicals. Overall and from an evolutionary perspective, the results suggest that receptors to aromatics emerged first and have been more conserved during the evolution of Lepidoptera, whereas receptors to terpenes and the aliphatic compounds emerged more recently and evolved faster (especially aliphatic receptors, which include pheromone receptors) [127]. These properties correlate well with the ecological needs of herbivorous and nectar-feeding insects, such as moths, since aromatics and monoterpenes are the major constituents of plant odors emitted by flowers and leaves. Sex pheromone receptors and a large part of ORs tuned to terpenes and short chain acetates appear to belong to later lineages with a higher rate of evolution.

5. Sex Pheromone Autodetection

The phenomenon of a sex pheromone producer insect capable of detecting its conspecific sex pheromone components is termed autodetection [131]. One of the first cases recorded on autodetection was reported by Schneider et al. [132] in which electroantennographic responses to both pheromone components released by females of *Panaxia quadripunctaria* Poda (Lepidoptera, Arctiidae) were equally detected and with similar amplitude by both sexes. It revealed that not only males could detect the female-produced pheromones, but females were not 'anosmic' (unable to detect their conspecific sex pheromone) for their own attractant. In some cases, the pheromone only attracts males, thus acting as a sex pheromone, but may induce other behavioral effects on females, e.g., a repellent effect, an advance or delay in calling initiation, or an increase in calling frequency, among others [133]. Sex pheromones are secreted by one sex and cause an intraspecific attractant response and mating in individuals of the opposite sex, but in some rare cases, the pheromone attracts both sexes, functioning more like an aggregation pheromone. The aggregation pheromones are emitted by insects of one sex and cause individuals of both sexes to join for feeding and reproduction. In sex pheromone autodetection, female aggregation may increase the possibilities of mating success or induce dispersal at high population levels.

Until 2015, most of the autodetection studies were focused on Lepidoptera and Coleoptera, of which 28 cases (67%) involved species in the order Lepidoptera, 12 (29%) in the Coleoptera, one (2%) in the Blattodea, and one (2%) in the Diptera [133]. Among Lepidoptera, responses of 7 families (Tortricidae, Noctuidae, Arctiidae, Cossidae, Sesiidae, Yponomeuta, and Pyralidae) out 11 families were detected positive to their sex pheromone by Electroantennography (EAG) and Single Sensillum Recording (SSR) of female antennae, in at least one species within a given family, and only the Saturniidae, Bombycidae, and Geometridae lepidopteran families showed no response to their pheromones.

Bakthavatsalam and coworkers [134] also proved that gravid females of *H. armigera* (Hubner) respond to their pheromone blend (mixture of (Z)-11-hexadecenal and (Z)-9-hexadecenal in 97:3 ratio) eliciting stronger responses than unmated males. However, virgin or gravid females showed poor response in wind-tunnel studies, and an oviposition bioassay where gravid females were allowed to oviposit in the presence and absence of pheromone odors indicated that there was no difference in the number of eggs laid. Although morphological differences in antennal size and complexity might correlate with differential pheromone detection ability between sexes in some families, such as Saturniidae, many other families such as Tortricidae and Noctuidae were not morphologically different [117]. Moreover, comparing studies of antennae morphology and pheromone sensitivity among various types of sensilla confirmed that sex pheromone detection is not directly related with the gross morphology of antennae [135]. The proteins of PBPs and PRs are crucial in pheromone detection. At least one species in each of the 9 Lepidoptera and Coleoptera families tested contained these proteins (or precursors) [133]. In comparison to males, female antennae exhibited dramatically fewer PBPs and PRs.

Contradictory results were reported in the autodetection of *G. molesta* females. Kuhns et al. [136] found lower mating success after exposure to pheromone but not in calling behavior, while Stelinski et al. [137] reported no effect of pheromone pre-exposure on mating success. Also, Stelinski and coworkers [138] previously reported that females elicited significantly higher EAG responses than solvent-treated controls as well as an advance of the onset of female calling. These contradictory results from two different studies on the same insect highlight variations in behavioral responses, depending possibly on the assay conditions.

As the first report of an insect of the family Gelechiidae exhibiting autodetection, the tomato leafminer *Tuta absoluta* Meyrick (Lepidoptera: Gelechiidae) females exhibited electrophysiological responses to their own pheromone. However, the elicited EAG responses were much lower than those induced by males [139]. The depolarizations displayed by virgin females when stimulated with the binary mixture were significantly higher than those displayed by mated females, although detection of the individual pheromone compounds was generally higher in mated females but not significantly.

A new example of autodetection was found recently in female *Contarinia nasturtii* (Diptera: Cecidomyiidae) 140]. It represents a second family of Diptera exhibiting a case of pheromone autodetection. In both laboratory and field experiments, females exposed to stereospecific and racemic three-component pheromone blends called more frequently and for longer periods than midges in control treatments. Additionally, pre-exposure to stereospecific and racemic pheromone component blends reduced subsequent matings on females (42% and 35%, respectively) vs. 68% of female midges mated under control conditions. The authors concluded that in pheromone-treated fields while more frequent callings may increase the probability of females being detected by males, a reduction in females predisposition to mate would enhance the efficacy of mating disruption experiments [140].

Although the combined ecological and molecular data available suggested that sex pheromones had been detected only in adult moths, Poivet et al. [141] provided a strong evidence that caterpillars can detect and be attracted by the adult sex pheromone of *S. littoralis*. The authors showed that larvae walk towards a sex pheromone source, since their antennae house olfactory receptor neurons (ORNs) that respond to the pheromone and express the PBPs already identified in adults. Moreover, the larvae significantly preferred a food source with the major pheromone component to other lacking the chemical. This unexpected larval behavior may open new expectations in pest control strategies [141]. The responses of olfactory sensilla on the larval antennae of *Heliothis virescens* to specific sex pheromone components were also later reported by single sensillum recordings (SSRs) [142]. Two pheromone receptors HR6 and HR13 were found to be expressed in two and three candidate pheromone response cells, respectively, and other cells expressed PBP1 and PBP2. The results suggested that the responsiveness of larval sensilla to the female sex pheromone is based on similar molecular machinery as in adult male antennae.

As noticed above, the behaviors of autodetection are diverse, and some are even contradictory, such as repelling/attraction or dispersal/aggregation. This suggests that a complete understanding of the autodetection behavior of a species is necessary before applying sex pheromone traps in monitoring insect pest populations and mating disruption strategies.

6. Resistance of Insects to Sex Pheromones

To date, there are only a few reports on the development of resistance of insects to sex pheromones. The first report on the potential for evolution of resistance to pheromones was described for the pink bollworm moth, *P. gossypiella* (Saunders) (Lepidoptera: Gelechiidae) by Haynes and coworkers [143]. After an extensive examination of the release rates and blend ratios of pheromonal components emitted by field-collected *P. gossypiella* females, the authors found no evidence of resistance to pheromones applied to cotton fields to disrupt mating. Haynes et al. [143] theorized that while resistance to the *P. gosspiella* pheromone is still an opportunity when used profoundly in managing insect pests as a mating disruptant, there are modern alternative IPM options to prevent the development of resistance to sex pheromones. On the other hand, since mating disruption using synthetic pheromones did not cause any insect mortality, resistance was not likely to emerge. Haynes and Baker [144] stated that a slight change in pheromone emission in insects would represent effective resistance to disruptant pheromones. They found that female *P. gossypiella* from the desert cotton-growing areas of southern California emitted a significantly higher pheromone (20%) in 1984 and 1985 than in 1982 and 1983. They hypothesized that this upsurge could result from selection pressure provided by the continuous application of mating disruptants for population control. However, for blend quality [145] and mating disruption (MD) [146], higher quantity of pheromones have been used than the previous studies.

Evenden and Haynes [147] suggested that MD experiments using the same pheromone blend along the years might affect certain pheromone phenotypes, and therefore modify the chemical communication channels of the insect. This could lead to generation of resistance to MD. In contrast, other authors reported that resistance to MD may be a function of a genetically-based change in the output and reaction to pheromone components of the target pest [148,149].

The first example of resistance to MD was documented on one of Japan's major tea pests, the smaller tea leafroller moth, *Adoxophyes honmai* Yasuda (Lepidoptera: Tortricidae) [150]. The authors noticed that using (Z)-11-tetradecenyl acetate as mating disruptant for four years induced a disruption percentage of pheromone trap catches of 96%, but 14–16 years later the percentage of catches became less than 50%. Moreover, the application of the compound to other previously untreated tea fields induced a disruption higher than 99%. These results brought the authors to term "resistance" of the leafroller to (Z)-11-tetradecenyl acetate. In new experiments, application of the 4-component pheromone blend of the insect (63:31:4:2 mixture of (Z)-9-tetradecenyl acetate, (Z)-11-tetradecenyl acetate, (E)-11-tetradecenyl acetate, and 10-methyldodecyl acetate) to disrupt the resistant population induced a 99% disruption and a decrease in larval development [150]. Moreover, resistant males supposedly utilize the widespread odor of (Z)-11-tetradecenyl acetate released from MD devices installed in the tea canopy for orientation, while they trace a directional plume of (Z)-9-tetradecenyl acetate emitted from the lure [151].

Tabata et al. [152] noticed that resistant males were attracted to lures with significantly deviated ratios of the pheromone components: 72% responded to the mixture lacking (Z)-11-tetradecenyl acetate, which is indispensable for the response behavior of the susceptible males. To avoid this resistance, the authors developed a new strain of resistant insects by rearing field-collected resistant individuals with the pheromone for more than 70 generations [153]. The new males could respond and copulate with their conspecifics even in the presence of 1 mg L^{-1} of the disruptant. In addition, the composition of the sex pheromone blend produced and emitted by females was not changed compared with those of females sensitive to MD.

The continuous exposure or pre-exposure to high concentrations of sex pheromone as in mating disruption experiments can elicit habituation or desensitization [154]. This mechanism implies a reduction of response when pest species are treated with species-specific dispensers [155]. Adaptation of the peripheral neurons system to the sex pheromone has been shown to occur in most moths, but not in all cases where it has been examined, including *G. molesta* [154], *C. pomonella* [156], or *E. postvittana* [157]. In an interesting work, the possible different effects of behavioral interference with high pheromone loadings on four orders of pest insects were studied [157]. The tests were implemented on the light brown apple moth, *E. postvittana* (Lepidoptera: Tortricidae), the citrophilous mealybug, *Pseudococcus calceolariae* (Homoptera: Pseudococcidae), the apple leaf curling midge, *Dasineura mali* (Diptera: Cecidomyiidae), and the Argentine ant, *Linepithema humile* (Hymenoptera: Formicidae). Only in the tortricid moth, pre-exposure of male moths to the main sex pheromone, (*E*)-11-tetradecenyl acetate, significantly reduced their subsequent behavioral responses to the pheromone stimulus compared with the untreated insects, like in usual habituation experiments on pheromone pre-exposure [158]. The insects in the three other orders showed no evidence for habituation to behaviorally active amounts of synthetic pheromone. The authors concluded that the range of mechanisms opened to use in pest management for the three order insects is potentially reduced, and at the same time it raises questions about the adaptive benefit of habituation in Lepidoptera [157].

7. Application of Insect Sex Pheromones

Many of the conventional methods using hazardous chemicals for insect pest management have been banned because of their adverse effect on the environment and human health. To overcome the negative effect of synthetic chemicals, many researchers have emphasized the need to develop and formulate eco-friendly and more specific agricultural practices for IPM [159]. The specific characteristics of insect pheromones determine their potential uses as environmental-friendly behavioral regulators in agriculture. On one hand, sex pheromones are species specific compounds with a potential use in population monitoring and mass trapping; on the other hand, pheromones are chemical messengers between the male and female adults of an insect species, with which one sex communicates their sex partners the predisposition and willingness to mating, pointing to a potential use in mating disruption. As early as in the 1970–80s, research studies were undertaken to confirm the possible practical application of pheromones in pest populations monitoring, mass trapping, mating disruption and push-pull strategies [7,9–11].

7.1. Interactions between Pheromones and Insects Biological Control Agents

The review article by Sharma et al. [160] has covered the most recent literature on the interactions between insect biological control agents and semiochemicals (predominantly pheromones). Therefore, we will discuss the work done in the last two years.

7.1.1. Pheromones vs Entomopathogenic Fungi

It is believed that entomopathogenic fungi act slowly and take time to cause mortality in insects. On the other hand, the combined use of entomopathogenic fungi with chemical insecticides [161] or other entomopathogenic fungi [162] improves microbial control agents' efficacy. In the same approach, entomopathogenic fungi and pheromones can be used to increase the effectiveness (additive or synergistic) against the target insects. Thus, the simultaneous application of *Metarhizium anisopliae* var. acridum and the aggregation pheromone phenylacetonitrile to *Schistocerca gregaria* Forsskål (Orthoptera: Acrididae) fifth-instar nymphs exhibited an additive interaction [163]. Synergistic interaction was also demonstrated when nymphs were exposed to pheromone first and then treated with *M. anisopliae* var. *acridum*. Thus, Gutiérrez-Cárdenas et al. [164] disclosed the potential of entomopathogenic fungi for the management of adults of *Spodoptera frugiperda* using the auto-dissemination technique. Three isolates of *Beauveria bassiana* (Balsamo-Crivelli) Vuillemin and six isolates of *Metarhizium anisopliae* (Metschnikoff) (Hypocreales: Clavicipitaceae) were reported to be pathogenic to the insect, and the isolate *M. anisopliae* Ma-San Rafel-2 was used for auto-dissemination. The fungus and the synthetic sex pheromone attracted, infected, and killed males 5–8 days after dissemination, paving the way for the potential management of *S. frugiperda* [164]. In the same way, Akutse et al. [165] evaluated the effect of 22 entomopathogenic fungi isolates on the fall armyworm *S. frugiperda*, and their compatibility with the pheromone FALLTRACK lure. All fungal isolates screened were pathogenic to the moths, particularly *Beauveria bassiana* ICIPE 621 and *M. anisopliae* ICIPE 7, which caused 100% mortality of the moths. Both isolates were also found compatible with FALLTRACT lure, as the lure did not affect the conidial germination in the laboratory [165]. These results suggest that ICIPE 7 and ICIPE 621 could be used combined with *S. frugiperda* pheromone as an "attract and infect" strategy for sustainable management of the fall armyworm. Also, Akutse et al. [166] observed high efficacy of *M. anisopliae* isolates ICIPE 18, ICIPE 20, and ICIPE 665 against both adult and fourth instar larvae of the tomato pest *T. absoluta*, when used in combination with the *Tuta* pheromone, TUA-Optima®, for mass trapping and auto-dissemination. The results suggest them as promising and appropriate applicants as fungal-based biopesticides against *T. absoluta* and other solanaceous pests.

7.1.2. Pheromones and Bacteria

Only a few new studies are available on pheromones and bacteria's combined use against crop insect pests. For example, Sammani et al. [167] studied the effects of the sex pheromone components (Z,E)-9,12-tetradecadienyl acetate and (Z)-9-tetradecen-1-yl acetate of *Cadra cautella* (Walker) (Lepidoptera: Pyralidae) in the presence of botanical oils on insect mating, and the burrowing ability of *C. cautella* larvae in different types of flour treated with spinosad (*Saccharopolyspora spinosa*), a bacterial organism isolated from soil. The mating success was higher with botanical oils alone but declined with pheromone exposure either alone or combined with botanical oils [167]. These studies indicated that the mating and burrowing of *C. cautella* is influenced by its pheromone and by exposure to botanicals and spinosad.

For the first time, Ren et al. [168] highlighted the influence of microbial symbionts on insect pheromones and provide an example of direct bacterial production of pheromones in insects. The authors used *Bactrocera dorsalis* (Hendel) (Diptera: Tephritidae) as a model system and demonstrated that *Bacillus* sp. in the rectum of male *B. dorsalis* plays a pivotal role in sex pheromones production. They also showed that 2,3,5-trimethylpyrazine and 2,3,5,6-tetramethylpyrazine are sex pheromones produced in the male rectum.

7.2. Monitoring

One of the most widespread and successful applications of sex pheromones is in the detection and monitoring of pest populations [11]. Monitoring systems are based on the relationship between trap captures and the pest population or damage induced by the pest species. The number of male catches is used to establish thresholds for making decisions on when it is advisable to take treatment actions. Sex pheromones are very useful for evaluating trap catches because they are highly sensitive when detecting low insect population levels and are species-specific. These features also allow for the detection and survey of invasive species, and permit growers to perform timely insecticidal applications, thereby reducing economic and environmental costs [2].

Factors including trap design (e.g., type, color, height), trap location, type of dispenser, pheromone dose and purity, and environmental conditions during the trapping period can influence male catches by pheromone-baited traps [169–172]. In addition, it is essential to have a good knowledge of the pest biology and geographical distribution of the pest [11]. For monitoring, pheromone traps are now available for a wide range of insect pests, mostly Lepidoptera, although some are also for Coleoptera and Diptera. Some examples of using sex pheromones for monitoring Lepidoptera and Coleoptera follow.

For monitoring *Coleophora deauratella* Leinig and Zeller (Lepidoptera: Coleophoridae), an invasive pest of the red clover in Canada, field experiments were conducted to optimize several pheromone-baited trap features [169]. The type of substrate used to release the pheromone and lure age did not affect trap catches. However, moth capture in non-saturating green Unitraps was significantly higher than other traps and the authors recommended their use with either gray or red rubber septa lures.

In studies directed to monitor the presence of the banana root borer *Cosmopolites sordidus* (Germar) (Coleoptera: Curculionidae) in banana plantations of the island of Guam (USA), the male-produced aggregation pheromone (sordidin) was deployed in lure packs containing 90 mg of pheromone [170]. Ground traps were found to be superior to ramp and pitfall traps, and brown traps were more attractive to insects than other colored traps. In addition, ground traps located in the shade of the canopy were more effective than those placed in sunlight.

To optimize the monitoring of the sweetpotato weevil, *Cylas formicarius* (Fabricius) (Coleoptera: Brentidae) with the sex pheromone (Z)-3-dodecenyl-(E)-2-butenoate, several parameters affecting trap catches were evaluated [171]. Each lure contained 10 mg of pheromone and emitted the active ingredient at <0.01 mg/day. Pherocon unitraps caught higher numbers of males than ground, funnel water, and delta traps, with medium-sized traps being the most effective. The weevil preferred red over other colors, particularly light red. For optimum catches, traps should be positioned 50 cm above the crop canopy [171].

In monitoring studies of the pea leaf weevil, *Sitona lineatus* (L.) (Coleoptera: Curculionidae), several factors that potentially could affect capture rates of pheromone-baited traps were considered [172]. Lures contained 10 mg of pheromone and emitted the active ingredient at 0.1 mg/day. Pheromone-baited pitfall and ramp traps caught significantly more adults than ground and delta traps and pitfall traps containing gray rubber septa captured more adults than traps baited with membrane formulations.

In a dose-response field test for monitoring the processionary moth *T. pityocampa* (Denis and Schiffermüller) (Lepidoptera: Thaumetopoeidae), the number of male captures increased significantly with the dosage of the pheromone to a plateau at ca. 10 mg [173,174]. Among several trap designs, plate sticky traps showed the highest trapping efficiency and the mean trap captures correlated well with the nest density. The utilization of four plate sticky traps containing 0.2 mg of pityolure proved to be a cost-effective tool for monitoring population densities of the pest [174].

Recently, we have investigated the effects of air flow on the host location of adults of citrus psyllid, Diaphorina citri, and the trapping efficiency of perforated yellow trapping plate (Zeng, unpublished). It was found that the airflow influenced the movement of psyllid adults and the catching rate of the trap. The catching number with the perforated yellow plate was 45.41%, significantly higher than that of the non-perforated yellow trap plate (21.67%). These results indicated that it is important to use an air-flowable trap device baited with sex pheromones for the monitoring of certain species of insect pests.

7.3. Mass Trapping

Mass trapping is a direct control strategy that uses a large number of pheromone traps to reduce the population density of the target species and/or pest damage [175,176]. Compared to mating disruption, mass trapping is more efficient when both control methods have an equal number of pheromone sources [177]. This is because mating disruption only delays the searching sex, whereas traps delay them indefinitely. Mass trapping has been particularly successful in controlling large weevils in tropical crops, such as oil palms, palmito palms, plantains, and bananas. The key biological factors for success appear to be the relatively long life and slow reproductive rate of the weevils, and the fact that the aggregation pheromones attract both sexes. In the case of sex pheromones that only attract one sex, mass trapping is generally less effective despite some larger projects collecting billions of individuals [177]. Several factors, such as trap design and density, population level, biology of the target pest, isolation, and risk of immigration can influence the success of the application [10,175,178]. The pheromone composition of the target pest and the cost for its production may be also crucial for an economically feasible control by mass trapping. This is the case of the economically important click beetle *Melanotus communis* Gyllenhal (Coleoptera: Elateridae) whose female-produced pheromone is a single, readily synthesized chemical, 13-tetradecenyl acetate [179]. In addition, the cost of traps and the manual labor required to deploy them should be economically competitive to other control methods. Another factor that can affect the feasibility of mass trapping is the need for using a large quantity of pheromone that may not be cost-effective. Traps must catch 80–95% of the males to effectively reduce population [180].

Mass trapping has been mainly used against the following pests: *C. pomonella* (L.), *Zeuzera pyrina* (L.), and *Cossus cossus* (L.) in orchards; *S. littoralis* (Boisduval) and *P. gossipiella* (Saunders) in cotton and oilseed; bark beetles; palm weevils; corn rootworms; *Ephestia* spp. and *Plodia interpunctella* (Hb.) in stored products and food industries; or the gypsy moth *Lymantria dispar* (L.), and boll weevil *Anthonomus grandis* (Boheman) as invasive species [10,178,180].

Mass trapping of *T. absoluta* Meyrick (Lepidoptera: Gelechiidae), a major threat of tomato worldwide, has been implemented using homemade traps (translucent plastic cylinders) or water traps lured with a sex pheromone dispenser [181,182]. The traps are most effective when placed near ground level and, ideally, they should be loaded with 0.5 mg of pheromone. The use of 48 traps/ha reduced leaf damage more efficiently than conventional insecticide treatment.

Mass trapping of *C. sordidus* (Germar) (Coleoptera: Curculionidae) with pheromone-baited pitfall traps and *Metamasius hemipterus* (Coleoptera: Curculionidae) with pheromone-sugarcane-baited traps were conducted in commercial banana plots [183]. Capture rates of *C. sordidus* and *M. hemipterus* declined by >75 % and corn damage decreased by 61–64% in trapping plots. In addition, banana bunch weights increased 23% relative to control plots. The authors estimated that the increase in the yield value obtained was about USD $4240 per year per hectare, while the cost of the experiment was approximately USD $185 per year per hectare [183].

Control of the processionary moth by mass trapping has been pursued over the years due to the economic impact of the moth on pine forests and woodlands. In 2015, Martin [184] reported that a minimum of 4 funnel traps baited with 1 mg of the pheromone (Z)-13-hexadecen-11-ynyl acetate was necessary to be effective in a small site and 6 traps per hectare in large sites. These results were consistent with those obtained by Trematerra et al. [185] in central Italy, who reported a reduction of 53% and 79% of adults caught in 2016 and 2017 in the mass trapping parcel in comparison to the control parcel. Consequently, the average number of winter nests per tree in the trapping parcel was reduced by 88% after one year and 94% after two years, when compared to the nest reduction of 43% and 80% observed in the control parcel.

7.4. Mating Disruption

Mating disruption (MD) is a strategy based on the permeation of the crop with synthetic sex pheromone to disrupt chemical communication between sexes and, thus, preventing mating. To date, MD is the most developed pheromone-based technology for the direct control of moth pests [186]. The species-specificity and low toxicity of pheromone applications have led to consider MD a reliable tool for use in area-wide programs to control insect pests and manage invasive species. Microencapsulation, hand application, aerial dispensers, and matrix formulations (SPLAT, Specialized Pheromone and Lure Application Technology), have been used for pheromone emission [34]. Ideally, the dispensers should release pheromones at a constant rate, should be mechanically applicable, completely biodegradable, and made from renewable and cheap organic materials, be economically cheap, and eco-toxicologically inert [187]. The application of this technology has increased almost exponentially in the last 30 years, and it was calculated that the surface of crops being controlled for specific pests amounted to 770,000 ha in 2010 [11,188].

The most successful cases of pest control by MD are the gypsy moth *L. dispar* [186]; the codling moth *C. pomonella* [189]; the grapevine moth *Lobesia botrana* [190]; the oriental fruit moth *G. molesta* (Busck) [191]; the raisin moth *E. cautella* (Walker), the Mediterranean flour moth *Ephestia kuehniella* Zeller, and the Indian meal moth *P. interpunctella* (Hübner) [192], and the carpenter moth *Cossus insularis* (Staudinger) (Lepidoptera: Cossidae) [193].

MD has also been helpful to control plant-feeding midges that cause important crop losses in forestry and horticultural and fruit crops [2]. For instance, application of the pheromone of the swede midge *Contarinia nasturtii* Kieffer (Diptera: Cecidomyiidae) on fields of Brussels sprouts, broccoli, and cauliflower resulted in a 59–91% reduction in damage [2]. More recently, Hodgdon and coworkers [140] tested in the laboratory and in the field whether exposure to synthetic pheromone influenced female calling and the subsequent propensity to mate. In some species, females possess pheromone receptors and autodetection of their own pheromones induce them to alter their mating behavior (see Sex pheromone autodetection above). This was the case for the swede midge; while 68% of female midges mated under control conditions only 42% and 35% of females mated when pre-exposed to stereospecific and racemic three-component blends, respectively [140].

MD on tomato greenhouses has been implemented for control of *T. absoluta*. The use of 30 g/ha of pheromone can be sufficiently effective in high-containment glasshouses, with façades closed by insect-proof nets, to control the moth populations for 4 months and reduce the percentage of damaged fruits, but not in the open field or unscreened greenhouses [194,195]. The insects' ability to undergo parthenogenesis or multiple matings stresses that immigration of mated females into greenhouses should be prevented in order to improve effectiveness of MD.

The sex pheromone of the European grape vine moth *L. botrana* in Ecoflex fibers, a cheap organic co-polyester and completely biodegradable within half a year, has been tested in MD experiments in Southwest Germany with promising results [187]. After 7 weeks of treatment, disruption effects of ca. 95% were obtained. The use of suitable mesofibers is protected by European and US patents. In this line, the authors later developed Electrospun mesofibers, novel biodegradable pheromone dispensers with diameters ranging from 0.6 to 3.5 μm. The dispensers are biodegradable and harmless to non-target organisms [196]. More recently, Luchi and coworkers [197] evaluated the efficacy of the MD products Isonet® L TT and the biodegradable Isonet® L TT BIO in reducing *L. botrana* damage on grapevine. Experiments were conducted in Central and Northern Italy over three years. The trials allowed a reliable control of the three generations of *L. botrana* during the whole grape-growing season.

A MD approach using pheromone puffer dispensers were considered to control *C. deauratella* at three red clover seed production fields in Alberta, Canada [198]. In all plots, aerosol-emitting pheromone puffers were able to reduce male *C. deauratella* orientation to traps by 60.7% to 93.7% compared with control plots. However, there was no corresponding decrease in larval numbers or increase in seed yield. Important challenges of this experimentation appeared to be the immigration of mated females and high population densities [198].

Univoltine species, like the processionary moth, might be ideal for MD since a single annual application might lead to population suppression without need for another application. However, the timing of application is always critical. In MD trials conducted in Aosta Valley (NW Italy) in 2016–2017, the number of males collected was significantly lower in the plots where MD was performed in comparison to control plots [199]. In addition, the total number of nests recorded per tree was significantly lower in MD plots. The technique appears to be the most appropriate control strategy for the processionary moth, and address that repeated annual applications of MD could dramatically reduce population densities below the economic injury level [199].

7.5. Push-Pull Strategy

The push-pull strategy, the simultaneous use of an attractant and repellent stimulus to divert pests, is an increasingly employed sustainable alternative to traditional pesticides. This strategy aims at reducing crop injury by modifying pest distribution using repellent stimuli to 'push' the insect pest away from the crop, and at the same time attractant stimuli to 'pull' the pest to other areas out of the crop. The development of push-pull strategies has been mainly directed to agricultural systems to manage insecticide resistance threats or diminish the use of insecticides. This strategy requires knowledge of insect biology, chemical ecology, and interaction between host plants and natural enemies [200]. Although there is a large variety of 'push' and 'pull' components, such as synthetic repellents, host- and non-host volatiles, host-derived semiochemicals, antifeedants, oviposition stimulants, and oviposition deterrents, among others, we will concentrate only on those strategies dealing with sex pheromones.

Sex pheromones can contribute in push-pull experiments to establish the timing of introduction of the stimuli and other population-decreasing actions [200]. Sex and aggregation pheromones attraction to herbivores can be reinforced by the synergistic action of host plant volatiles [201,202]. In push-pull trials against aphids, nepetalactone, one of the aphid sex pheromone component, and (Z)-jasmone a host-plant volatile that attracts aphid parasitoids, may be used to pull parasitoids into the crop [203]. In addition, the parasitoids can be pushed to the field from nearby locations by the action of tricosane and pentacosane, the lady beetle pheromone [204].

In a three years experiment in alfalfa fields, slow-release formulations of the aphid sex pheromone components (4a*S*,7*S*,7a*R*)-nepetalactone and (1*R*,4a*S*,7*S*,7a*R*)-nepetalactol significantly decreased population of the pea aphid, *Acyrthosiphon pisum* Harris [205]. At the same time, parasitism by the aphid parasitoids *Aphidius ervi* Haliday and *Praon barbatum* Mackauer was significantly increased, while no pheromone effects were detected for predators. These results show that the aphid sex pheromone can attract aphid parasitoids and enhance their ability to suppress aphid abundance in the field [205].

8. Conclusions

The excessive use of synthetic chemical insecticides causes pollution, pest resurgence, and resistance problems. Sex pheromones are natural insect behavior regulators that serve as suitable chemical agents in sustainable agriculture. They have the advantages over the hazardous chemicals of not killing the pest, but reducing the number of male adults, their reproduction rate, and guiding the timely application of insecticides. In the last two decades, studies have been mainly focused on the identification of new sex pheromones, characterization of sex pheromone perception mechanisms, and integration of these new advances in pheromone research to IPM programs. Mating disruption is probably the semiochemical-based technique most successfully used in IPM [137]. So far, studies on MD mechanisms have been focused on male moths almost exclusively, but studies on pheromone autodetection by females have determined that modeling MD mechanisms will increase in complexity [206]. The actual effects of the different female behaviors on MD largely remain to be understood, but their knowledge should prove useful for evaluating the potential of this strategy in pest control. From the molecular point of view, significant research is being performed on the discovery of new olfactory receptors, pheromone binding proteins, sensory binding proteins and chemical signal transduction mechanisms involved in sex pheromone communication. Advances in sequence technologies, genomics and transcriptomics of insect olfactory system will help develop new technologies for more sustainable pest management strategies, thereby reducing the use of synthetic insecticides.

Author Contributions: Writing of initial draft and literature search, S.A.H.R.; writing of the manuscript and review, G.V.P.R., J.G., X.Z., and A.G.; editing and supervision, A.G.; funding acquisition, X.Z., G.V.P.R., and J.G.; final revision, X.Z., G.V.P.R., J.G., and A.G. All authors have read and agreed to the published version of the manuscript.

Funding: Thanks to the National Key R&D Program of China (2018YFD0201102-4) and the Department of Science and Technology of Guangdong Province (2015B090903076) for financial support. This work is also partially supported by USDA-ARS Research Project# 6066-22000-084-00D-Insect Control and Resistance Management in Corn, Cotton, Sorghum, Soybean, and Sweet Potato, and Alternative Approaches to Tarnished Plant Bug Control in the Southern United States.

Institutional Review Board Statement: Not applicable.

Acknowledgments: We are grateful to Russell Jurenka (Iowa State University, USA) for suggestion on the topics included in this manuscript.

Conflicts of Interest: The authors declare no conflict of interest.

References

1. Seybold, S.J.; Bentz, B.J.; Fettig, C.J.; Lundquist, J.E.; Progar, R.A.; Gillette, N.E. Management of western North American bark beetles with semiochemicals. *Annu. Rev. Entomol.* **2018**, *63*, 407–432. [CrossRef] [PubMed]
2. Smart, L.; Aradottir, G.; Bruce, T. Role of semiochemicals in integrated pest management. In *Integrated Pest Management*; Abrol, D.P., Ed.; Academic Press: Amsterdam, The Netherlands, 2014; pp. 93–109. [CrossRef]
3. Butenandt, V.A. Uber den sexsual-lockstoff des seidenspinners Bombyx mori. *Reindarstellung und Konstitution. Z. Naturforsch. B* **1959**, *14*, 283.
4. Karlson, P.; Lüscher, M. 'Pheromones': A new term for a class of biologically active substances. *Nature* **1959**, *183*, 55–56. [CrossRef]
5. Petkevicius, K.; Löfstedt, C.; Borodina, I. Insect sex pheromone production in yeasts and plants. *Curr. Opin. Biotechnol.* **2020**, *65*, 259–267. [CrossRef] [PubMed]
6. Larsson, M.C. Pheromones and other semiochemicals for monitoring rare and endangered species. *J. Chem. Ecol.* **2016**, *42*, 853–868. [CrossRef] [PubMed]
7. Tewari, S.; Leskey, T.C.; Nielsen, A.L.; Piñero, J.C.; Rodriguez-Saona, C.R. Use of pheromones in insect pest management, with special attention to weevil pheromones. In *Integrated Pest Management*; Abrol, D.P., Ed.; Academic Press: Amsterdam, The Netherlands, 2014; pp. 141–168. [CrossRef]
8. Yew, J.Y.; Chung, H. Insect pheromones: An overview of function, form, and discovery. *Prog. Lipid Res.* **2015**, *59*, 88–105. [CrossRef] [PubMed]
9. Reddy, G.V.; Guerrero, A. New pheromones and insect control strategies. *Vitam. Horm.* **2010**, *83*, 493–519. [CrossRef] [PubMed]
10. Trematerra, P. Advances in the use of pheromones for stored-product protection. *J. Pest Sci.* **2012**, *85*, 285–299. [CrossRef]
11. Witzgall, P.; Kirsch, P.; Cork, A. Sex pheromones and their impact on pest management. *J. Chem. Ecol.* **2010**, *36*, 80–100. [CrossRef] [PubMed]
12. Hostachy, C.; Couzi, P.; Portemer, G.; Hanafi-Portier, M.; Murmu, M.; Deisig, N.; Dacher, M. Exposure to conspecific and heterospecific sex-pheromones modulates gustatory habituation in the moth Agrotis ipsilon. *Front. Physiol.* **2019**, *10*, 1518. [CrossRef] [PubMed]
13. Molander, M.A.; Eriksson, B.; Winde, I.B.; Zou, Y.; Millar, J.G.; Larsson, M.C. The aggregation-sex pheromones of the cerambycid beetles *Anaglyptus mysticus* and *Xylotrechus antilope* ssp. *antilope*: New model species for insect conservation through pheromone-based monitoring. *Chemoecology* **2019**, *29*, 111–124. [CrossRef]
14. El-Sayed, A.M. The Pherobase: Database of Pheromones and Semiochemicals. 2021. Available online: https://www.pherobase.com (accessed on 19 January 2021).
15. Ma, T.; Liu, Z.T.; Zhang, Y.Y.; Sun, Z.H.; Li, Y.Z.; Wen, X.J.; Chen, X.Y. Electrophysiological and behavioral responses of *Diaphania glauculalis* males to female sex pheromone. *Environ. Sci. Pollut. Res. Int.* **2015**, *22*, 15046–15054. [CrossRef]
16. El-Sayed, A.M.; Gibb, A.R.; Mitchell, V.J.; Lee-Anne, M.; Manning, L.-A.M.; Revell, J.; Thistleton, B.; Suckling, D.M. Identification of the sex pheromone of *Conogethes pluto*: A pest of Alpinia. *Chemoecology* **2013**, *23*, 93–101. [CrossRef]
17. Peng, C.L.; Gu, P.; Li, J.; Chen, Q.Y.; Feng, C.H.; Luo, H.H.; Du, Y.J. Identification and field bioassay of the sex pheromone of *Trichophysetis cretacea* (Lepidoptera: Crambidae). *J. Econ. Entomol.* **2012**, *105*, 1566–1572. [CrossRef]
18. Symmes, E.J.; Dewhirst, S.Y.; Birkett, M.A.; Campbell, C.A.; Chamberlain, K.; Pickett, J.A.; Zalom, F.G. The sex pheromones of mealy plum (*Hyalopterus pruni*) and leaf-curl plum (*Brachycaudus helichrysi*) aphids: Identification and field trapping of male and gynoparous aphids in prune orchards. *J. Chem. Ecol.* **2012**, *38*, 576–583. [CrossRef] [PubMed]
19. Levi-Zada, A.; Ben-Yehuda, S.; Dunkelblum, E.; Gindin, G.; Fefer, D.; Protasov, A.; Kuznetsowa, T.; Manulis-Sasson, S.; Mendel, Z. Identification and field bioassays of the sex pheromone of the yellow-legged clearwing *Synanthedon vespiformis* (Lepidoptera: Sesiidae). *Chemoecology* **2011**, *21*, 227–233. [CrossRef]
20. El-Sayed, A.M.; Mitchell, V.J.; Manning, L.A.; Suckling, D.M. New sex pheromone blend for the lightbrown apple moth, *Epiphyas postvittana*. *J. Chem. Ecol.* **2011**, *37*, 640–646. [CrossRef]
21. Acín, P.; Rosell, G.; Guerrero, A.; Quero, C. Sex pheromone of the Spanish population of the beet armyworm *Spodoptera exigua*. *J. Chem. Ecol.* **2010**, *36*, 778–786. [CrossRef]
22. Eliyahu, D.; Nojima, S.; Santangelo, R.G.; Carpenter, S.; Webster, F.X.; Kiemle, D.J.; Gemeno, C.; Leal, W.S.; Schal, C. Unusual macrocyclic lactone sex pheromone of Parcoblatta lata, a primary food source of the endangered red-cockaded woodpecker. *Proc. Natl. Acad. Sci. USA* **2012**, *109*, E490–E496. [CrossRef]
23. Guerrero, A.; Ramos, V.E.; López, S.; Alvarez, J.M.; Domínguez, A.; Coca-Abia, M.M.; Bosch, M.P.; Quero, C. Enantioselective synthesis and activity of all diastereoisomers of (E)-phytal, a pheromone component of the Moroccan locust, *Dociostaurus maroccanus*. *J. Agric. Food Chem.* **2019**, *67*, 72–80. [CrossRef]
24. Vacas, S.; Navarro, I.; Marzo, J.; Navarro-Llopis, V.; Primo, J. Sex pheromone of the invasive mealybug citrus pest, *Delottococcus aberiae* (Hemiptera: Pseudococcidae). A new monoterpenoid with a necrodane skeleton. *J. Agric. Food Chem.* **2019**, *67*, 9441–9449. [CrossRef]

25. Bouwer, M.C.; Slippers, B.; Degefu, D.; Wingfield, M.J.; Lawson, S.; Rohwer, E.R. Identification of the sex pheromone of the tree infesting cossid moth *Coryphodema tristis* (Lepidoptera: Cossidae). *PLoS ONE* **2015**, *10*, e0118575. [CrossRef]

26. Būda, V.; Blažytė-Čereškienė, L.; Radžiutė, S.; Apšegaitė, V.; Stamm, P.; Schulz, S.; Aleknavičius, D.; Mozūraitis, R. Male-produced (−)-δ-heptalactone, pheromone of fruit fly *Rhagoletis batava* (Diptera: Tephritidae), a sea buckthorn berries pest. *Insects* **2020**, *11*, 138. [CrossRef]

27. Gago, R.; Allyson, J.D.; McElfresh, J.S.; Haynes, K.F.; McKenney, J.; Guerrero, A.; Millar, J. A tetraene aldehyde as the major sex pheromone component of the promethea moth (*Callosamia promethea* (Drury)). *J. Chem. Ecol.* **2013**, *39*, 1263–1272. [CrossRef]

28. Weiss, I.; Rössler, T.; Hofferberth, J.; Brummer, M.; Ruther, J.; Stökl, J. A nonspecific defensive compound evolves into a competition avoidance cue and a female sex pheromone. *Nat. Commun.* **2013**, *4*, 2767. [CrossRef] [PubMed]

29. Kong, X.B.; Liu, K.W.; Wang, H.B.; Zhang, S.F.; Zhang, Z. Identification and behavioral evaluation of sex pheromone components of the Chinese pine caterpillar moth, *Dendrolimus tabulaeformis*. *PLoS ONE* **2012**, *7*, e33381. [CrossRef] [PubMed]

30. Shibasaki, H.; Yamamoto, M.; Yan, Q.; Naka, H.; Suzuki, T.; Ando, T. Identification of the sex pheromone secreted by a nettle moth, *Monema flavescens*, using gas chromatography/fourier transform infrared spectroscopy. *J. Chem. Ecol.* **2013**, *39*, 350–357. [CrossRef]

31. De Oliveira, M.W.; Borges, M.; Andrade, C.K.; Laumann, R.A.; Barrigossi, J.A.; Blassioli-Moraes, M.C. Zingiberenol, (1R,4R,1′S)-4-(1′,5′-dimethylhex-4′-enyl)-1-methylcyclohex-2-en-1-ol, identified as the sex pheromone produced by males of the rice stink bug *Oebalus poecilus* (Heteroptera: Pentatomidae). *J. Agric. Food Chem.* **2013**, *61*, 7777–7785. [CrossRef] [PubMed]

32. Blassioli-Moraes, M.C.; Khrimian, A.; Michereff, M.F.F.; Magalhães, D.M.; Hickel, E.; de Freitas, T.F.S.; Barrigossi, J.A.F.; Laumann, R.A.; Silva, A.T.; Guggilapu, S.D.; et al. Male-produced sex pheromone of *Tibraca limbativentris* revisited: Absolute configurations of zingiberenol stereoisomers and their influence on chemotaxis behavior of conspecific females. *J. Chem. Ecol.* **2020**, *46*, 1–9. [CrossRef]

33. Millar, J.G.; McElfresh, J.S.; Romero, C.; Vila, M.; Marí-Mena, N.; Lopez-Vaamonde, C. Identification of the sex pheromone of a protected species, the Spanish moon moth *Graellsia isabellae*. *J. Chem. Ecol.* **2010**, *36*, 923–932. [CrossRef] [PubMed]

34. Welter, S.; Pickel, C.; Millar, J.; Cave, F.; Van Steenwyk, R.; Dunley, J. Pheromone mating disruption offers selective management options for key pests. *Calif. Agric.* **2005**, *59*, 16–22. [CrossRef]

35. Löfstedt, C.; Wahlberg, N.; Millar, J.M. Evolutionary patterns of pheromone diversity in Lepidoptera. In *Pheromone Communication in Moths: Evolution, Behavior and Application*; Allison, J.D., Cardé, R.T., Eds.; University of California Press: Berkeley, CA, USA, 2016; pp. 43–78.

36. Ando, T.; Inomata, S.I.; Yamamoto, M. Lepidopteran sex pheromones. In *The Chemistry of Pheromones and Other Semiochemicals I, Topics in Current Chemistry*; Schulz, S., Ed.; Springer: Berlin/Heidelberg, Germany, 2004; Volume 239, pp. 51–96.

37. Naka, H.; Fujii, T. Chemical divergences in the sex pheromone communication systems in moths. In *Insect Sex Pheromone Research and Beyond*; Ishikawa, Y., Ed.; Springer: Singapore, 2020; pp. 3–17. [CrossRef]

38. Allison, J.D.; Cardé, R.T. *Pheromone Communication in Moths: Evolution, Behavior, and Application*, 1st ed.; University of California Press: Oakland, CA, USA, 2016; p. 416.

39. Jurenka, R. Insect pheromone biosynthesis. *Top. Curr. Chem.* **2004**, *239*, 97–132. [PubMed]

40. Jurenka, R. Regulation of pheromone biosynthesis in moths. *Curr. Opin. Insect Sci.* **2017**, *24*, 29–35. [CrossRef]

41. Tillman, J.A.; Seybold, S.J.; Jurenka, R.A.; Blomquist, G.J. Insect pheromones–an overview of biosynthesis and endocrine regulation. *Insect Biochem. Mol. Biol.* **1999**, *29*, 481–514. [CrossRef]

42. Blomquist, G.J.; Jurenka, R.; Schal, C.; Tittiger, C. Pheromone production: Biochemistry and molecular biology. In *Insect Endocrinology, Gilbert, L.*, Ed.; Academic London: London, UK, 2012. [CrossRef]

43. Park, H.Y.; Kim, M.S.; Paek, A.; Jeong, S.E.; Knipple, D.C. An abundant acyl-CoA (Delta9) desaturase transcript in pheromone glands of the cabbage moth, *Mamestra brassicae*, encodes a catalytically inactive protein. *Insect Biochem. Mol. Biol.* **2008**, *38*, 581–595. [CrossRef]

44. Ding, B.J.; Löfstedt, C. Analysis of the *Agrotis segetum* pheromone gland transcriptome in the light of sex pheromone biosynthesis. *BMC Genom.* **2015**, *16*, 711. [CrossRef]

45. Moto, K.I.; Suzuki, M.G.; Hull, J.J.; Kurata, R.; Takahashi, S.; Yamamoto, M.; Okano, K.; Imai, K.; Ando, T.; Matsumoto, S. Involvement of a bifunctional fatty-acyl desaturase in the biosynthesis of the silkmoth, *Bombyx mori*, sex pheromone. *Proc. Natl. Acad. Sci. USA* **2004**, *101*, 8631–8636. [CrossRef] [PubMed]

46. Ding, B.J.; Hofvander, P.; Wang, H.L.; Durrett, T.P.; Stymne, S.; Löfstedt, C. A plant factory for moth pheromone production. *Nat. Commun.* **2014**, *5*, 1–7. [CrossRef] [PubMed]

47. Yin, C.; Shen, G.; Guo, D.; Wang, S.; Ma, X.; Xiao, H.; Liu, J.; Zhang, Z.; Liu, Y.; Zhang, Y.; et al. InsectBase: A resource for insect genomes and transcriptomes. *Nucleic Acids Res.* **2016**, *44*, 801–807. [CrossRef] [PubMed]

48. Tupec, M.; Buček, A.; Valterová, I.; Pichová, I. Biotechnological potential of insect fatty acid-modifying enzymes. *Z. Naturforsch. C* **2017**, *72*, 387–403. [CrossRef] [PubMed]

49. Liu, W.; Ma, P.W.; Marsella-Herrick, P.; Rosenfield, C.L.; Knipple, D.C.; Roelofs, W. Cloning and functional expression of a cDNA encoding a metabolic acyl-CoA delta 9-desaturase of the cabbage looper moth, *Trichoplusia ni*. *Insect Biochem. Mol. Biol.* **1999**, *29*, 435–443. [CrossRef]

50. Zhang, Y.N.; Zhang, X.Q.; Zhu, G.H.; Zheng, M.Y.; Yan, Q.; Zhu, X.Y.; Xu, J.W.; Zhang, Y.Y.; He, P.; Sun, L.; et al. A Δ9 desaturase (SlitDes11) is associated with the biosynthesis of ester sex pheromone components in *Spodoptera litura*. *Pest. Biochem. Physiol.* **2019**, *156*, 152–159. [CrossRef]
51. Fujii, T.; Yasukochi, Y.; Rong, Y.; Matsuo, T.; Ishikawa, Y. Multiple Δ11-desaturase genes selectively used for sex pheromone biosynthesis are conserved in *Ostrinia* moth genomes. *Insect Biochem. Mol. Biol.* **2015**, *61*, 62–68. [CrossRef]
52. Hagström, Å.K.; Albre, J.; Tooman, L.K.; Thirmawithana, A.H.; Corcoran, J.; Löfstedt, C.; Newcomb, R.D. A novel fatty acyl desaturase from the pheromone glands of *Ctenopseustis obliquana* and *C. herana* with specific Z5-desaturase activity on myristic acid. *J. Chem. Ecol.* **2014**, *40*, 63–70. [CrossRef]
53. Wang, H.L.; Liénard, M.A.; Zhao, C.H.; Wang, C.Z.; Löfstedt, C. Neofunctionalization in an ancestral insect desaturase lineage led to rare Δ6 pheromone signals in the Chinese tussah silkworm. *Insect Biochem. Mol. Biol.* **2010**, *40*, 742–751. [CrossRef] [PubMed]
54. Liénard, M.A.; Lassance, J.M.; Wang, H.L.; Zhao, C.H.; Piskur, J.; Johansson, T.; Löfstedt, C. Elucidation of the sex pheromone biosynthesis producing 5,7-dodecadienes in *Dendrolimus punctatus* (Lepidoptera: Lasiocampidae) reveals Delta 11- and Delta 9-desaturases with unusual catalytic properties. *Insect Biochem. Mol. Biol.* **2010**, *40*, 440–452. [CrossRef]
55. Hao, G.; Liu, W.; O'Connor, M.; Roelofs, W. Acyl-CoA Z9- and Z10-desaturase genes from a New Zealand leafroller moth species, *Planotortrix octo*. *Insect Biochem. Mol. Biol.* **2002**, *32*, 961–966. [CrossRef]
56. Roelofs, W.L.; Rooney, A.P. Molecular genetics and evolution of pheromone biosynthesis in Lepidoptera. *Proc. Natl. Acad. Sci. USA* **2003**, *100*, 9179–9184. [CrossRef]
57. Ding, B.J.; Liénard, M.A.; Wang, H.L.; Zhao, C.H.; Löfstedt, C. Terminal fatty-acyl-CoA desaturase involved in sex pheromone biosynthesis in the winter moth (*Operophtera brumata*). *Insect Biochem. Mol. Biol.* **2011**, *41*, 715–722. [CrossRef] [PubMed]
58. Xia, Y.H.; Zhang, Y.N.; Ding, B.J.; Wang, H.L.; Löfstedt, C. Multi-functional desaturases in two *Spodoptera* moths with Δ11 and Δ12 desaturation activities. *J. Chem. Ecol.* **2019**, *45*, 378–387. [CrossRef] [PubMed]
59. Serra, M.; Piña, B.; Abad, J.L.; Camps, F.; Fabriàs, G. A multifunctional desaturase involved in the biosynthesis of the processionary moth sex pheromone. *Proc. Natl. Acad. Sci. USA* **2007**, *104*, 16444–16449. [CrossRef]
60. Zhang, Y.N.; Zhang, L.W.; Chen, D.S.; Sun, L.; Li, Z.Q.; Ye, Z.F.; Zheng, M.Y.; Li, J.B.; Zhu, X.Y. Molecular identification of differential expression genes associated with sex pheromone biosynthesis in *Spodoptera exigua*. *Mol. Genet. Genom.* **2017**, *292*, 795–809. [CrossRef]
61. Bai, Y.; McCoy, J.G.; Levin, E.J.; Sobrado, P.; Rajashankar, K.R.; Fox, B.G.; Zhou, M. X-ray structure of a mammalian stearoyl-CoA desaturase. *Nature* **2015**, *524*, 252–256. [CrossRef] [PubMed]
62. Matsumoto, S. Molecular mechanisms underlying sex pheromone production in moths. *Biosci. Biotechnol. Biochem.* **2010**, *74*, 223–231. [CrossRef]
63. Hagström, A.K.; Liénard, M.A.; Groot, A.T.; Hedenström, E.; Löfstedt, C. Semi-selective fatty acyl reductases from four heliothine moths influence the specific pheromone composition. *PLoS ONE* **2012**, *7*, e37230. [CrossRef] [PubMed]
64. Antony, B.; Fujii, T.; Moto, K.; Matsumoto, S.; Fukuzawa, M.; Nakano, R.; Tatsuki, S.; Ishikawa, Y. Pheromone-gland-specific fatty-acyl reductase in the adzuki bean borer, *Ostrinia scapulalis* (Lepidoptera: Crambidae). *Insect Biochem. Mol. Biol.* **2009**, *39*, 90–95. [CrossRef]
65. Antony, B.; Ding, B.J.; Moto, K.; Aldosari, S.A.; Aldawood, A.S. Two fatty acyl reductases involved in moth pheromone biosynthesis. *Sci. Rep.* **2016**, *6*, 29927. [CrossRef] [PubMed]
66. Liénard, M.A.; Hagstroem, A.K.; Lassance, J.-M.; Löfstedt, C. Evolution of multicomponent pheromone signals in small ermine moths involves a single fatty-acyl reductase gene. *Proc. Natl. Acad. Sci. USA* **2010**, *107*, 10955–10960. [CrossRef] [PubMed]
67. Günther, C.S.; Chervin, C.; Marsh, K.B.; Newcomb, R.D.; Souleyre, E.J. Characterisation of two alcohol acyltransferases from kiwifruit (*Actinidia* spp.) reveals distinct substrate preferences. *Phytochemistry* **2011**, *72*, 700–710. [CrossRef]
68. Shalit, M.; Katzir, N.; Tadmor, Y.; Larkov, O.; Burger, Y.; Shalekhet, F.; Lastochkin, E.; Ravid, U.; Amar, O.; Edelstein, M.; et al. Acetyl-CoA: Alcohol acetyltransferase activity and aroma formation in ripening melon fruits. *J. Agric. Food Chem.* **2001**, *49*, 794–799. [CrossRef]
69. Ding, B.J.; Lager, I.; Bansal, S.; Durrett, T.P.; Stymne, S.; Löfstedt, C. The yeast ATF1 acetyltransferase efficiently acetylates insect pheromone alcohols: Implications for the biological production of moth pheromones. *Lipids* **2016**, *51*, 469–475. [CrossRef]
70. Yang, C.Y.; Lee, S.; Choi, K.S.; Jeon, H.Y.; Boo, K.S. Sex pheromone production and response in Korean populations of the diamondback moth, *Plutella xylostella*. *Entomol. Exp. Appl.* **2007**, *124*, 293–298. [CrossRef]
71. Chen, D.S.; Dai, J.Q.; Han, S.C. Identification of the pheromone biosynthesis genes from the sex pheromone gland transcriptome of the diamondback moth, *Plutella xylostella*. *Sci. Rep.* **2017**, *7*, 16255. [CrossRef] [PubMed]
72. Vogel, H.; Heidel, A.J.; Heckel, D.G.; Groot, A.T. Transcriptome analysis of the sex pheromone gland of the noctuid moth *Heliothis virescens*. *BMC Genom.* **2010**, *11*, 29. [CrossRef]
73. Hansson, B.S.; Stensmyr, M.C. Evolution of insect olfaction. *Neuron* **2011**, *72*, 698–711. [CrossRef] [PubMed]
74. Hansson, B.S. *Insect Olfaction*; Springer: Berlin/Heidelberg, Germany, 2013; p. 458. [CrossRef]
75. Leal, W.S. Odorant reception in insects: Roles of receptors, binding proteins, and degrading enzymes. *Annu. Rev. Entomol.* **2013**, *58*, 373–391. [CrossRef]
76. Cardé, R.T.; Willis, M.A. Navigational strategies used by insects to find distant, wind-borne sources of odor. *J. Chem. Ecol.* **2008**, *34*, 854–866. [CrossRef] [PubMed]
77. De Bruyne, M.; Baker, T.C. Odor detection in insects: Volatile codes. *J. Chem. Ecol.* **2008**, *34*, 882–897. [CrossRef]

78. Fleischer, J.; Pregitzer, P.; Breer, H.; Krieger, J. Access to the odor world: Olfactory receptors and their role for signal transduction in insects. *Cell. Mol. Life Sci.* **2018**, *75*, 485–508. [CrossRef] [PubMed]
79. Montagné, N.; de Fouchier, A.; Newcomb, R.D.; Jacquin-Joly, E. Advances in the identification and characterization of olfactory receptors in insects. *Prog. Mol. Biol. Transl. Sci.* **2015**, *130*, 55–80. [CrossRef]
80. Jeanne, J.M.; Fişek, M.; Wilson, R.I. The organization of projections from olfactory glomeruli onto higher-order neurons. *Neuron* **2018**, *98*, 1198–1213.e6. [CrossRef] [PubMed]
81. Zhao, Z.; McBride, C.S. Evolution of olfactory circuits in insects. *J. Comp. Physiol. A* **2020**, *206*, 353–367. [CrossRef] [PubMed]
82. Fialho, M.D.C.Q.; Guss-Matiello, C.P.; Zanuncio, J.C.; Campos, L.A.O.; Serrão, J.E. A comparative study of the antennal sensilla in corbiculate bees. *J. Apic. Res.* **2014**, *53*, 392–403. [CrossRef]
83. Tsang, T.K.; Bushong, E.A.; Boassa, D.; Hu, J.; Romoli, B.; Phan, S.; Dulcis, D.; Su, C.Y.; Ellisman, M.H. High-quality ultrastructural preservation using cryofixation for 3D electron microscopy of genetically labeled tissues. *eLife* **2018**, *7*. [CrossRef]
84. Zhang, Q.; Lee, W.A.; Paul, D.L.; Ginty, D.D. Multiplexed peroxidase-based electron microscopy labeling enables simultaneous visualization of multiple cell types. *Nat. Neurosci.* **2019**, *22*, 828–839. [CrossRef]
85. Schmidt, H.R.; Benton, R. Molecular mechanisms of olfactory detection in insects: Beyond receptors. *Open Biol.* **2020**, *10*, 200252. [CrossRef]
86. Wang, G.; Carey, A.F.; Carlson, J.R.; Zwiebel, L.J. Molecular basis of odor coding in the malaria vector mosquito *Anopheles gambiae*. *Proc. Natl. Acad. Sci. USA* **2010**, *107*, 4418–4423. [CrossRef]
87. Zhang, J.; Liu, Y.; Walker, W.B.; Dong, S.L.; Wang, G.R. Identification and localization of two sensory neuron membrane proteins from *Spodoptera litura* (Lepidoptera: Noctuidae). *Insect Sci.* **2015**, *22*, 399–408. [CrossRef] [PubMed]
88. Sato, K.; Pellegrino, M.; Nakagawa, T.; Nakagawa, T.; Vosshall, L.B.; Touhara, K. Insect olfactory receptors are heteromeric ligand-gated ion channels. *Nature* **2008**, *452*, 1002–1006. [CrossRef] [PubMed]
89. Wicher, D.; Schäfer, R.; Bauernfeind, R.; Stensmyr, M.C.; Heller, R.; Heinemann, S.H.; Hansson, B.S. Drosophila odorant receptors are both ligand-gated and cyclic-nucleotide-activated cation channels. *Nature* **2008**, *452*, 1007–1011. [CrossRef]
90. Missbach, C.; Dweck, H.K.; Vogel, H.; Vilcinskas, A.; Stensmyr, M.C.; Hansson, B.S.; Grosse-Wilde, E. Evolution of insect olfactory receptors. *eLife* **2014**, *3*, e02115. [CrossRef]
91. Sakurai, T.; Namiki, S.; Kanzaki, R. Molecular and neural mechanisms of sex pheromone reception and processing in the silkmoth *Bombyx mori*. *Front. Physiol.* **2014**, *5*, 125. [CrossRef] [PubMed]
92. Brito, N.F.; Moreira, M.F.; Melo, A.C. A look inside odorant-binding proteins in insect chemoreception. *J. Insect Physiol.* **2016**, *95*, 51–65. [CrossRef] [PubMed]
93. Jiang, X.J.; Guo, H.; Di, C.; Yu, S.; Zhu, L.; Huang, L.Q.; Wang, C.Z. Sequence similarity and functional comparisons of pheromone receptor orthologs in two closely related Helicoverpa species. *Insect Biochem. Mol. Biol.* **2014**, *48*, 63–74. [CrossRef] [PubMed]
94. Cassau, S.; Krieger, J. The role of SNMPs in insect olfaction. *Cell Tissue Res.* **2021**, *383*, 21–33. [CrossRef] [PubMed]
95. Gomez-Diaz, C.; Bargeton, B.; Abuin, L.; Bukar, N.; Reina, J.H.; Bartoi, T.; Graf, M.; Ong, H.; Ulbrich, M.H.; Masson, J.F.; et al. A CD36 ectodomain mediates insect pheromone detection via a putative tunnelling mechanism. *Nat. Commun.* **2016**, *7*, 11866. [CrossRef] [PubMed]
96. Liu, S.; Chang, H.; Liu, W.; Cui, W.; Liu, Y.; Wang, Y.; Ren, B.; Wang, G. Essential role for SNMP1 in detection of sex pheromones in *Helicoverpa armigera*. *Insect Biochem. Mol. Biol.* **2020**, *127*, 103485. [CrossRef]
97. Zhao, Y.; Ding, J.; Zhang, Z.; Liu, F.; Zhou, C.; Mu, W. Sex- and tissue-specific expression profiles of odorant binding protein and chemosensory protein genes in *Bradysia odoriphaga* (Diptera: Sciaridae). *Front. Physiol.* **2018**, *9*, 107. [CrossRef]
98. Zhang, Y.N.; Qian, J.L.; Xu, J.W.; Zhu, X.Y.; Li, M.Y.; Xu, X.X.; Liu, C.X.; Xue, T.; Sun, L. Identification of chemosensory genes based on the transcriptomic analysis of six different chemosensory organs in *Spodoptera exigua*. *Front. Physiol.* **2018**, *9*, 432. [CrossRef]
99. Laughlin, J.D.; Ha, T.S.; Jones, D.N.M.; Smith, D.P. Activation of pheromone-sensitive neurons is mediated by conformational activation of pheromone-binding protein. *Cell* **2008**, *133*, 1255–1265. [CrossRef]
100. Tian, Z.; Zhang, Y. Molecular characterization and functional analysis of pheromone binding protein 1 from *Cydia pomonella* (L.). *Insect Mol. Biol.* **2016**, *25*, 769–777. [CrossRef]
101. Oliveira, D.S.; Brito, N.F.; Franco, T.A.; Moreira, M.F.; Leal, W.S.; Melo, A.C.A. Functional characterization of odorant binding protein 27 (RproOBP27) from *Rhodnius prolixus* antennae. *Front. Physiol.* **2018**, *9*, 1175. [CrossRef] [PubMed]
102. Chang, H.; Liu, Y.; Yang, T. Pheromone binding proteins enhance the sensitivity of olfactory receptors to sex pheromones in *Chilo suppressalis*. *Sci. Rep.* **2015**, *5*, 13093. [CrossRef] [PubMed]
103. Wen, M.; Li, E.; Li, J.; Chen, Q.; Zhou, H.; Zhang, S.; Li, K.; Ren, B.; Wang, Y.; Yin, J. Molecular characterization and key binding sites of sex pheromone-binding proteins from the meadow moth, *Loxostege sticticalis*. *J. Agric. Food Chem.* **2019**, *67*, 12685–12695. [CrossRef] [PubMed]
104. Jing, D.; Zhang, T.; Bai, S.; He, K.; Prabu, S.; Luan, J.; Wang, Z. Sexual-biased gene expression of olfactory-related genes in the antennae of *Conogethes pinicolalis* (Lepidoptera: Crambidae). *BMC Genom.* **2020**, *21*, 244. [CrossRef]
105. Liu, N.Y.; Yang, F.; Yang, K.; He, P.; Niu, X.H.; Xu, W.; Anderson, A.; Dong, S.L. Two subclasses of odorant-binding proteins in *Spodoptera exigua* display structural conservation and functional divergence. *Insect Mol. Biol.* **2015**, *24*, 167–182. [CrossRef]
106. Yang, H.; Su, T.; Yang, W.; Yang, C.-P.; Chen, Z.-M.; Lu, L.; Liu, Y.-L.; Tao, Y.-Y. Molecular characterization, expression pattern and ligand-binding properties of the pheromone-binding protein gene from *Cyrtotrachelus buqueti*. *Physiol. Entomol.* **2017**, *42*, 369–378. [CrossRef]

107. Chen, X.L.; Li, G.W.; Xu, X.L.; Wu, J.X. Molecular and functional characterization of odorant binding protein 7 from the oriental fruit moth *Grapholita molesta* (Busck) (Lepidoptera: Tortricidae). *Front. Physiol.* **2018**, *9*, 1762. [CrossRef]
108. Barrozo, R.B.; Gadenne, C.; Anton, S. Switching attraction to inhibition: Mating-induced reversed role of sex pheromone in an insect. *J. Exp. Biol.* **2010**, *213*, 2933–2939. [CrossRef]
109. Kromann, S.H.; Saveer, A.M.; Binyameen, M.; Bengtsson, M.; Birgersson, G.; Hansson, B.S.; Schlyter, F.; Witzgall, P.; Ignell, R.; Becher, P.G. Concurrent modulation of neuronal and behavioural olfactory responses to sex and host plant cues in a male moth. *Proc. Biol. Sci.* **2015**, *282*, 20141884. [CrossRef]
110. Hatano, E.; Saveer, A.M.; Borrero-Echeverry, F.; Strauch, M.; Zakir, A.; Bengtsson, M.; Ignell, R.; Anderson, P.; Becher, P.G.; Witzgall, P.; et al. Herbivore-induced plant volatile interferes with host plant and mate location in moths through suppression of olfactory signalling pathways. *BMC Biol.* **2015**, *13*, 75. [CrossRef]
111. Jarriault, D.; Gadenne, C.; Rospars, J.P.; Anton, S. Quantitative analysis of sex-pheromone coding in the antennal lobe of the moth *Agrotis ipsilon*: A tool to study network plasticity. *J. Exp. Biol.* **2009**, *212*, 1191–1201. [CrossRef]
112. Barrozo, R.B.; Jarriault, D.; Simeone, X.; Gaertner, C.; Gadenne, C.; Anton, S. Mating-induced transient inhibition of responses to sex pheromone in a male moth is not mediated by octopamine or serotonin. *J. Exp. Biol.* **2010**, *213*, 1100–1106. [CrossRef]
113. Deisig, N.; Kropf, J.; Vitecek, S.; Pevergne, D.; Rouyar, A.; Sandoz, J.C.; Lucas, P.; Gadenne, C.; Anton, S.; Barrozo, R. Differential interactions of sex pheromone and plant odour in the olfactory pathway of a male moth. *PLoS ONE* **2012**, *7*, e33159. [CrossRef] [PubMed]
114. Chaffiol, A.; Dupuy, F.; Barrozo, R.B.; Kropf, J.; Renou, M.; Rospars, J.-P.; Anton, S. Pheromone modulates plant odor responses in the antennal lobe of a moth. *Chem. Senses* **2014**, *39*, 451–463. [CrossRef]
115. Hoffmann, A.; Bourgeois, T.; Munoz, A.; Anton, S.; Gevar, J.; Dacher, M.; Renou, M. Plant volatile alters the perception of sex pheromone blend ratios in a moth. *J. Comp. Physiol. A* **2020**, *206*, 553–570. [CrossRef]
116. Ian, E.; Kirkerud, N.H.; Galizia, C.G.; Berg, B.G. Coincidence of pheromone and plant odor leads to sensory plasticity in the heliothine olfactory system. *PLoS ONE* **2017**, *12*, e0175513. [CrossRef] [PubMed]
117. Borrero-Echeverry, F.; Bengtsson, M.; Nakamuta, K.; Witzgall, P. Plant odor and sex pheromone are integral elements of specific mate recognition in an insect herbivore. *Evolution* **2018**, *72*, 2225–2233. [CrossRef]
118. Ammagarahalli, B.; Gemeno, C. Interference of plant volatiles on pheromone receptor neurons of male *Grapholita molesta* (Lepidoptera: Tortricidae). *J. Insect Physiol.* **2015**, *81*, 118–128. [CrossRef]
119. Renou, M.; Party, V.; Rouyar, A.; Anton, S. Olfactory signal coding in an odor background. *BioSystems* **2015**, *136*, 35–45. [CrossRef] [PubMed]
120. Trona, F.; Anfora, G.; Balkenius, A.; Bengtsson, M.; Tasin, M.; Knight, A.; Janz, N.; Witzgall, P.; Ignell, R. Neural coding merges sex and habitat chemosensory signals in an insect herbivore. *Proc. R. Soc. B* **2013**, *280*, 20130267. [CrossRef] [PubMed]
121. Galizia, C.G.; Rössler, W. Parallel olfactory system in insects: Anatomy and function. *Annu. Rev. Entomol.* **2010**, *55*, 399–420. [CrossRef] [PubMed]
122. Varela, N.; Avilla, J.; Gemeno, C.; Anton, S. Ordinary glomeruli in the antennal lobe of male and female tortricid moth *Grapholita molesta* (Busck) (Lepidoptera: Tortricidae) process sex pheromone and host-plant volatiles. *J. Exp. Biol.* **2011**, *214*, 637–645. [CrossRef]
123. Trona, F.; Anfora, G.; Bengtsson, M.; Witzgall, P.; Ignell, R. Coding and interaction of sex pheromone and plant volatile signals in the antennal lobe of the codling moth *Cydia pomonella*. *J. Exp. Biol.* **2010**, *213*, 4291–4303. [CrossRef]
124. Anton, S.; Rössler, W. Plasticity and modulation of olfactory circuits in insects. *Cell Tissue Res.* **2021**, *383*, 149–164. [CrossRef] [PubMed]
125. Renou, M.; Anton, S. Insect olfactory communication in a complex and changing world. *Curr. Opin. Insect Sci.* **2020**, *42*, 1–7. [CrossRef]
126. Murmu, M.S.; Hanoune, J.; Choi, A.; Bureau, V.; Renou, M.; Dacher, M.; Deisig, N. Modulatory effects of pheromone on olfactory learning and memory in moths. *J. Insect Physiol.* **2020**, *127*, 104159. [CrossRef]
127. De Fouchier, A.; Walker III, W.B.; Montagné, N.; Steiner, C.; Binyameen, M.; Schlyter, F.; Chertemps, T.; Maria, A.; François, M.C.; Monsempes, C.; et al. Functional evolution of Lepidoptera olfactory receptors revealed by deorphanization of a moth repertoire. *Nat. Commun.* **2017**, *8*, 15709. [CrossRef] [PubMed]
128. Larsson, M.C.; Domingos, A.I.; Jones, W.D.; Chiappe, M.E.; Amrein, H.; Vosshall, L.B. Or83b encodes a broadly expressed odorant receptor essential for *Drosophila* olfaction. *Neuron* **2004**, *43*, 703–714. [CrossRef]
129. Hallem, E.A.; Ho, M.G.; Carlson, J.R. The molecular basis of odor coding in the *Drosophila* antenna. *Cell* **2004**, *117*, 965–979. [CrossRef]
130. Legeai, F.; Malpel, S.; Montagné, N.; Monsempes, C.; Cousserans, F.; Merlin, C.; François, M.-C.; Maïbèche-Coisné, M.; Gavory, F.; Poulain, J.; et al. An expressed sequence tag collection from the male antennae of the noctuid moth *Spodoptera littoralis*: A resource for olfactory and pheromone detection research. *BMC Genom.* **2011**, *12*, 86. [CrossRef]
131. Ochieng, S.A.; Anderson, P.; Hansson, B.S. Antennal lobe projection patterns of olfactory receptor neurons involved in sex pheromone detection in *Spodoptera littoralis* (Lepidoptera: Noctuidae). *Tissue Cell* **1995**, *27*, 221–232. [CrossRef]
132. Schneider, D.; Schulz, S.; Priesner, E.; Ziesmann, J.; Francke, W. Autodetection and chemistry of female and male pheromone in both sexes of the tiger moth *Panaxia quadripunctaria*. *J. Comp. Physiol. A* **1998**, *182*, 153–161. [CrossRef]

133. Holdcraft, R.; Rodriguez-Saona, C.; Stelinski, L.L. Pheromone autodetection: Evidence and implications. *Insects* **2016**, *7*, 17. [CrossRef] [PubMed]

134. Bakthavatsalam, N.; Vinutha, J.; Ramakrishna, P.; Raghavendra, A.; Ravindra, K.; Verghese, A. Autodetection in *Helicoverpa armigera* (Hubner). *Curr. Sci.* **2016**, 2261–2267. [CrossRef]

135. Hillier, N.; Kleineidam, C.; Vickers, N.J. Physiology and glomerular projections of olfactory receptor neurons on the antenna of female *Heliothis virescens* (Lepidoptera: Noctuidae) responsive to behaviorally relevant odors. *J. Comp. Physiol. A* **2006**, *192*, 199–219. [CrossRef] [PubMed]

136. Kuhns, E.H.; Pelz-Stelinski, K.; Stelinski, L.L. Reduced mating success of female tortricid moths following intense pheromone auto-exposure varies with sophistication of mating system. *J. Chem. Ecol.* **2012**, *38*, 168–175. [CrossRef] [PubMed]

137. Stelinski, L.; Holdcraft, R.; Rodriguez-Saona, C. Female moth calling and flight behavior are altered hours following pheromone autodetection: Possible implications for practical management with mating disruption. *Insects* **2014**, *5*, 459–473. [CrossRef]

138. Stelinski, L.; Il'Ichev, A.; Gut, L. Antennal and behavioral responses of virgin and mated oriental fruit moth (Lepidoptera: Tortricidae) females to their sex pheromone. *Ann. Entomol. Soc. Am.* **2006**, *99*, 898–904. [CrossRef]

139. Domínguez, A.; López, S.; Bernabé, A.; Guerrero, A.; Quero, C. Influence of age, host plant and mating status in pheromone production and new insights on perception plasticity in *Tuta absoluta*. *Insects* **2019**, *10*, 256. [CrossRef] [PubMed]

140. Hodgdon, E.A.; Hallett, R.H.; Heal, J.D.; Swan, A.E.; Chen, Y.H. Synthetic pheromone exposure increases calling and reduces subsequent mating in female *Contarinia nasturtii* (Diptera: Cecidomyiidae). *Pest Manag. Sci.* **2021**, *77*, 548–556. [CrossRef] [PubMed]

141. Poivet, E.; Rharrabe, K.; Monsempes, C.; Glaser, N.; Rochat, D.; Renou, M.; Marion-Poll, F.; Jacquin-Joly, E. The use of the sex pheromone as an evolutionary solution to food source selection in caterpillars. *Nat. Commun.* **2012**, *3*, 1047. [CrossRef] [PubMed]

142. Zielonka, M.; Gehrke, P.; Badeke, E.; Sachse, S.; Breer, H.; Krieger, J. Larval sensilla of the moth *Heliothis virescens* respond to sex pheromone components. *Insect Mol. Biol.* **2016**, *25*, 666–678. [CrossRef] [PubMed]

143. Haynes, K.; Gaston, L.; Pope, M.M.; Baker, T.C. Potential for evolution of resistance to pheromones: Interindividual and interpopulational variation in chemical communication system of pink bollworm moth. *J. Chem. Ecol.* **1984**, *10*, 1551–1565. [CrossRef] [PubMed]

144. Haynes, K.F.; Baker, T.C. Potential for evolution of resistance to pheromones. *J. Chem. Ecol.* **1988**, *14*, 1547–1560. [CrossRef]

145. Doane, C.C.; Brooks, T.W. Research and development of pheromones for insect control with emphasis on the pink bollworm. In *Management of Insect Pests with Semiochemicals*; Mitchell, E.R., Ed.; Springer: Boston, MA, USA, 1981; pp. 285–303. [CrossRef]

146. Collins, R.; Cardé, R. Variation in and heritability of aspects of pheromone production in the pink bollworm moth, *Pectinophora gossypiella* (Lepidoptera: Gelechiidae). *Ann. Entomol. Soc. Am.* **1985**, *78*, 229–234. [CrossRef]

147. Evenden, M.; Haynes, K. Potential for the evolution of resistance to pheromone-based mating disruption tested using two pheromone strains of the cabbage looper, *Trichoplusia ni*. *Entomol. Exp. Appl.* **2001**, *100*, 131–134. [CrossRef]

148. Löfstedt, C. Moth pheromone genetics and evolution. *Philos. Trans. R. Soc. Lond. B Biol. Sci.* **1993**, *340*, 167–177.

149. Phelan, P. Evolution of sex pheromones and the role of asymmetric tracking. In *Insect Chemical Ecology: An Evolutionary Approach*; Roitberg, B.D., Isman, M.B., Eds.; Chapman and Hall: New York, NY, USA, 1992; pp. 265–314.

150. Mochizuki, F.; Fukumoto, T.; Noguchi, H.; Sugie, H.; Morimoto, T.; Ohtani, K. Resistance to a mating disruptant composed of (Z)-11-tetradecenyl acetate in the smaller tea tortrix, *Adoxophyes honmai* (Yasuda) (Lepidoptera: Tortricidae). *Appl. Entomol. Zool.* **2002**, *37*, 299–304. [CrossRef]

151. Mochizuki, F.; Noguchi, H.; Sugie, H.; Tabata, J.; Kainoh, Y. Sex pheromone communication from a population resistant to mating disruptant of the smaller tea tortrix, *Adoxophyes honmai Yasuda* (Lepidoptera: Tortricidae). *Appl. Entomol. Zool.* **2008**, *43*, 293–298. [CrossRef]

152. Tabata, J.; Noguchi, H.; Kainoh, Y.; Mochizuki, F.; Sugie, H. Behavioral response to sex pheromone-component blends in the mating disruption-resistant strain of the smaller tea tortrix, *Adoxophyes honmai* Yasuda (Lepidoptera: Tortricidae), and its mode of inheritance. *Appl. Entomol. Zool.* **2007**, *42*, 675–683. [CrossRef]

153. Tabata, J.; Noguchi, H.; Kainoh, Y.; Mochizuki, F.; Sugie, H. Sex pheromone production and perception in the mating disruption-resistant strain of the smaller tea leafroller moth, *Adoxophyes honmai*. *Entomol. Exp. Appl.* **2007**, *122*, 145–153. [CrossRef]

154. Stelinski, L.L.; Vogel, K.J.; Gut, L.J.; Miller, J.R. Seconds-long pre-exposures to pheromone from rubber septum or polyethelene tube dispensers alters subsequent behavioral responses of male *Grapholita molesta* (Lepidoptera:Tortricidae) in a sustained-flight tunnel. *Environ. Entomol.* **2005**, *34*, 696–704. [CrossRef]

155. Miller, J.R.; Gut, L.J. Mating disruption for the 21st century: Matching technology with mechanism. *Environ. Entomol.* **2015**, *44*, 427–453. [CrossRef]

156. Judd, G.J.R.; Gardiner, M.G.T.; DeLury, N.C.; Karg, G. Reduced antennal sensitivity, behavioural response, and attraction of male codling moths, *Cydia pomonella*, to their pheromone (E,E)-8,10-dodecadien-1-ol following various pre-exposure regimes. *Entomol. Exp. Appl.* **2005**, *114*, 65–78. [CrossRef]

157. Suckling, D.M.; Stringer, L.D.; Jiménez-Pérez, A.; Walter, G.H.; Sullivan, N.; El-Sayed, A.M. With or without pheromone habituation: Possible differences between insect orders? *Pest Manag. Sci.* **2018**, *74*, 1259–1264. [CrossRef] [PubMed]

158. Bartell, R.J.; Lawrence, L.A. Reduction in sexual responsiveness of male light-brown apple moth [*Epiphyas postvittana* (Wlk.)] following previous brief pheromonal exposure is concentration dependent. *J. Aust. Entomol. Soc.* **1976**, *15*, 236. [CrossRef]

159. Sarles, L.; Verhaeghe, A.; Francis, F.; Verheggen, F.J. Semiochemicals of *Rhagoletis* fruit flies: Potential for integrated pest management. *Crop Prot.* **2015**, *78*, 114–118. [CrossRef]
160. Sharma, A.; Sandhi, R.K.; Reddy, G.V. A review of interactions between insect biological control agents and semiochemicals. *Insects* **2019**, *10*, 439. [CrossRef] [PubMed]
161. Bitsadze, N.; Jaronski, S.; Khasdan, V.; Abashidze, E.; Abashidze, M.; Latchininsky, A.; Samadashvili, D.; Sokhadze, I.; Rippa, M.; Ishaaya, I. Joint action of *Beauveria bassiana* and the insect growth regulators diflubenzuron and novaluron, on the migratory locust, *Locusta migratoria*. *J. Pest Sci.* **2013**, *86*, 293–300. [CrossRef]
162. Reddy, G.V.; Zhao, Z.; Humber, R.A. Laboratory and field efficacy of entomopathogenic fungi for the management of the sweetpotato weevil, *Cylas formicarius* (Coleoptera: Brentidae). *J. Invertebr. Pathol.* **2014**, *122*, 10–15. [CrossRef]
163. Abdellaoui, K.; Miladi, M.; Mkhinini, M.; Boughattas, I.; Hamouda, A.B.; Hajji-Hedfi, L.; Tlili, H.; Acheuk, F. The aggregation pheromone phenylacetonitrile: Joint action with the entomopathogenic fungus *Metarhizium anisopliae* var. acridum and physiological and transcriptomic effects on *Schistocerca gregaria* nymphs. *Pest. Biochem. Physiol.* **2020**, *167*, 104594. [CrossRef]
164. Gutiérrez-Cárdenas, O.G.; Cortez-Madrigal, H.; Malo, E.A.; Gómez-Ruíz, J.; Nord, R. Physiological and pathogenical characterization of *Beauveria bassiana* and *Metarhizium anisopliae* isolates for management of adult *Spodoptera frugiperda*. *Southwest. Entomol.* **2019**, *44*, 409–421. [CrossRef]
165. Akutse, K.S.; Khamis, F.M.; Ambele, F.C.; Kimemia, J.W.; Ekesi, S.; Subramanian, S. Combining insect pathogenic fungi and a pheromone trap for sustainable management of the fall armyworm, *Spodoptera frugiperda* (Lepidoptera: Noctuidae). *J. Invertebr. Pathol.* **2020**, *177*, 107477. [CrossRef]
166. Akutse, K.S.; Subramanian, S.; Khamis, F.M.; Ekesi, S.; Mohamed, S.A. Entomopathogenic fungus isolates for adult *Tuta absoluta* (Lepidoptera: Gelechiidae) management and their compatibility with *Tuta* pheromone. *J. Appl. Entomol.* **2020**, *144*, 777–787. [CrossRef]
167. Sammani, A.M.; Dissanayaka, D.M.; Wijayaratne, L.K.; Bamunuarachchige, T.C.; Morrison, W.R. Effect of pheromones, plant volatiles and spinosad on mating, male attraction and burrowing of *Cadra cautella* (Walk.) (Lepidoptera: Pyralidae). *Insects* **2020**, *11*, 845. [CrossRef] [PubMed]
168. Ren, L.; Yingao, M.; Xie, M.; Lu, Y.; Cheng, D. Rectal bacteria produce sex pheromones in the male oriental fruit fly. *bioRxiv* **2020**. [CrossRef]
169. Mori, B.A.; Evenden, M.L. Factors affecting pheromone-baited trap capture of male *Coleophora deauratella*, an invasive pest of clover in Canada. *J. Econ. Entomol.* **2013**, *106*, 844–854. [CrossRef] [PubMed]
170. Reddy, G.V.; Cruz, Z.T.; Guerrero, A. Development of an efficient pheromone-based trapping method for the banana root borer *Cosmopolites sordidus*. *J. Chem. Ecol.* **2009**, *35*, 111–117. [CrossRef]
171. Reddy, G.V.; Gadi, N.; Taianao, A.J. Efficient sex pheromone trapping: Catching the sweetpotato weevil, *Cylas formicarius*. *J. Chem. Ecol.* **2012**, *38*, 846–853. [CrossRef]
172. Reddy, G.V.; Shrestha, G.; Miller, D.A.; Oehlschlager, A.C. Pheromone-trap monitoring system for pea leaf weevil, *Sitona lineatus*: Effects of trap type, lure type and trap placement within fields. *Insects* **2018**, *9*, 75. [CrossRef]
173. Athanassiou, C.G.; Kavallieratos, N.G.; Gakis, S.F.; Kyrtsa, L.A.; Mazomenos, B.E.; Gravanis, F.T. Influence of trap type, trap colour, and trapping location on the capture of the pine moth, *Thaumetopoea pityocampa*. *Entomol. Exp. Appl.* **2007**, *122*, 117–123. [CrossRef]
174. Jactel, H.; Menassieu, P.; Vétillard, F.; Barthélémy, B.; Piou, D.; Frérot, B.; Rousselet, J.; Goussard, F.; Branco, M.; Battisti, A. Population monitoring of the pine processionary moth (Lepidoptera: Thaumetopoeidae) with pheromone-baited traps. *For. Ecol. Manag.* **2006**, *235*, 96–106. [CrossRef]
175. Jones, O.T. Practical applications of pheromones and other semiochemicals. In *Insect Pheromones and Their Use in Pest Management*; Howse, P., Stevens, I., Jones, O., Eds.; Chapman and Hall: London, UK, 1998; pp. 261–355.
176. Trematerra, P.; Colacci, M. Recent advances in management by pheromones of *Thaumetopoea* moths in urban parks and woodland recreational areas. *Insects* **2019**, *10*, 395. [CrossRef]
177. Byers, J.A. Modelling female mating success during mass trapping and natural competitive attraction of searching males or females. *Entomol. Exp. Appl.* **2012**, *145*, 228–237. [CrossRef]
178. El-Sayed, A.M.; Suckling, D.M.; Wearing, C.H.; Byers, J.A. Potential of mass trapping for long-term pest management and eradication of invasive species. *J. Econ. Entomol.* **2006**, *99*, 1550–1564. [CrossRef] [PubMed]
179. Williams, L.; Serrano, J.M.; Johnson, P.J.; Millar, J.G. 13-Tetradecenyl acetate, a female-produced sex pheromone component of the economically important click beetle *Melanotus communis* (Gyllenhal) (Coleoptera: Elateridae). *Sci. Rep.* **2019**, *9*, 16197. [CrossRef] [PubMed]
180. Hegazi, E.; Khafagi, W.E.; Konstantopoulou, M.; Raptopoulos, D.; Tawfik, H.; El-Aziz, G.M.A.; El-Rahman, S.M.A.; Atwa, A.; Aggamy, E.; Showeil, S. Efficient Mass-trapping method as an alternative tactic for suppressing populations of leopard moth (Lepidoptera: Cossidae). *Ann. Entomol. Soc. Am.* **2009**, *102*, 809–818. [CrossRef]
181. Abbes, K.; Chermiti, B. Comparison of two marks of sex pheromone dispensers commercialized in Tunisia for their efficiency to monitor and to control by mass-trapping Tuta absoluta under greenhouses. *Tunis. J. Plant Prot.* **2011**, *6*, 133–148.
182. Lobos, E.; Occhionero, M.; Werenitzky, D.; Fernandez, J.; Gonzalez, L.M.; Rodriguez, C.; Calvo, C.; Lopez, G.; Oehlschlager, A.C. Optimization of a trap for *Tuta absoluta* Meyrick (Lepidoptera: Gelechiidae) and trials to determine the effectiveness of mass trapping. *Neotrop. Entomol.* **2013**, *42*, 448–457. [CrossRef]

183. Alpizar, D.; Fallas, M.; Oehlschlager, A.C.; Gonzalez, L.M. Management of *Cosmopolites sordidus* and *Metamasius hemipterus* in banana by pheromone-based mass trapping. *J. Chem. Ecol.* **2012**, *38*, 245–252. [CrossRef]
184. Martin, J.C. Development of environment-friendly strategies in the management of processionary moths. In *Processionary Moths and Climate Change: An Update*; Roques, A., Ed.; Springer: Dordrecht, The Netherlands, 2015. [CrossRef]
185. Trematerra, P.; Colacci, M.; Sciarretta, A. Mass-trapping trials for the control of pine processionary moth in a pine woodland recreational area. *J. Appl. Entomol.* **2019**, *143*, 129–136. [CrossRef]
186. Lance, D.R.; Leonard, D.S.; Mastro, V.C.; Walters, M.L. Mating disruption as a suppression tactic in programs targeting regulated lepidopteran pests in US. *J. Chem. Ecol.* **2016**, *42*, 590–605. [CrossRef] [PubMed]
187. Hummel, H.E.; Langner, S.S.; Eisinger, M.T. Pheromone dispensers, including organic polymer fibers, described in the crop protection literature: Comparison of their innovation potential. *Commun. Agric. Appl. Biol. Sci.* **2013**, *78*, 233–252. [PubMed]
188. Ioriatti, C.; Lucchi, A. Semiochemical strategies for tortricid moth control in apple orchards and vineyards in Italy. *J. Chem. Ecol.* **2016**, *42*, 571–583. [CrossRef] [PubMed]
189. Witzgall, P.; Stelinski, L.; Gut, L.; Thomson, D. Codling moth management and chemical ecology. *Annu. Rev. Entomol.* **2008**, *53*, 503–522. [CrossRef] [PubMed]
190. Gordon, D.; Zahavi, T.; Anshelevich, L.; Harel, M.; Ovadia, S.; Dunkelblum, E.; Harari, A.R. Mating disruption of *Lobesia botrana* (Lepidoptera: Tortricidae): Effect of pheromone formulations and concentrations. *J. Econ. Entomol.* **2005**, *98*, 135–142. [CrossRef]
191. Stelinski, L.L.; Miller, J.R.; Ledebuhr, R.; Siegert, P.; Gut, L.J. Season-long mating disruption of *Grapholita molesta* (Lepidoptera: Tortricidae) by one machine application of pheromone in wax drops (SPLAT-OFM). *J. Pest Sci.* **2007**, *80*, 109–117. [CrossRef]
192. Trematerra, P.; Athanassiou, C.; Stejskal, V.; Sciarretta, A.; Kavallieratos, N.; Palyvos, N. Large-scale mating disruption of *Ephestia* spp. and *Plodia interpunctella* in Czech Republic, Greece and Italy. *J. Appl. Entomol.* **2011**, *135*, 749–762. [CrossRef]
193. Hoshi, H.; Takabe, M.; Nakamuta, K. Mating disruption of a carpenter moth, *Cossus insularis* (Lepidoptera: Cossidae) in apple orchards with synthetic sex pheromone, and registration of the pheromone as an agrochemical. *J. Chem. Ecol.* **2016**, *42*, 606–611. [CrossRef]
194. Cocco, A.; Deliperi, S.; Delrio, G. Control of *Tuta absoluta* (Meyrick) (Lepidoptera: Gelechiidae) in greenhouse tomato crops using the mating disruption technique. *J. Appl. Entomol.* **2013**, *137*, 16–28. [CrossRef]
195. Vacas, S.; Alfaro, C.; Primo, J.; Navarro-Llopis, V. Studies on the development of a mating disruption system to control the tomato leafminer, *Tuta absoluta* Povolny (Lepidoptera: Gelechiidae). *Pest Manag. Sci.* **2011**, *67*, 1473–1480. [CrossRef] [PubMed]
196. Hummel, H.E.; Langner, S.S.; Breuer, M. Electrospun mesofibers, a novel biodegradable pheromone dispenser technology, are combined with mechanical deployment for efficient IPM of *Lobesia botrana* in vineyards. *Commun. Agric. Appl. Biol. Sci.* **2015**, *80*, 331–341. [PubMed]
197. Lucchi, A.; Ladurner, E.; Iodice, A.; Savino, F.; Ricciardi, R.; Cosci, F.; Conte, G.; Benelli, G. Eco-friendly pheromone dispensers-a green route to manage the European grapevine moth? *Environ. Sci. Pollut. Res. Int.* **2018**, *25*, 9426–9442. [CrossRef] [PubMed]
198. Mori, B.A.; Evenden, M.L. Challenges of mating disruption using aerosol-emitting pheromone puffers in red clover seed production fields to control *Coleophora deauratella* (Lepidoptera: Coleophoridae). *Environ. Entomol.* **2015**, *44*, 34–43. [CrossRef] [PubMed]
199. Ferracini, C.; Saitta, V.; Pogolotti, C.; Rollet, I.; Vertui, F.; Dovigo, L. Monitoring and management of the pine processionary moth in the North-Western Italian Alps. *Forests* **2020**, *11*, 1253. [CrossRef]
200. Cook, S.M.; Khan, Z.R.; Pickett, J.A. The use of push-pull strategies in integrated pest management. *Annu. Rev. Entomol.* **2007**, *52*, 375–400. [CrossRef]
201. Dickens, J. Plant volatiles moderate response to aggregation pheromone in Colorado potato beetle. *J. Appl. Entomol.* **2006**, *130*, 26–31. [CrossRef]
202. Reddy, G.V.; Guerrero, A. Interactions of insect pheromones and plant semiochemicals. *Trends Plant Sci.* **2004**, *9*, 253–261. [CrossRef]
203. Powell, W.; Pickett, J.A. Manipulation of parasitoids for aphid pest management: Progress and prospects. *Pest Manag. Sci.* **2003**, *59*, 149–155. [CrossRef]
204. Nakashima, Y.; Birkett, M.A.; Pye, B.J.; Pickett, J.A.; Powell, W. The role of semiochemicals in the avoidance of the seven-spot ladybird, *Coccinella septempunctata*, by the aphid parasitoid, *Aphidius ervi*. *J. Chem. Ecol.* **2004**, *30*, 1103–1116. [CrossRef] [PubMed]
205. Nakashima, Y.; Ida, T.Y.; Powell, W.; Pickett, J.A.; Birkett, M.A.; Taki, H.; Takabayashi, J. Field evaluation of synthetic aphid sex pheromone in enhancing suppression of aphid abundance by their natural enemies. *BioControl* **2016**, *61*, 485–496. [CrossRef]
206. Miller, J.R.; Gut, L.J.; de Lame, F.M.; Stelinski, L.L. Differentiation of competitive vs. non-competitive mechanisms mediating disruption of moth sexual communication by point sources of sex pheromone (part I): Theory. *J. Chem. Ecol.* **2006**, *10*, 2089–2114. [CrossRef] [PubMed]